Studies in Systems, Decision and Control

Volume 467

Series Editor

Janusz Kacprzyk, Systems Research Institute, Polish Academy of Sciences, Warsaw, Poland

The series "Studies in Systems, Decision and Control" (SSDC) covers both new developments and advances, as well as the state of the art, in the various areas of broadly perceived systems, decision making and control–quickly, up to date and with a high quality. The intent is to cover the theory, applications, and perspectives on the state of the art and future developments relevant to systems, decision making, control, complex processes and related areas, as embedded in the fields of engineering, computer science, physics, economics, social and life sciences, as well as the paradigms and methodologies behind them. The series contains monographs, textbooks, lecture notes and edited volumes in systems, decision making and control spanning the areas of Cyber-Physical Systems, Autonomous Systems, Sensor Networks, Control Systems, Energy Systems, Automotive Systems, Biological Systems, Vehicular Networking and Connected Vehicles, Aerospace Systems, Automation, Manufacturing, Smart Grids, Nonlinear Systems, Power Systems, Robotics, Social Systems, Economic Systems and other. Of particular value to both the contributors and the readership are the short publication timeframe and the world-wide distribution and exposure which enable both a wide and rapid dissemination of research output.

Indexed by SCOPUS, DBLP, WTI Frankfurt eG, zbMATH, SCImago.

All books published in the series are submitted for consideration in Web of Science.

Didier Theilliol · Józef Korbicz · Janusz Kacprzyk
Editors

Recent Developments in Model-Based and Data-Driven Methods for Advanced Control and Diagnosis

Springer

Editors
Didier Theilliol
Faculté des Sciences et Techniques
University of Lorraine
Nancy, France

Janusz Kacprzyk
Systems Research Institute
Polish Academy of Sciences
Warsaw, Poland

Józef Korbicz
Institute Control and Computation Engineering
University of Zielona Góra
Zielona Gora, Poland

ISSN 2198-4182 ISSN 2198-4190 (electronic)
Studies in Systems, Decision and Control
ISBN 978-3-031-27539-5 ISBN 978-3-031-27540-1 (eBook)
https://doi.org/10.1007/978-3-031-27540-1

This Springer imprint is published by the registered company Springer Nature Switzerland AG
The registered company address is: Gewerbestrasse 11, 6330 Cham, Switzerland

Preface

This book contains selected papers presented at the 16th European Workshop on Advanced Control and Diagnosis, ACD 2022 for short, held in Nancy (France) on November 16–18, 2022.

The conference ACD 2022 (https://acd2022.cran.univ-lorraine.fr/) was organized by European Advanced Intelligent Control and Diagnosis working group with the support of the Centre of Research in Automatic Control (CRAN) and Ecole Polytechnique (Polytech Nancy), Universite de Lorraine. ACD 2022 was technically co-sponsored by IFAC (the International Federation of Automatic Control) Technical Committee 6.4. Fault Detection, Supervision & Safety of Technical Processes—SAFEPROCESS of which Société d'Automatique, de Génie Industriel et deProductique SAGIP is the National Member Organization for France.

The series of ACD conferences has been an opportunity for high-profile scientists and experts in the subject of advanced control, diagnosis, and health monitoring of systems to present recent results on focused on advanced fault-tolerant control strategies, health-aware control design strategies, advanced control approaches, deep learning-based methods for control and diagnosis, reinforcement learning-based approaches for advanced control, diagnosis & prognosis techniques applied to industrial problems, Industry 4.0 as well as instrumentation and sensors.

The series of conferences is an excellent forum for the exchange of knowledge and experience and sharing solutions in the academic and industrial environment. An important task of this forum is the integration of scientists and engineers from various industries, as well as producers of hardware and software for computer control and diagnostic systems.

This book is divided into six parts:

I. Observers and Estimation,
II. Diagnosis and Prognosis,
III. Advanced Control,
IV. Nonlinear Systems, Localization, and FDI,
V. Machine Learning, Artificial Intelligence, and Data-Driven Methods, and
VI. Robust and Fault-Tolerant Control.

We sincerely thank all the participants and the reviewers of the articles from the International Program Committee for their personal scientific contributions to the conference. I extend special appreciation to the authors of the accepted articles that are published in this collective book by Springer, as well as to the speakers of the plenary:

- Prof. Vicenç Puig: *Cyberphysical Security of Critical Infrastructures*
- Universitat Politècnica de Catalunya, BarcelonaTech (UPC)—Spain
- Dr. Lorenzo Fagiano (Italy) *Dealing with uncertainty in learning-based control and optimization: the Set membership paradigm*
- Politecnico di Milano, Milan—Italy
- Dr. Jean Baptiste Mouret (France) *Damage compensation in robotics without diagnosis* INRIA, Nancy—France

We believe that this collective book will become a great reference tool for scientists and researchers working in the area of fault diagnosis and fault tolerant control. Readers are kindly encouraged to contact the corresponding authors for further details concerning their research and presented results.

Nancy, France
Zielona Gora, Poland
Warsaw, Poland
November 2022

Editor Chair:
Didier Theilliol
Editor co-chairs:
Józef Korbicz
Janusz Kacprzyk

In collaboration with
General Chair: J. C. Ponsart
IPC Chair: M. S. Jha
IPC Co-Chair: H. Schulte

Acknowledgments We are grateful to the members of the ACD 2022 National Organizing Committee, especially to General Chair J. C. Ponsart, International Program Committee Chair M. S. Jha and Co-Chair H. Schulte for supervising many technical matters, to P. Weber and all National organizing committee members for organizational and administrative support, and to L. Suel for the secretariat.

As always, only the great synergistic effort of the organizers makes the conference a successful scientific event—especially in such difficult times of pandemic and war.

Contents

Observers and Estimation

H_∞ Stochastic State-Multiplicative Uncertain Systems-Robust Luenberger Filters

E. Gershon

Abstract Linear, continuous-time state-multiplicative noisy systems are considered. The problem of H_∞ Luenberger filtering for either deterministic norm-bounded or polytopic-type uncertain systems are solved via a simple LMI(s) condition. In this problem, a cost function is defined which is the expected value of the standard H_∞ performance cost with respect to the stochastic parameters. An illustrative example is given that demonstrates the tractability of our solution method in the robust uncertain case.

Keywords H_∞ Luenberger Filters · Stochastic systems · Polytopic uncertainty

1 Introduction

We address the problem of H_∞ Luenberger-type filtering of continuous-time, state-multiplicative noisy linear systems. The multiplicative noise appears in the system model in both the dynamic matrix and in the measurement matrix and the system may encounter either norm-bounded or polytopic-type uncertainties.

The field of control and estimation of stochastic state-multiplicative noisy systems has been greatly developed since the onset of the H_∞ control theory in the early 80's (see [1, 2] and the references therein). Within a span of more than three decades, many approaches to the study of the various stochastic control and filtering problems, including those that ensure a worst case performance bound in the H_∞ sense, have been derived for both: delay-free systems [1–9] and state-delayed, linear, stochastic systems [10–14].

Delay-free systems with parameter uncertainties that are modeled as white noise processes in a linear setting have been treated in [1–9] for both the continuous-time and the discrete-time cases. Such models of uncertainty are encountered in many areas of applications such as: nuclear fission and heat transfer, population models

E. Gershon (✉)
Holon Institute of Technology, Holon, Israel
e-mail: eli_g@hit.ac.il

D. Theilliol et al. (eds.), *Recent Developments in Model-Based and Data-Driven Methods for Advanced Control and Diagnosis*, Studies in Systems, Decision and Control 467,
https://doi.org/10.1007/978-3-031-27540-1_1

and immunology. In control theory, such models are encountered in gain scheduling when the scheduling parameters are corrupted with measurement noise.

The problem of H_∞ filtering of the above systems has been addressed via the application of a Luenberger filter type or via the general-type filter where in the latter case, the solution of the corresponding robust problem has been solved for both types of the above uncertainties. The former **robust** case where the simple Luenberger filter is applied has not been tackled before, hence the rational for addressing this type of filtering. Although the use of the general-type filter accounts also for the Luenberger filter type (being a special case), the unique solution of the Luenberger-type filtering may serve as a template in the predictive control case (say, state-feedback control) where the predictor may itself be of the Luenberger type form. In this study, following the problem formulation, we bring the BRL (Bounded Real Lemma) solution for stochastic systems as a preliminary result needed for the solution of the robust Luenberger type filter. We then treat the case of norm-bounded uncertainties followed by the solution of the robust polytopic case. The theory developed is demonstrated via a simple example that shows the tractability of our solution method.

Notation: Throughout the paper the superscript 'T' stands for matrix transposition, $\mathcal{R}^n$ denotes the n dimensional Euclidean space and $\mathcal{R}^{n\times m}$ is the set of all $n \times m$ real matrices. For a symmetric $P \in \mathcal{R}^{n\times n}$, $P > 0$ means that it is positive definite. We denote expectation by $\mathcal{E}\{\cdot\}$ and we provide all spaces $\mathcal{R}^k,\ k \geq 1$ with the usual inner product $< \cdot, \cdot >$ and with the standard Euclidean norm $|| \cdot ||$. The space of vector functions that are square integrable over $[0 \quad \infty)$ is denoted by $\mathcal{L}_2$ and $col\{a,\ b\}$ implies $[a^T\ b^T]^T$. We denote by $L^2(\Omega, \mathcal{R}^k)$ the space of square-integrable $\mathcal{R}^k-$ valued functions on the probability space $(\Omega, \mathcal{F}, \mathcal{P})$, where Ω is the sample space, $\mathcal{F}$ is a σ algebra of a subset of Ω called events and $\mathcal{P}$ is the probability measure on $\mathcal{F}$. By $(\mathcal{F}_t)_{t>0}$ we denote an increasing family of σ-algebras $\mathcal{F}_t \subset \mathcal{F}$. We also denote by $\tilde{L}^2([0, T); \mathcal{R}^k)$ the space of nonanticipative stochastic process $f(\cdot) = (f(t))_{t\in[0,T]}$ in $\mathcal{R}^k$ with respect to $(\mathcal{F}_t)_{t\in[0,T)}$ satisfying

$$||f(\cdot)||^2_{\tilde{L}_2} = \mathcal{E}\{\int_0^T ||f(t)||^2 dt\} = \int_0^T \mathcal{E}\{||f(t)||^2\}dt < \infty.$$

Stochastic differential equations will be interpreted to be of $It\hat{o}$ type [1].

2 Problem Formulation

We consider the following linear system:

$$\begin{aligned} dx(t) &= Ax(t)dt + Dx(t)d\nu(t) + B_1 w(t)dt, \\ dy(t) &= C_2 x(t)dt + Fx(t)d\zeta(t) + D_{21} w(t)dt \end{aligned} \qquad (1a, b)$$

with the objective vector

$$z(t) = C_1 x(t), \tag{2}$$

where $x(t) \in \mathcal{R}^n$ is the system state vector, $w(t) \in \mathcal{R}^p$ is the exogenous disturbance signal, $y(t) \in \mathcal{R}^m$ is the measured output and $z(t) \in \mathcal{R}^r$ is the state combination (objective function signal) to be estimated. The variables $\nu(t)$ and $\zeta(t)$ are zero-mean real scalar Wiener processes that satisfy:

$$\mathcal{E}\{d\nu(t)\} = 0,\ \mathcal{E}\{d\zeta(t)\}=0,\ \mathcal{E}\{d\nu(t)^2\}=dt,$$

$$\mathcal{E}\{d\zeta(t)^2\}=dt,\ \ \mathcal{E}\{d\nu(t)d\zeta(t)\}= 0.$$

3 Preliminaries

The solution of the robust Luenberger filter is achieved in the sequel by resorting to the BRL of stochastic state-multiplicative continuous-time linear systems. This BRL was obtained by several groups (see, for example, [1], Chap. 2) and it is brought here as a preliminary result. We consider the system of (1a) and (2) and the following index of performance;

$$J_B \stackrel{\Delta}{=} ||z(t)||^2_{\tilde{L}_2} - \gamma^2[||w(t)||^2_{\tilde{L}_2}. \tag{3}$$

we obtain the following result:

Lemma 1 *Consider the system of (1a) and (2). For a prescribed scalar $\gamma > 0$, a necessary and sufficient condition for J_B of (3) to be negative for all nonzero $w(t) \in \tilde{L}^2([0,\infty); \mathcal{R}^p)$, is that there exist $0 < P \in \mathcal{R}^{n \times n}$ such that the following LMI holds:*

$$\begin{bmatrix} A^T P + PA + D^T PD & PB_1 & C_1^T \\ * & -\gamma^2 I & 0 \\ * & * & -I \end{bmatrix} < 0. \tag{4}$$

In the sequel we will solve the estimation problem under study by applying the above Lemma 1 to the estimation error as given in the next section.

4 Robust Luenberger-Type Estimator

In this section we treat the problem of designing a robust norm-bounded Luenberger-type estimator which will be utilized in latter sections. The nominal counterpart of the latter problem, can be found in [1]. Denoting

$$\Delta A = A - \bar{A} \quad \text{and} \quad \Delta C_2 = C_2 - \bar{C}_2$$

where $\bar{A}$ and $\bar{C}_2$ are the nominal values for A and C_2, respectively, we consider:

$$\begin{aligned} &d\hat{x}(t) = \bar{A}\hat{x}(t)dt + L(dy(t) - \bar{C}_2\hat{x}(t)dt) = \\ &\bar{A}\hat{x}(t)dt + L\bar{C}_2e(t)dt + L\Delta C_2x(t)dt + LFx(t)d\zeta + LD_{21}w(t)dt, \\ &\hat{z}(t) = \bar{C}_1\hat{x}(t), \end{aligned} \qquad (5a, b)$$

where

$$e(t) \stackrel{\Delta}{=} x(t) - \hat{x}(t).$$

Denoting $\bar{z}(t) = z(t) - \hat{z}(t)$, we consider the following cost function:

$$J_F \stackrel{\Delta}{=} ||\bar{z}(t)||^2_{\tilde{L}_2} - \gamma^2[||w(t)||^2_{\tilde{L}_2}. \qquad (6)$$

Given $\gamma > 0$, we seek an estimate $\bar{C}_1\hat{x}(t)$ of $C_1x(t)$ over the infinite time horizon $[0, \infty)$ such that J_F is negative for all nonzero $w(t) \in \tilde{L}^2([0, \infty); \mathcal{R}^p)$.

It is readily found that

$$de(t) = Ax(t)dt + Dx(t)d\nu(t) + B_1w(t)dt - \bar{A}\hat{x}(t)dt$$

$$-L\bar{C}_2e(t)dt - L\Delta C_2x(t)dt - LFx(t)d\zeta(t) - LD_{21}w(t)dt$$

or

$$de(t) = (\bar{A} - L\bar{C}_2)e(t)dt + \Delta Ax(t)dt - L\Delta C_2x(t)dt$$

$$+ Dx(t)d\nu(t) - LFx(t)d\zeta(t) + (B_1 - LD_{21})w(t)dt. \qquad (7)$$

Denoting $\eta(t) = col\{x(t),\ e(t)\}$ we obtain:

$$d\eta(t)=\tilde{A}\eta(t)dt+\tilde{D}\eta(t)d\nu(t)+\tilde{F}\eta(t)d\zeta(t)+\tilde{B}_1w(t)dt$$

where

$$\tilde{A}=\begin{bmatrix} \bar{A}+\Delta A & 0 \\ \Delta A-L\Delta C_2 & \bar{A}-L\bar{C}_2 \end{bmatrix},\ \tilde{D}=\begin{bmatrix} D & 0 \\ D & 0 \end{bmatrix}, \qquad (8a\text{-}e)$$

$$\tilde{F}=\begin{bmatrix} 0 & 0 \\ -LF & 0 \end{bmatrix},\ \tilde{B}_1=\begin{bmatrix} B_1 \\ B_1-LD_{21} \end{bmatrix}$$

and

$$\bar{z}(t) = \tilde{C}_1\eta(t),$$

where $\Delta C_1 = C_1 - \bar{C}_1$ and $\tilde{C}_1 = [\Delta C_1 \ \ \bar{C}_1]$. We thus arrive at the following result.

Theorem 1 *Consider the system of (1a, b) and (2). For a prescribed scalar $\gamma > 0$, a necessary and sufficient condition for J_F of (6) to be negative for all nonzero $w(t) \in \tilde{L}^2([0, \infty); \mathcal{R}^p)$, is that there exist $0 < P \in \mathcal{R}^{2n \times 2n}$, $L \in \mathcal{R}^{n \times m}$ such that the following inequality holds:*

$$\begin{bmatrix} \tilde{A}^T P + P\tilde{A} + \tilde{D}^T P \tilde{D} & P\tilde{B}_1 & \tilde{C}_1^T & \tilde{F}^T P \\ * & -\gamma^2 I & 0 & 0 \\ * & * & -I & 0 \\ * & * & * & -P \end{bmatrix} < 0. \tag{9}$$

Remark 1 In the case where there is no uncertainty, $A = \bar{A}$, $C_2 = \bar{C}_2$ and $C_1 = \bar{C}_1$, the standard result of ([1], see p. 31) is retrieved. We note that in order to extract the filter gain L, one must put a structure on the $2n \times 2n$ Lyapunov function P, as appears in the sequel. In case of uncertainty one may consider both types: norm-bounded and polytopic-type uncertainties.

In the norm-bounded case we assume that:

$$A = \bar{A} + \Delta A, \quad C_2 = \bar{C}_2 + \Delta C_2$$

$$\text{and } [\Delta A,\ \Delta C_2] = H\delta(x, t)[E_1\ \ E_2],\ \ ||\delta(x, t)|| < 1. \qquad \text{(10a-d)}$$

We then have:

$$\tilde{A} = \begin{bmatrix} \bar{A} & 0 \\ 0 & \bar{A} - L\bar{C}_2 \end{bmatrix} + \begin{bmatrix} H \\ H \end{bmatrix} \delta(x, t) \begin{bmatrix} E_1 & 0 \end{bmatrix} +$$

$$\begin{bmatrix} 0 \\ -LH \end{bmatrix} \delta(x, t) \begin{bmatrix} E_2 & 0 \end{bmatrix}.$$

Using Young's inequality ([2]) and (10d) the LMI solution readily follows. Note that for simplicity we took $\Delta C_1 = 0$.

Uncertainty of the polytopic type can also be solved for, around nominal values of A and C_2 (i.e. taking $\Delta A = 0, \quad \Delta C_2 = 0$). We assume that the system matrices in (1a, b), (2) lie within the following polytope:

$$\bar{\Omega} = Co\{\bar{\Omega}_1, \bar{\Omega}_2, ..., \bar{\Omega}_N\}, \tag{11}$$

where

$$\bar{\Omega}_i \triangleq \begin{bmatrix} A^{(i)} & B_1^{(i)} & C_1^{(i)} & C_2^{(i)} & D_{21}^{(i)} \end{bmatrix} \tag{12}$$

and where N is the number of vertices. In other words:

$$\bar{\Omega} = \sum_{i=1}^{N} \bar{\Omega}_i f_i, \quad \sum_{i=1}^{N} f_i = 1 \quad , f_i \geq 0. \tag{13}$$

Denoting, for the uncertain polytopic case,

$$\hat{A}^{(i)} = \begin{bmatrix} \bar{A}^{(i)} & 0 \\ 0 & \bar{A}^{(i)} - L\bar{C}_2^{(i)} \end{bmatrix}, \quad \hat{D}^{(i)} = \begin{bmatrix} D^{(i)} & 0 \\ D^{(i)} & 0 \end{bmatrix},$$

$$\hat{F}^{(i)} = \begin{bmatrix} 0 & 0 \\ -LF^{(i)} & 0 \end{bmatrix}, \quad \hat{B}_1^{(i)} = \begin{bmatrix} B_1^{(i)} \\ B_1^{(i)} - LD_{21}^{(i)} \end{bmatrix},$$

$i = 1, 2, ..., N$, we obtain the following result—the so-called 'quadratic' solution—by applying a single Lyapunov function over the whole uncertainty polytope:

Theorem 2 *Consider the system of (1a, b) and (2), where the system matrices lie within the polytope of (11). For a prescribed scalar $\gamma > 0$, a necessary and sufficient condition for J_F of (6) to be negative for all nonzero $w(t) \in \tilde{L}^2([0, \infty); \mathcal{R}^p)$, is that there exist $0 < P \in \mathcal{R}^{2n \times 2n}$, $L \in \mathcal{R}^{n \times m}$ such that the following set of LMIs holds:*

$$\begin{bmatrix} \Upsilon_{11}^{(i)} & P\hat{B}_1^{(i)} & \tilde{C}_1^{(i),T} & \hat{F}^{(i),T}P \\ * & -\gamma^2 I & 0 & 0 \\ * & * & -I & 0 \\ * & * & * & -P \end{bmatrix} < 0. \tag{14}$$

where

$$\Upsilon_{11}^{(i)} = \hat{A}^{(i),T}P + P\tilde{A}^{(i)} + \hat{D}^{(i),T}P\hat{D}^{(i)}$$

and where $i = 1, 2, N$. As in the case of Theorem 1 for the nominal case, in order to extract the filter gain L one must apply a special structure to the above Lyapunov function P. Taking the following structure:

$$P = \begin{bmatrix} \bar{P} & \alpha\hat{P} \\ \alpha\hat{P} & \hat{P} \end{bmatrix}, \quad \alpha > 0,$$

we obtain the following result, denoting $Y = \hat{P}L$,

Lemma 2 *Consider the system of (1a, b) and (2) where the system matrices lie within the polytope of (11). For a prescribed scalar $\gamma > 0$ and a given tuning parameter $\alpha > 0$, J_F of (6) is negative for all nonzero $w(t) \in \tilde{L}^2([0, \infty); \mathcal{R}^p)$, if there exist $0 < \bar{P} \in \mathcal{R}^{n \times n}$, $0 < \hat{P} \in \mathcal{R}^{n \times n}$, $Y \in \mathcal{R}^{n \times m}$ such that the following set of LMIs hold:*

$$\begin{bmatrix} \hat{\Upsilon}_{11}^{(i)} & P\hat{B}_1^{(i)} & \tilde{C}_1^{(i),T} & \hat{F}^{(i),T}P & \hat{D}^{(i),T}P \\ * & -\gamma^2 I & 0 & 0 & 0 \\ * & * & -I & 0 & 0 \\ * & * & * & -P & 0 \\ * & * & * & * & -P \end{bmatrix} < 0 \tag{15}$$

for $i = 1, 2, ...N$ *where* $\hat{\Upsilon}_{11}^{(i)} =$

$$\begin{bmatrix} \bar{P}\bar{A}^{(i)} + \bar{A}^{(i),T}\bar{P} & \alpha[\hat{P}\bar{A}^{(i)} - Y\bar{C}_2^{(i)} - \bar{A}^{(i),T}\hat{P}] \\ * & \hat{P}\bar{A}^{(i)} - Y\bar{C}_2^{(i)} + \bar{A}^{(i),T}\hat{P} - \bar{C}_2^{(i),T}Y^T \end{bmatrix}.$$

5 Example

We consider the system of (1a, b) and (2) with

$$A = \begin{bmatrix} 0 & 1 \pm a \\ -1 & 0.4 \end{bmatrix}, \quad D = \begin{bmatrix} 0 & 0.3 \\ 0 & -0.12 \end{bmatrix},$$

$$B_1 = \begin{bmatrix} 1 \\ -1 \end{bmatrix} \text{ and } C_1 = \begin{bmatrix} -0.5 & 1 \\ 0 & 0 \end{bmatrix},$$

where $C_2 = [0.1 \ \ 0.1]$ and $a = 0.2$. A Luenberger-type filter is sought that stabilizes the system and achieves a negative J_F for a minimum γ. Considering the nominal case (i.e. $a = 0$) and applying the result of Lemma 2, where a single vertex is solved, we find that the error system can be stabilized for $\alpha = 0.1$ for a near minimum of $\gamma = 5.72$, where the following block matrices of the Lyapunov function are obtained:

$$\bar{P} = \begin{bmatrix} 3.43 & 0.52 \\ 0.52 & 2.94 \end{bmatrix}, \quad \hat{P} = \begin{bmatrix} 2.78 & 0.26 \\ 0.26 & 1.71 \end{bmatrix}.$$

Solving for the robust uncertain case, we apply the result of Lemma 2 (for two vertices) and obtain a near minimum of $\gamma = 7.54$.

6 Conclusions

The theory of linear H_∞ robust Luenberger-type filtering is developed for continuous-time systems, where the stochastic uncertainty is encountered in both the dynamic and the measurement matrices of the state space model of the system. Two uncertain strategies are accounted for where the matrices uncertainties are either norm-bounded or of the polytopic-type nature. In both cases, the solution is obtained by applying the BRL result of stochastic state-multiplicative noisy systems to the error system. In the norm-bounded case a single LMI is obtained, whereas a simple and readily tractable set of LMIs is obtained for the polytopic case.

An inherent over-design is admitted to the polytopic-type solution due to the special structure imposed on the Lyapunov function. The numerical example demonstrates the efficiency and tractability of the proposed results.

References

1. Gershon, E., Shaked, U., Yaesh, I.: H_∞ control and Estimation of State-multiplicative linear systems. Lecture Notes in Control and Information Sciences, vol. 318. LNCIS, Springer (2005)
2. Gershon, E., Shaked, U.: Advanced topics in control and estimation of state-multiplicative noisy systems. Lecture Notes in Control and Information Sciences, vol. 439. LNCIS, Springer (2013)
3. Chen, B.S., Zhang, W.: Stochastic H_2/H_∞ control with state-dependent noise. IEEE Trans. Autom. Control **49**(1), 45–57 (2004)
4. Costa, O.L.V., Kubrusly, C.S.: State-feedback H_∞ -control for discrete-time infinite-dimensional stochastic bilinear systems. J. Math. Syst. Estim. Control **6**, 1–32 (1996)
5. Willems, J.L., Willems, J.C.: Feedback stabilizability for stochastic systems with state and control dependent noise. Automatica **12**, 277–283 (1976)
6. Dragan, V., Morozan, T., Halanay, A.: Optimal stabilizing compensator for linear systems with state-dependent noise. Stoch. Anal. Appl. **10**, 557–572 (1992)
7. Dragan, V., Halanay, A., Stoica, A.: A small gain theorem for linear stochastic system. Syst. Control Lett. **30**, 243–251 (1997)
8. Hinriechsen, D., Pritchard, A.J.: Stochastic H_∞. SIAM J. Control Optim. **36**(5), 1504–1538 (1998)
9. Dragan, V., Stoica, A.: A γ attenuation problem for discrete-time time-varying stochastic systems with multiplicative noise. Reprint series of the Institute of Mathematics of the Romanian Academy, vol. 10 (1998)
10. Florchinger, P., Verriest, E.I.: Stabilization of nonlinear stochastic systems with delay feedback. In: Proceedings of the 32nd IEEE Conferences on Decision and Control, San Antonio, TX, pp. 859–860, Dec. 1993
11. Chen, W.H., Guan, Z., Lu, X.: Delay-dependent exponential stability of uncertain stochastic systems with multiple delays: an LMI approach. Syst. Control Lett. **54**, 547–555 (2005)
12. Gershon, E., Shaked, U.: Robust H_∞ output-feedback control of state-multiplicative stochastic systems with delay. Int. J. Robust Nonlinear Control **21**(11), 1283–1296 (2011)
13. Xu, S., Lam, J., Chen, T.: Robust H_∞ control for uncertain discrete stochastic time-delay systems. Syst. Control Lett. **51**, 203–215 (2004)
14. Yue, D., Tian, E., Zhang, Y.: Stability analysis of discrete systems with stochastic delay and its applications. Int. J. Innov. Comput. Inf. Control **5**(8), 2391–2403 (2009)

Anticipating the Loss of Unknown Input Observability for Sampled LPV Systems

Gustave Bainier, Jean-Christophe Ponsart, and Benoît Marx

Abstract Given a continuous-time Linear Parameter-Varying (LPV) system with a sampled scheduling parameter and subject to an unknown input, this paper provides—under some Lipschitz assumptions—an exact discretization of an extended system which translates the sampled-data unknown input estimation problem into a discrete-time LPV observer design problem with norm-bounded uncertainties. The bounds developed in this process account for the inter-sample behavior of the scheduling parameter, and allow for an estimation of some near-future observability Gramians, from which it is possible to lower bound the number of samples for which the unknown input is guaranteed to remain observable.

1 Introduction

The continuous-time Linear Parameter-Varying (LPV) representation is a powerful tool to study a large class of nonlinear systems using a linear-like framework, which generally facilitates controller and observer synthesis [16, 18]. It relies on the construction of a scheduling parameter θ typically accounting for the non-linearities of the initial system, and which is assumed to be continuously known or estimated in real time. However, in practice, the values of θ are generally recovered at a fixed sampling rate and are assumed to be held constant between two samples [20]. This zero-order hold (ZOH) assumption made on the values of θ creates a sampled-data problem which has already been discussed in the LPV literature [8, 15, 19].

The present paper considers a continuous-time LPV system subject to an unknown input. Taking into consideration the sampled-data problem described above, an exact discretization of an extended version of the system is given with consideration for the uncertainties due to the ZOH assumption made on θ. This discretization effectively translates the sampled-data unknown input estimation problem into a discrete-time LPV observer design problem with norm-bounded uncertainties. To the authors'

G. Bainier (✉) · J.-C. Ponsart · B. Marx
Université de Lorraine, CNRS, CRAN, 54000 Nancy, France
e-mail: gustave.bainier@univ-lorraine.fr

D. Theilliol et al. (eds.), *Recent Developments in Model-Based and Data-Driven Methods for Advanced Control and Diagnosis*, Studies in Systems, Decision and Control 467,
https://doi.org/10.1007/978-3-031-27540-1_2

knowledge, this is a new approach in the sampled-data literature. The exhibited bounds on the uncertainties rely on some Lipschitz assumptions, which are similar to the usual bounds put on the variations of θ that are typically found in the literature of LPV systems. These bounds constitute a generalization of the ones found in [1], the idea behind their proofs being similar.

Following the ideas of [1], the norm-bounded uncertainties of the exactly discretized extended-system allow for some real-time structural analysis. Specifically, from the last known value of θ, the observability Gramian of the extended discrete-time LPV system can be estimated between two future instants, as if the values of θ were known in advance. In practice, this Gramian-estimation can be used to evaluate in real time a number of samples for which the unknown input is guaranteed to remain observable. To the authors' knowledge, this is an uninvestigated question of the LPV literature. Some results on the robust observability of discrete time varying systems could arguably be leveraged to obtain similar bounds [14, 17], but the approach of this paper is more straightforward, the uncertainties being treated here as "yet to be known" values.

The paper is organised as follows: Sect. 2 introduces the notations used throughout the paper. Section 3 provides technical results, allowing for a smoother reading of the next two sections. Section 4 translates the sampled-data unknown input estimation problem into a discrete-time LPV observer design problem with norm-bounded uncertainties. Section 5 estimates the near-future observability Gramians of the previously constructed system in order to lower-bound the number of samples left until the unknown input observability of the initial system may be lost. Section 6 finally provides an illustrative example. Some perspectives are discussed in Sect. 7 to conclude the paper.

2 Notations

$\mathbb{N}$ is the set of natural numbers. $\mathbb{K}$ stands for $\mathbb{R}$ or $\mathbb{C}$, resp. the set of real and complex numbers. $\mathbb{K}^{p\times q}$ denotes the set of matrices with p rows, q columns and coefficients in $\mathbb{K}$. $\|\cdot\|$ stands for the spectral norm. $\mu : \mathbb{K}^{p\times p} \to \mathbb{R}$ denotes the logarithmic norm induced by the spectral norm. Given a matrix $E \in \mathbb{K}^{p\times p}$, E^* denotes its conjugate transpose, E positive-definite is denoted $E \succ 0$, and by definition $\mu(E) = \lim_{h\to 0^+} \frac{\|I_p+hE\|-1}{h}$ where $I_p \in \mathbb{K}^{p\times p}$ stands for the identity matrix [6]. Given the following continuous-time system $\dot{x}(t) = A(t)x(t)$ with $A : \mathbb{R} \to \mathbb{K}^{n\times n}$ a continuous function, the state transition matrix from t_1 to t_2 of such a system is often abstractly denoted $\Phi(t_2, t_1)$. However, this paper considers the state-transition matrix to be a product integral, which is denoted in the following way [1, 4, 21]:

$$\Phi(t_2, t_1) = \prod_{t_1}^{t_2} e^{A(s)\mathrm{d}s} \tag{1}$$

3 Preliminary Results

Among the many properties of the logarithmic norm and of the product integral, several of them play a crucial role in this document.

Lemma 1 (Duhamel's Formula) *Let $E, F : \mathbb{R} \to \mathbb{K}^{p\times p}$ be two continuous functions. For all $t_1, t_2 \in \mathbb{R}$*

$$\prod_{t_1}^{t_2} e^{E(s)ds} - \prod_{t_1}^{t_2} e^{F(s)ds} = \int_{t_1}^{t_2} \left(\prod_{u}^{t_2} e^{F(s)ds}\right)(E(u) - F(u))\left(\prod_{t_1}^{u} e^{E(s)ds}\right) du \quad (2)$$

Proof See Theorem 5.1 of [4]. □

Lemma 2 *Let $E : \mathbb{R} \to \mathbb{K}^{p\times p}$ be a continuous function. For all $t_1, t_2 \in \mathbb{R}$ such that $t_1 \leq t_2$*

$$\left\|\prod_{t_1}^{t_2} e^{E(s)ds}\right\| \leq e^{\int_{t_1}^{t_2} \mu(E(s))ds} \quad (3)$$

Proof This result is a consequence of Grönwall's lemma [3]. □

Lemma 3 *If $E : \mathbb{R} \to \mathbb{K}^{p\times p}$ is a L_E-Lipschitz function, then $\mu(E(\cdot))$ is also a L_E-Lipschitz function.*

Proof For $t_1, t_2 \in \mathbb{R}$

$$\begin{aligned} \mu(E(t_1)) &= \mu(E(t_2) + E(t_1) - E(t_2)) \\ &\leq \mu(E(t_2)) + \mu(E(t_1) - E(t_2)) \\ \Rightarrow \mu(E(t_1)) - \mu(E(t_2)) &\leq \mu(E(t_1) - E(t_2)) \leq \|E(t_1) - E(t_2)\| \end{aligned} \quad (4)$$

interverting the role of t_1 and t_2 in the previous equations yields the same upper-bound. Since E is L_E-Lipschitz, combining (4) with its interverted counterpart provides

$$|\mu(E(t_1)) - \mu(E(t_2))| \leq \|E(t_1) - E(t_2)\| \leq L_E\,|t_1 - t_2| \quad (5)$$

which concludes the proof. □

Lemma 4 *Let $E : \mathbb{R} \to \mathbb{K}^{p\times p}$ be a L_E-Lipschitz function with for all $t \in \mathbb{R}$, $\mu(E(t)) \leq \sigma$ and let τ be defined by $\tau = t_0 + \frac{\sigma - \mu(E(t_0))}{L_E}$. For all $t_0, t_1, t_2 \in \mathbb{R}$ such that $t_0 \leq t_1 \leq t_2$*

$$\left\|\prod_{t_1}^{t_2} e^{E(s)ds}\right\| \leq \begin{cases} e^{(t_2-t_1)\sigma} & \text{if } \tau \leq t_1 \\ e^{t_2\sigma - t_1(\mu(E(t_0)) - \frac{L_E}{2}(2t_0 - t_1)) - \frac{1}{2L_E}(\sigma - \mu(E(t_0)) + L_E t_0)^2} & \text{if } \tau \in (t_1, t_2) \\ e^{(t_2-t_1)\mu(E(t_0)) + \frac{1}{2}L_E\left(t_2^2 - t_1^2 + 2t_0(t_1 - t_2)\right)} & \text{if } \tau \geq t_2 \end{cases} \quad (6)$$

Proof Lemma 3 provides

$$\int_{t_1}^{t_2} \mu(E(s))\mathrm{d}s \leq \int_{t_1}^{t_2} \min(\mu(E(t_0) + L_E(s - t_0), \sigma)\mathrm{d}s \tag{7}$$

and since $s \mapsto \mu(E(t_0)) + L_E(s - t_0)$ is increasing, it follows

$$\begin{aligned}(7) &\leq \min_{s\in(t_1,t_2)} (s - t_1)\mu(E(t_0)) + \frac{1}{2}L_E(s^2 - 2t_0 s - t_1^2 + 2t_0 t_1) + (t_2 - s)\sigma \\ &= \min_{s\in(t_1,t_2)} as^2 + bs + c\end{aligned} \tag{8}$$

The minimum of $as^2 + bs + c$ being reached for $\tau = -b/2a$, the sharpest upper-bound of (7) depends on whether τ belongs to (t_1, t_2) or not. The results are then applied to Lemma 2, providing (6) and thus concluding the proof. □

Remark 1 The first and last upper-bounds of (6) stay true for all values of τ, contrary to the second upper-bound which only holds when $\tau \in (t_1, t_2)$.

Lemma 5 *If $E : \mathbb{R} \to \mathbb{K}^{p\times q}$ and $F : \mathbb{R} \to \mathbb{K}^{p\times r}$ are L_E and L_F-Lipschitz functions respectively, then for all $k \in \mathbb{N}$, the following function is $(L_E + L_F)$-Lipschitz*

$$\begin{aligned} G_k : \mathbb{R} &\to \mathbb{K}^{(q+r+k)\times(q+r+k)} \\ t &\mapsto \begin{bmatrix} E(t) & F(t) & 0 \\ 0 & 0 & I_k \\ 0 & 0 & 0 \end{bmatrix}\end{aligned} \tag{9}$$

Proof Given $t_1, t_2 \in \mathbb{R}$, the following inequalities hold

$$\begin{aligned} &\|G_k(t_1) - G_k(t_2)\| = \left\|G_k^*(t_1) - G_k^*(t_2)\right\| = \sup_{\|x\|=1} \left\| \begin{bmatrix} E^*(t_1) - E^*(t_2) & 0 \\ F^*(t_1) - F^*(t_2) & 0 \\ 0 & 0 \end{bmatrix} x \right\| \\ &\leq \sup_{\|x_1\|=1} \left\|\begin{bmatrix} E^*(t_1) - E^*(t_2) & 0 \end{bmatrix} x_1\right\| + \sup_{\|x_2\|=1} \left\|\begin{bmatrix} F^*(t_1) - F^*(t_2) & 0 \end{bmatrix} x_2\right\| \\ &= \|E(t_1) - E(t_2)\| + \|F(t_1) - F(t_2)\| \leq (L_E + L_F)\,|t_1 - t_2| \end{aligned} \tag{10}$$

which concludes the proof. □

Lemma 6 *Let $E : \mathbb{R} \to \mathbb{K}^{p\times p}$ be a L_E-Lipschitz function with for all $t \in \mathbb{R}$, $\mu(E(t)) \leq \sigma$. For all $t_0, t_1, t_2 \in \mathbb{R}$ such that $t_0 \leq t_1 \leq t_2$.*

$$\left\| \prod_{t_1}^{t_2} e^{E(s)ds} - e^{(t_2-t_1)E(t_0)} \right\| \leq \min \left\{ \begin{array}{c} \frac{1}{2}L_E\left(t_2^2 - t_1^2 + 2t_0(t_1 - t_2)\right) e^{(t_2-t_1)\sigma}, \\ \left(e^{\frac{1}{2}L_E\left(t_2^2 - t_1^2 + 2t_0(t_1 - t_2)\right)} - 1\right) e^{(t_2-t_1)\mu(E(t_0))} \end{array} \right\} \tag{11}$$

Proof The proof of the first inequality can be found in [1]. The second inequality only relies on the Lipschitz assumption on E. Given $t_0, t_1, t_2 \in \mathbb{R}$ such that $t_0 \leq t_1 \leq t_2$, Lemma 1 provides

$$\begin{aligned}
\prod_{t_1}^{t_2} e^{E(s)\mathrm{d}s} - e^{(t_2-t_1)E(t_0)} &= \prod_{t_1}^{t_2} e^{E(s)\mathrm{d}s} - \prod_{t_1}^{t_2} e^{E(t_0)\mathrm{d}s} \\
&= \int_{t_1}^{t_2} \left(\prod_{u}^{t_2} e^{E(t_0)\mathrm{d}s}\right) (E(u) - E(t_0)) \left(\prod_{t_1}^{u} e^{E(s)\mathrm{d}s}\right) \mathrm{d}u \\
&= \int_{t_1}^{t_2} e^{(t_2-u)E(t_0)} (E(u) - E(t_0)) \left(\prod_{t_1}^{u} e^{E(s)\mathrm{d}s}\right) \mathrm{d}u
\end{aligned} \tag{12}$$

then, by submultiplicativity of the operator norm and the Lipschitz asumption on E

$$\| (12) \| \leq L_E \int_{t_1}^{t_2} e^{(t_2-u)\mu(E(t_0))} (u - t_0) \left\| \prod_{t_1}^{u} e^{E(s)\mathrm{d}s} \right\| \mathrm{d}u \tag{13}$$

the third upper-bound of Lemma 4 is finally applied, providing

$$\begin{aligned}
\| (12) \| &\leq L_E \left(\int_{t_1}^{t_2} (u - t_0) e^{\frac{1}{2} L_E (u^2 - 2t_0 u)} \mathrm{d}u \right) e^{-\frac{1}{2} L_E (t_1^2 - 2t_0 t_1) + (t_2 - t_1)\mu(E(t_0))} \\
&= L_E \left(\frac{1}{L_E} e^{\frac{1}{2} L_E t_2 (t_2 - 2t_0)} - \frac{1}{L_E} e^{\frac{1}{2} L_E t_1 (t_1 - 2t_0)} \right) e^{-\frac{1}{2} L_E t_1 (t_1 - 2t_0) + (t_2 - t_1)\mu(E(t_0))} \\
&= \left(e^{\frac{1}{2} L_E (t_2^2 - t_1^2 + 2t_0 (t_1 - t_2))} - 1 \right) e^{(t_2 - t_1)\mu(E(t_0))}
\end{aligned} \tag{14}$$

which concludes the proof. □

Remark 2 If $t_1 < \tau < t_2$, with τ given in Lemma 4, a sharper (but very involved) upper-bound to (11) can be found.

4 An Exact Discretization of Sampled LPV Systems Subject to an Unknown Input

This section deals with the state and unknown input estimation of a sampled LPV system of the following form

$$\begin{aligned}
\dot{x}(t) &= A(\theta(t))x(t) + F_1(\theta(t)) f(t) \\
y(t) &= C(t)x(t) + F_2(t) f(t)
\end{aligned} \tag{15}$$

where $x(t) \in \mathbb{K}^{n_x}$ is the state of the system, $y(t) \in \mathbb{K}^{n_y}$ is its output, $f(t) \in \mathbb{K}^{n_f}$ is its unknown input and $\theta(t) \in \mathbb{K}^{n_\theta}$ is a sampled scheduling parameter measured in real time. Note that the coefficients of $C(t)$ and $F_2(t)$ are assumed to be known at all time t, hence, these matrices are not θ-dependent. This section provides a discrete representation of this system which takes into account the intrinsic uncertainties that come with the sampling of θ.

Polynomial Unknown Input Assumption: f is a polynomial function of degree $q-1$ (hence for all t, the q-th derivative of $f(t)$ is null).

A common way [5, 9, 23] to reconstruct the unknown input $f(t)$ of (15) consists in finding an observer for the augmented LPV system

$$\begin{aligned} \dot{z}(t) &= \overline{A}(\theta(t))z(t) \\ y(t) &= \overline{C}(t)z(t) \end{aligned} \tag{16}$$

$$\text{with} \quad z(t) = \begin{bmatrix} x(t) \\ f(t) \\ f^{(1)}(t) \\ \vdots \\ f^{(q-1)}(t) \end{bmatrix} \quad \overline{A}(\theta(t)) = \begin{bmatrix} A(\theta(t)) & F_1(\theta(t)) & 0 & \dots & 0 \\ 0 & 0 & I_{n_f} & \dots & 0 \\ \vdots & \vdots & \vdots & \ddots & \vdots \\ 0 & 0 & 0 & \dots & I_{n_f} \\ 0 & 0 & 0 & \dots & 0 \end{bmatrix}, \tag{17}$$

$$\overline{C}(t) = \begin{bmatrix} C(t) & F_2(t) & 0 \dots 0 \end{bmatrix} \tag{18}$$

If the evolution of θ were exactly known between t_1 and t_2, it would be possible to perfectly find $z(t_2)$ based on the values of $z(t_1)$ and θ through the relation $z(t_2) = \left(\prod_{t_1}^{t_2} e^{\overline{A}(\theta(s))\mathrm{d}s}\right) z(t_1)$.

Sampling Assumption: In practice, θ is measured at a sampling period T_s, and its values are not only unknown in the future, but also unknown in-between two samples.

For all $k, m, n \in \mathbb{N}$ with $n \geq m$, $\theta(kT_s)$ now represents the last known value of θ. The exact discretization of (16) accounting for these future and inter-sample uncertainties is given by

$$\begin{aligned} z_{k+n} &= \left(e^{(n-m)T_s\overline{A}(\theta_k)} + \Delta_{k,n,m}\right) z_{k+m} \\ y_{k+m} &= \overline{C}_{k+m} z_{k+m} \end{aligned} \tag{19}$$

where $\Delta_{k,n,m} \in \mathbb{K}^{n_z \times n_z}$ stands for a norm-bounded matrix of uncertainties (with $n_z = n_x + qn_f$), and where for all $p \in \mathbb{N}$, $z_p = z(pT_s)$, $y_p = y(pT_s)$, $\theta_p = \theta(pT_s)$ and $\overline{C}_p = \overline{C}(pT_s)$.

Remark 3 As a side note, notice how the peculiar structure of $\overline{A}(\theta_k)$ provides for all $s \in \mathbb{R}$

$$e^{s\overline{A}(\theta_k)} = \begin{bmatrix} e^{sA(\theta_k)} & \int_0^s e^{(s-u)A(\theta_k)} \begin{bmatrix} F_1(\theta_k) & 0 \dots 0 \end{bmatrix} e^{uN} \mathrm{d}u \\ 0 & e^{sN} \end{bmatrix} \tag{20}$$

with $N \in \mathbb{K}^{qn_f \times qn_f}$ the nilpotent matrix $N = \begin{bmatrix} 0 & I_{(q-1)n_f} \\ 0 & 0 \end{bmatrix}$.

If $A(\theta(\cdot))$ and $F_1(\theta(\cdot))$ are assumed to be L_A-Lipschitz and L_{F_1}-Lipschitz respectively—which is a similar assumption to the usual bounds put on the variations of θ that are typically found in the literature of LPV systems—then the following bounds are easily deduced from Lemmas 5 and 6

$$\left\| \Delta_{k,n,m} \right\| \leq \left(e^{\frac{1}{2}\left(L_A + L_{F_1}\right)\left(n^2 - m^2 + 2k(m-n)\right)T_s^2} - 1 \right) e^{(n-m)T_s \mu(\overline{A}(\theta_k))} \tag{21}$$

Moreover, if the logarithmic norm of $\overline{A}(\theta(\cdot))$ is also upper-bounded by $\sigma \in \mathbb{R}$—which is a similar assumption to the usual bounds put on the values of θ (see Lemma 1 of [10])—then the following bounds are also verified

$$\left\| \Delta_{k,n,m} \right\| \leq \frac{1}{2} \left(L_A + L_{F_1} \right) \left(n^2 - m^2 + 2k(m-n) \right) T_s^2 e^{(n-m)T_s \sigma} \tag{22}$$

Depending on the context, one of the previous upper-bounds might be sharper than the other. One may also bound $\Delta_{k,n,m}$ inside a polytope, or more specifically between two matrices in a component-wise approach, which can be achieved by shifting the Lipschitz assumption from $A(\theta(\cdot))$ and $F_1(\theta(\cdot))$ to $\theta(\cdot)$ directly.

Finally, to reconstruct the unknown input $f(t)$ while taking into account the intersample behavior of θ, it is possible to build a robust observer [7, 11–13, 22] for the discrete-time uncertain LPV system (19) taken with $n = 1$ and $m = 0$.

5 Anticipating the Loss of Unknown Input Observability

In this section, the possible loss of unknown input observability of the system (15) is anticipated using a sufficient condition of finite-time observability for (19) in a near future. This sufficient condition is based on upper-bounding, in terms of spectral norm, the difference between the observability Gramian of (19) and the observability Gramian of (19) taken without its uncertainties (i.e. with $\Delta_{k,n,m} = 0$). These Gramians are denoted W_k^{Δ} and W_k respectively, where θ_k still stands for the last known value of the scheduling parameter. On one hand, W_k^{Δ} evaluated between $(k+m)T_s$ and $(k+n)T_s$ can be expressed by:

$$\begin{aligned} W_{k,n,m}^{\Delta} &= \sum_{r=m}^{n-1} \left(e^{(r-m)T_s \overline{A}^*(\theta_k)} + \Delta_{k,r,m}^* \right) \overline{C}_{k+r}^* \overline{C}_{k+r} \left(e^{(r-m)T_s \overline{A}(\theta_k)} + \Delta_{k,r,m} \right) \\ &= \sum_{r=m}^{n-1} \left(\prod_{(k+m)T_s}^{(k+r)T_s} e^{\overline{A}(\theta(s))\mathrm{d}s} \right)^* \overline{C}_{k+r}^* \overline{C}_{k+r} \left(\prod_{(k+m)T_s}^{(k+r)T_s} e^{\overline{A}(\theta(s))\mathrm{d}s} \right) \end{aligned} \tag{23}$$

hence, $W^{\Delta}_{k,n,m}$ can only be fully computed if $\theta(t)$ is perfectly known for all $t \in [(k+m)T_s, (k+n)T_s]$, that is to say: in the future, and without sampling. On the other hand, W_k evaluated between $(k+m)T_s$ and $(k+n)T_s$ can be easily computed using:

$$W_{k,n,m} = \sum_{r=m}^{n-1} e^{(r-m)T_s\overline{A}^*(\theta_k)} \overline{C}^*_{k+r} \overline{C}_{k+r} e^{(r-m)T_s\overline{A}(\theta_k)} \tag{24}$$

The augmented system (19) is observable between $(k+m)T_s$ and $(k+n)T_s$ if and only if $W^{\Delta}_{k,n,m} \succ 0$ [2]. Moreover, following the ideas of [1], if the error $e_{k,n,m} = \|(W^{\Delta} - W)_{k,n,m}\|$ is upper-bounded by a constant $M_{k,n,m} \in \mathbb{R}$, then

$$W_{k,n,m} - M_{k,n,m} I_{n_z} \succ 0 \Rightarrow W^{\Delta}_{k,n,m} \succ 0 \tag{25}$$

Hence, $W_{k,n,m} - M_{k,n,m} I_{n_z} \succ 0$ is a sufficient condition of unknown input observability for (15) between $(k+m)T_s$ and $(k+n)T_s$. The upper-bound $M_{k,n,m}$ is easily deduced from Lemmas 4 to 6, and the following inequality [1]:

$$e_{k,n,m} \leq \sum_{r=m+1}^{n-1} \left(\left\| \prod_{(k+m)T_s}^{(k+r)T_s} e^{\overline{A}(\theta(s))\mathrm{d}s} \right\| + e^{(r-m)T_s \mu(\overline{A}(\theta_k))} \right) \|\Delta_{k,r,m}\| \, \|\overline{C}_{k+r}\|^2 \tag{26}$$

This approach generalizes the upper-bound found in [1], and extends the resulting structural analysis [1] to unknown input observability of sampled LPV systems.

Following from the previous results, a lower-bound to the number of samples for which the unknown input is guatanteed to remain observable is given by

$$m^*(\theta_k) = \max\{m \in \mathbb{N} : \exists n > m,\, W_{k,n,m} - M_{k,n,m} I_{n_z} \succ 0\} \tag{27}$$

where θ_k is still assumed to be the last known value of θ.

6 Illustrative Example

The following second order unknown input LPV system is considered

$$\begin{aligned} \dot{x}(t) &= \begin{bmatrix} -10 & 0 \\ \theta_2(t) & -15 \end{bmatrix} x(t) + \begin{bmatrix} \theta_1(t) \\ 0 \end{bmatrix} f(t) \\ y(t) &= \begin{bmatrix} 4 & 1 \\ 1 & 5 \end{bmatrix} x(t) \end{aligned} \tag{28}$$

where $\theta \in [0, 60] \times [-60, 60]$ is assumed to be a scheduling vector (with $L_A = 2$, $L_{F_1} = 3$) sampled at $T_s = 0.03s$, and the unknown input is assumed to be constant, hence $f^{(1)}(t) = 0$. The values of $m^*(\theta_k)$ for (28) with $k = 0$ are plotted on Fig. 1.

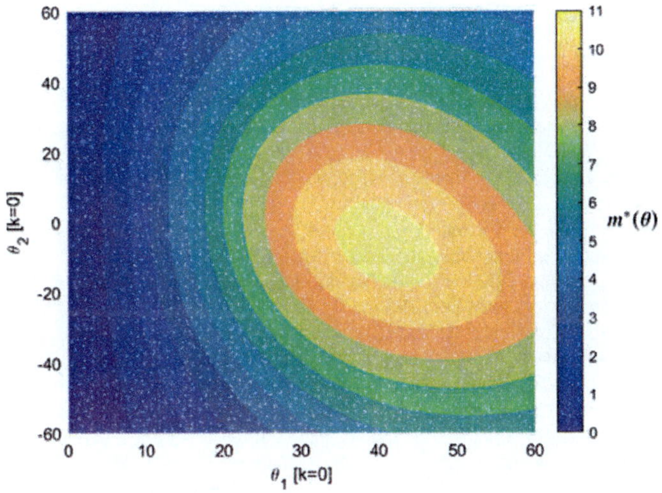

Fig. 1 $m^*(\theta)$, a lower-bound to the number of samples for which the unknown input observability of (28) is guatanteed

Despite the conservativeness of the bounds used to compute $m^*(\theta_k)$, some values of θ guarantee the observability of the unknown input for at least the next 11 samples. These results are encouraging since they were computed without consideration for the structure of the extended system $\overline{A}(\theta_k)$, which tends to have a large logarithmic norm $\mu(\overline{A}(\theta_k))$.

7 Conclusion and Perspectives

In this paper, given a continuous-time LPV system with a sampled scheduling parameter and an unknown input, and under a Lipschitz assumption, the sampled-data unknown input estimation problem has been translated into a discrete-time LPV robust observer design problem. The bounds that were developed in the process allowed for an estimation of some near-future observability Gramians, from which a lower-bound to the number of samples for which the unknown input is guatanteed to remain observable was exhibited. The obtained results could be further enhanced by taking into consideration the structure of the extended system $\overline{A}(\theta_k)$, in particular by using Eq. (20) in the computation of the Gramian estimation error. Moreover, the extension of this work to sampled systems with a non-constant sampling period remains to be investigated.

References

1. Bainier, G., Marx, B., Ponsart, J.-C.: Bounding the trajectories of continuous-time LPV systems with parameters known in real time. In: 5th IFAC Workshop on Linear Parameter Varying Systems (LPVS) (2022)
2. Callier, F.M., Desoer, C.A.: Controllability and Observability. The Discrete-Time Case, pp. 265–294. Springer, New York, NY (1991)
3. Desoer, C., Haneda, H.: The measure of a matrix as a tool to analyze computer algorithms for circuit analysis. IEEE Trans. Circuit Theory **19**(5), 480–486 (1972)
4. Dollard, J.D., Friedman, C.N.: Product Integration with Application to Differential Equations. Cambridge University Press, Dec. 1984
5. Gao, Z., Ding, S.X., Ma, Y.: Robust fault estimation approach and its application in vehicle lateral dynamic systems. Optim. Control Appl. Methods **28**(3), 143–156 (2007)
6. Germund, D.: Stability and error bounds in the numerical integration of ordinary differential equations. Kungl. tekniska hogskolans Handlingar. Almqvist & Wiksells, Uppsala (1959)
7. Han, W., Wang, Z., Shen, Y.: H-/L∞ fault detection observer for linear parameter-varying systems with parametric uncertainty. Int. J. Robust Nonlinear Control **29**(10), 2912–2926 (2019)
8. Hetel, L., Fiter, C., Omran, H., Seuret, A., Fridman, E., Richard, J.-P., Niculescu, S.I.: Recent developments on the stability of systems with aperiodic sampling: an overview. Automatica **76**, 309–335 (2017)
9. Ichalal, D., Marx, B., Ragot, J., Maquin, D.: Simultaneous state and unknown inputs estimation with PI and PMI observers for Takagi Sugeno model with unmeasurable premise variables. In: 2009 17th Mediterranean Conference on Control and Automation. IEEE, June 2009
10. Jungers, M., Deaecto, G.S., Geromel, J.C.: Bounds for the remainders of uncertain matrix exponential and sampled-data control of polytopic linear systems. Automatica **82**, 202–208 (2017)
11. Khan, A., Xie, W., Zhang, B., Liu, L.-W.: A survey of interval observers design methods and implementation for uncertain systems. J. Frankl. Inst. **358**(6), 3077–3126 (2021)
12. Li, J., Wang, Z., Shen, Y., Wang, Y.: Interval observer design for discrete-time uncertain Takagi-Sugeno fuzzy systems. IEEE Trans. Fuzzy Syst. **27**(4), 816–823 (2019)
13. Marx, B., Ichalal, D., Ragot, J.: Interval state estimation for uncertain polytopic systems. Int. J. Control **93**(11), 2564–2576 (2019)
14. Moheimani, S.R., Savkin, A., Petersen, I.: Necessary and sufficient conditions for robust observability of a class of discrete-time uncertain systems. In: Proceedings of 35th IEEE Conference on Decision and Control. IEEE
15. Ramezanifar, A., Mohammadpour, J., Grigoriadis, K.: Sampled-data control of LPV systems using input delay approach. In: IEEE Conference on Decision and Control. IEEE, Dec. 2012
16. Rugh, W.J., Shamma, J.S.: Research on gain scheduling. Automatica **36**(10), 1401–1425 (2000)
17. Seo, J., Chung, D., Park, C., Jee, G.-I.: The robustness of controllability and observability for discrete linear time-varying systems with norm-bounded uncertainty. IEEE Trans. Autom. Control **50**, 1039 – 1043, 08 (2005)
18. Shamma, J.S., Athans, M.: Guaranteed properties of gain scheduled control for linear parameter-varying plants. Automatica **27**(3), 559–564 (1991)
19. Tan, K., Grigoriadis, K., Wu, F.: Output-feedback control of LPV sampled-data systems. In: Proceedings of the American Control Conference. ACC. IEEE (2000)
20. Tóth, R., den Hof, P.V., Heuberger, P.: Discretisation of linear parameter-varying state-space representations. IET Control Theory Appl. **4**(10), 2082–2096 (2010)
21. Volterra, V., Hostinský, B.: Opérations infinitésimales linéaires, applications aux équations différentielles et fonctionnelles. Gauthier-Villars (1938)

22. Vu, V.-P., Wang, W.-J., Zurada, J.M., Chen, H.-C., Chiu, C.-H.: Unknown input method based observer synthesis for a discrete time uncertain T-S fuzzy system. IEEE Trans. Fuzzy Syst. **26**(2), 761–770 (2018)
23. Youssef, T., Karimi, H.R., Chadli, M.: Faults diagnosis based on proportional integral observer for TS fuzzy model with unmeasurable premise variable. In: 2014 IEEE International Conference on Fuzzy Systems (FUZZ-IEEE). IEEE, July 2014

Luenberger Observer Design for Robust Estimation of Battery State of Charge with Application to Lithium-Titanate Oxide Cells

Eero Immonen

Abstract The State of Charge (SoC) of a lithium-ion battery cannot be measured directly, and its estimation is one of the main challenges in electrification of transport, among others. In this article, we introduce a systematic methodology for designing Luenberger observers for on-line SoC estimation, based on a given second-order Equivalent Circuit Model (ECM) for the battery. We show how the proposed methodology can be augmented with the Linear Quadratic Regulator (LQR) optimal control framework, resulting in a computationally lightweight state estimator that also displays time optimal convergence. An application to Lithium-Titanate Oxide (LTO) battery cells is presented to illustrate the proposed approach.

Keywords Battery · Equivalent circuit model · Luenberger observer · Linear quadratic regulator · State of charge estimation · Lithium-titanate oxide cell

1 Introduction

Rechargeable lithium-ion batteries, typically composed of several interconnected cells, are instrumental for modern energy storage and mobility applications. The SoC of a battery cell is defined as the remaining capacity (Q_r) available relative to the capacity available when fully charged (Q_0), i.e. SoC $= Q_r/Q_0$. Thus:

$$\dot{\text{SoC}}(t) = \frac{d\text{SoC}(t)}{dt} = \eta \frac{I(t)}{Q_0} = \frac{I(t)}{Q_n}, \quad t > 0 \tag{1}$$

where $\text{SoC}(0) \in [0, 1]$, $I(t)$ is the current [A] through the battery (which by convention is negative in discharge conditions), $\eta \in (0, 1]$ is the efficiency and $Q_n = \frac{Q_0}{\eta}$ is the nominal (or effective) capacity [As]. Accurate knowledge of the battery SoC at

E. Immonen (✉)
Computational Engineering and Analysis Research Group, Turku University of Applied Sciences, 20520 Turku, Finland
e-mail: eero.immonen@turkuamk.fi

D. Theilliol et al. (eds.), *Recent Developments in Model-Based and Data-Driven Methods for Advanced Control and Diagnosis*, Studies in Systems, Decision and Control 467,
https://doi.org/10.1007/978-3-031-27540-1_3

all times is instrumental for efficient and safe operation, as well as for extending the lifetime of the battery [1, 2]. Unfortunately, SoC cannot be directly measured from the battery, but must be inferred from voltage and current measurements. The purpose of this article is to present a method, and its theoretical justification, for designing computationally lightweight Luenberger observers for on-line SoC estimation in the LQR optimal control framework.

A simple and well-known open-loop procedure called Coulomb counting relies on integrating Eq. (1) for estimating the SoC. However, this approach requires knowledge of the initial condition SoC(0), which is reasonable in practice only when the battery has been charged to full or discharged to empty. Moreover, in practice, the real efficiency $\eta < 1$, that results from thermal losses, is seldom known accurately and may change in time as the battery ages [3]. Further, the battery cell gradually loses capacity, i.e. Q_n decreases slowly, in both usage and storage [4]. Since Q_n is also slightly different for hot and cold temperatures [5], the SoC estimation problem becomes a non-trivial one, and requires methods that are robust to the above effects.

The present knowledge on battery SoC estimation is summarized in the many review articles, e.g. [2, 6–9]. Although sophisticated and remarkably accurate computational methods, including those based on Machine Learning [10], have been presented, the Battery Management System (BMS) hosting the SoC estimation algorithm in practice is often a relatively simple circuit board with limited memory and computing resources. Even the recurring matrix inversion operations required in (time-varying) Kalman filtering may thus be computationally too expensive in practice [11, 12]. Luenberger-type observers, attempting to estimate the joint state of a battery ECM (see Sect. 2) and the SoC model (1), have been suggested as computationally inexpensive alternatives in the literature [11, 13]. For strictly linear systems, the state estimate would then be obtained via a single stabilizing matrix gain, tuned offline, which is reasonable for low-resource BMS applications.

To address some controversy about the accuracy and need for trial-and-error tuning for Luenberger observers [14, 15], in this article, we aim to:

1. Present rigorous yet easy-to-verify sufficient conditions for the Luenberger observer gains to obtain an SoC estimate;
2. Show how tuning of the Luenberger gains can be carried out systematically in the LQR framework, resulting in time-optimal SoC estimates;
3. Demonstrate in an application to battery cells with the LTO chemistry that SoC estimation by the proposed method can be made accurate in the presence of small modeling errors and nonlinearities.

In the proposed approach, the designed observer gain is a vector of 3 SoC-dependent components, obtained offline from the solutions of a set of LQR problems. In contrast to Kalman filtering, during battery operation, the SoC estimate can thus be obtained by resolving the corresponding plant-observer dynamics without repeated matrix inversion at every time step (analogous to using a *steady state* Kalman filter). We demonstrate the proposed approach for LTO cells (cf. Sect. 4), which are important in many heavy-duty electric vehicle applications, among others [16].

2 Equivalent Circuit Model for the Battery

In this article, we consider batteries represented by second-order ECM systems, as illustrated in Fig. 1. This ECM system is driven by the dynamic current $[A]$ input $u(t) = I(t)$. The system output $V(t)$ models the battery terminal voltage $[V]$. It arises from the open-circuit voltage $V_0 > 0\ [V]$, the internal resistance $R_s > 0\ [\Omega]$ and $V_{12} = V_1 + V_2$, which is the voltage $[V]$ across the dashed subcircuit consisting of $R_1 > 0\ [\Omega]$, $C_1 > 0\ [F]$, $R_2 > 0\ [\Omega]$ and $C_2 > 0\ [F]$.

The following ECM-SoC system of equations represents that in Fig. 1:

$$\dot{\mathbf{x}}(t) = A\mathbf{x}(t) + Bu(t), \quad t > 0, \quad \mathbf{x}(0) = \mathbf{x}_0 \tag{2a}$$

$$\dot{\text{SoC}}(t) = \frac{u(t)}{Q_n}, \quad t > 0, \quad \text{SoC}(0) \in [0, 1] \tag{2b}$$

$$y(t) = C\mathbf{x}(t), \quad t \geq 0 \tag{2c}$$

$$V(t) = V_0(t) + R_s u(t) + y(t), \quad t \geq 0 \tag{2d}$$

with

$$\mathbf{x}(t) = \begin{pmatrix} V_1(t) \\ V_2(t) \end{pmatrix}, \quad u(t) = I(t), \quad y(t) = V_{12}(t) \tag{3}$$

and

$$A = \begin{pmatrix} -\frac{1}{R_1 C_1} & 0 \\ 0 & -\frac{1}{R_2 C_2} \end{pmatrix}, \quad B = \begin{pmatrix} \frac{1}{C_1} \\ \frac{1}{C_2} \end{pmatrix}, \quad C = \begin{pmatrix} 1 & 1 \end{pmatrix}, \quad \mathbf{x}_0 = \begin{pmatrix} V_1^0 \\ V_2^0 \end{pmatrix} \tag{4}$$

In general, the parameters (4) depend on SoC, i.e. $A = A(\text{SoC})$, $B = B(\text{SoC})$ and $\mathbf{x}_0 = \mathbf{x}_0(\text{SoC})$. Also the open-circuit voltage V_0 and internal resistance R_s depend on the battery SoC, i.e. $V_0 = V_0(\text{SoC})$ and $R_s = R_s(\text{SoC})$. We assume that all these dependencies on SoC are represented by continuously differentiable functions. In practice, see Sect. 4, this can be ensured by using a suitable interpolation scheme to

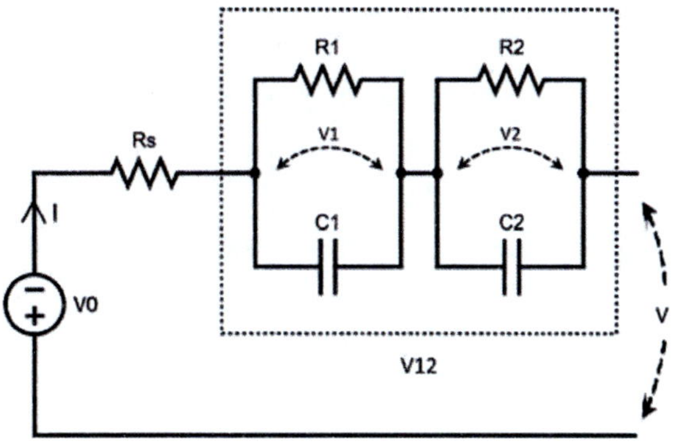

Fig. 1 Battery representation by a second-order ECM, with voltage output $V = V(t)$ and current input $I = I(t)$ [5, 17]

parameter data reported at discrete SoC points. We emphasize that these parameter dependencies on SoC can be identified off-line using standard dynamic test data (see e.g. [5, 17]), and we assume in the remainder of this article that they are known.

3 Luenberger Observer Design

3.1 *Preliminaries*

For Luenberger observer design, we have to make SoC explicit in the output of the dynamical system (2). To this end, without loss of generality, we propose to decompose the open-circuit voltage as:

$$V_0(\mathrm{SoC}) = V_0^a \cdot \mathrm{SoC} + V_0^b + V_0^{res}(\mathrm{SoC}) \tag{5}$$

where $V_0^a > 0$, $V_0^b > 0$ and, of course, the residual term $V_0^{res}(\mathrm{SoC})$ can be nonlinear. Further, we assume that the measured battery current $I_m(t)$ and the corresponding terminal voltage $V_m(t)$ are known at every $t \geq 0$. Then we set $u(t) = I_m(t)$.

If $\hat{s}(t)$ is an estimate of battery SoC at time t, then also the *synthetic measurement*

$$y_m(t) = V_m(t) - V_0^b - V_0^{res}(\hat{s}(t)) \tag{6}$$

can be immediately calculated at every $t \geq 0$. This estimate $\hat{s}(t)$ is obtained from a full state estimate for the system:

$$\dot{\mathbf{z}}(t) = \mathbf{A}\mathbf{z}(t) + \mathbf{B}u(t), \quad \mathbf{z}(0) = \mathbf{z}_0 \tag{7a}$$

$$v(t) = \mathbf{C}\mathbf{z}(t) + Du(t) \tag{7b}$$

with

$$\mathbf{A} = \begin{pmatrix} A & 0 \\ 0 & 0 \end{pmatrix}, \quad \mathbf{B} = \begin{pmatrix} B \\ \frac{1}{Q_n} \end{pmatrix}, \quad \mathbf{C} = \begin{pmatrix} C & V_0^a \end{pmatrix}, \quad D = R_s, \quad \mathbf{z}(t) = \begin{pmatrix} \mathbf{x}(t) \\ \mathrm{SoC}(t) \end{pmatrix} \tag{8}$$

Note that a state estimate $\hat{\mathbf{z}}(t)$ for the system (7)–(8) yields the SoC estimate:

$$\hat{s}(t) = (\mathbf{0} \;\; 1)\hat{\mathbf{z}}(t) \tag{9}$$

and, analogous to Eq. (2d), a prediction $V_p(t)$ for the battery voltage:

$$V_p(t) = \hat{v}(t) + V_0^b + V_0^{res}(\hat{s}(t)) = \mathbf{C}\hat{\mathbf{z}}(t) + Du(t) + V_0^b + V_0^{res}(\hat{s}(t)) \tag{10}$$

$$= \hat{V}_0(t) + R_s(\hat{s}(t))I_m(t) + \hat{V}_1(t) + \hat{V}_2(t) \tag{11}$$

If the state estimate was exact, i.e. $\hat{\mathbf{z}}(t) = \mathbf{z}(t)$, so that $\hat{\mathbf{x}}(t) = \mathbf{x}(t)$ and $\hat{s}(t) = \text{SoC}(t)$, and if the voltage model (2d) was accurate, i.e. $V(t) = V_m(t)$, then the predicted voltage would coincide with the measurement: $V_p(t) = V_0(t) + R_s I_m(t) + y(t) = V(t) = V_m(t)$. We propose to obtain the estimate $\hat{\mathbf{z}}(t)$ by utilizing a Luenberger state observer, as described in the following Subsections.

3.2 Main Results

Let $\mathbf{L} = (L_1\ L_2\ L_3)^T \in \mathbb{R}^{3\times 1}$. The Luenberger state observer for the system (7)–(8) is defined using the synthetic measurement (6) as:

$$\dot{\hat{\mathbf{z}}}(t) = \mathbf{A}\hat{\mathbf{z}}(t) + \mathbf{B}u(t) + \mathbf{L}(y_m(t) - \hat{v}(t)), \quad t > 0, \quad \hat{\mathbf{z}}(0) = \left(\hat{\mathbf{x}}(0)\ \hat{s}(0)\right)^T \tag{12a}$$

$$\hat{v}(t) = \mathbf{C}\hat{\mathbf{z}}(t) + Du(t), \quad t \geq 0 \tag{12b}$$

We begin by demonstrating that state estimation errors for the system (12) arise from modeling inaccuracies in $V(t)$ and nonlinearities in $V_0(\text{SoC})$.

Lemma 1 *If the voltage measurement is from* (2d), *i.e.* $V_m(t) \equiv V(t)$, *if the residual* $V_0^{res}(\text{SoC}) \equiv 0$ *and if* $\mathbf{L}$ *can be chosen such that* $\mathbf{A} - \mathbf{L}\mathbf{C} \in \mathbb{R}^{3\times 3}$ *is stable, i.e.*

$$\max\{\Re(\lambda) \mid \det\left[\lambda I - (A - LC)\right] = 0\} < 0 \tag{13}$$

then $\left\|\hat{z}(t) - z(t)\right\| \leq Me^{-\omega t}$ *for some* $M > 0, \omega > 0$ *and for all* $t \geq 0$.

Proof A direct calculation shows that, under the stated conditions, we have $y_m(t) - \hat{v}(t) = (V_1(t) - \hat{V}_1(t)) + (V_2(t) - \hat{V}_2(t)) + (V_0^a \cdot \text{SoC}(t) - V_0^a \cdot \hat{s}(t)) = \mathbf{C}(\mathbf{z}(t) - \hat{\mathbf{z}}(t))$. The state estimation error $\mathbf{e}(t) = \mathbf{z}(t) - \hat{\mathbf{z}}(t)$ satisfies the differential equation:

$$\begin{aligned}\dot{\mathbf{e}}(t) &= \dot{\mathbf{z}}(t) - \dot{\hat{\mathbf{z}}}(t) = \mathbf{A}\mathbf{z}(t) + \mathbf{B}u(t) - \mathbf{A}\hat{\mathbf{z}}(t) - \mathbf{B}u(t) - \mathbf{L}(y_m(t) - \hat{v}(t)) \\ &= \mathbf{A}(\mathbf{z}(t) - \hat{\mathbf{z}}(t)) - \mathbf{L}\mathbf{C}(\mathbf{z}(t) - \hat{\mathbf{z}}(t)) = (\mathbf{A} - \mathbf{L}\mathbf{C})\mathbf{e}(t)\end{aligned} \tag{14}$$

The eigenvalue condition (13) guarantees the existence of $M > 0, \omega > 0$ $\|\mathbf{e}(t)\| \leq Me^{-\omega t}$ for all $t \geq 0$.

Under the best-case conditions of Lemma 1, the battery SoC estimate converges at an exponential rate regardless of true SoC and the initial guess $\hat{s}(0)$. In practice, there are often modeling inaccuracies and nonlinearities in the voltage representations. We aim to demonstrate in a practical application (Sect. 4) that accurate state estimation is still possible. The following result shows that the condition (13) is easy to satisfy, and it is the basis for observer design in the proposed methodology.

Proposition 1 *If* $L_1 \geq 0$, $L_2 \geq 0$ *and* $L_3 > 0$, *then* $\mathbf{A} - \mathbf{L}\mathbf{C}$ *is stable.*

Proof Denote $A_1 = -\frac{1}{R_1C_1} < 0$ and $A_2 = -\frac{1}{R_2C_2} < 0$. A direct calculation shows that the first column of the Routh array for $\mathbf{A} - \mathbf{LC}$ consists of the following entries $R_i, i = 1, \ldots, 4$:

$$R_1 = 1 > 0, \quad R_2 = L_1 - A_1 - A_2 + L_2 + L_3 V_0^a > 0, \quad R_4 = A_1 A_2 L_3 V_0^a > 0,$$
$$R_3 = \frac{N}{R_2} \tag{15}$$

where the numerator

$$N = -A_1^2 A_2 + A_1^2 L_2 + A_1^2 L_3 V_0^a - A_1 A_2^2 + 2A_1 A_2 L_1 + 2A_1 A_2 L_2 + 2A_1 A_2 L_3 V_0^a$$
$$- A_1 L_1 L_2 - A_1 L_1 L_3 V_0^a - A_1 L_2^2 - 2A_1 L_2 L_3 V_0^a - A_1 L_3^2 V_0^{a^2}$$
$$+ A_2^2 L_1 + A_2^2 L_3 V_0^a - A_2 L_1^2 - A_2 L_1 L_2 - 2A_2 L_1 L_3 V_0^a - A_2 L_2 L_3 V_0^a - A_2 L_3^2 V_0^{a^2} \tag{16}$$

In the numerator expression (16), those terms with a minus sign are non-positive (hence non-negative when multiplied by the minus sign), and those terms with a plus sign are non-negative. Additionally, some terms, such as $-A_1^2 A_2$, are strictly positive. Hence $R_3 > 0$. The result follows from Routh's Test [18].

By Proposition 1, the exponential convergence of the SoC estimate does not depend on $\mathbf{B}$, and specifically on the battery nominal capacity Q_n. Consequently, as Q_n slowly decreases during battery use and storage, the same observer designed for a brand new battery still works, i.e. is robust with respect to Q_n.

Although they establish the theoretical foundations for using Luenberger observers for ECM-based SoC estimation, strictly speaking, the above results guarantee convergence of the state estimate only in the absence of parameter variations in the battery ECM. However, in practice, the ECM parameters only vary slowly relative to the rate of convergence of the state estimate. This is demonstrated in Sect. 4 in successful SoC state estimation for Lithium-Titanate Oxide battery cells. We aim to complement these results in the future by considering linear time-varying systems directly; see e.g. [19].

3.3 Designing the Observer Gain in the LQR Framework

Let $Q \in \mathbb{R}^{3\times3}$ be Hermitian and let $R > 0$. Clearly the pair $(\mathbf{A}, \mathbf{C})$ in the system (7)–(8) is observable, and we can thus design the gain $\mathbf{L}$ by considering the dynamical system:

$$\dot{\mathbf{r}}(t) = \mathbf{A}^T \mathbf{r}(t) + \mathbf{C}^T u(t), \quad t > 0, \quad \mathbf{r}(0) = \mathbf{r}_0 \tag{17}$$

and the associated infinite-horizon LQR objective:

$$J(\mathbf{r}) = \int_0^\infty \left[\mathbf{r}(s)^T Q\mathbf{r}(s) + Ru^2(s)\right]\mathrm{d}s \tag{18}$$

It is well-known (see e.g. [20]) that the state feedback control for the system (17) that minimizes the value of the cost (18) is $u(t) = -K\mathbf{r}(t)$ where $K = R^{-1}\mathbf{C}X$ and X is the unique solution of the Continuous-time Algebraic Riccati Equation (CARE):

$$\mathbf{A}X + X\mathbf{A}^T - X\mathbf{C}^T R^{-1}\mathbf{C}X + Q = 0 \tag{19}$$

By duality, we can choose $\mathbf{L} = K^T$ for the Luenberger state observer (12). Then also $\mathbf{A} - \mathbf{LC}$ is stable. Note that the only parameters chosen by hand in this approach would be Q and R that place emphasis on different parts of the optimal closed loop response through the objective (18).

3.4 Observer Design and Operation

The steps in the proposed Luenberger observer design procedure are as follows:

1. Identify the ECM model (2) parameter values (4) at a chosen set of points $\mathrm{SoC}_i \in [0, 1]$.
2. Establish continuously differentiable functions $V_0 = V_0(\mathrm{SoC})$, $R_s = R_s(\mathrm{SoC})$, $R_1 = R_1(\mathrm{SoC})$, $R_2 = R_2(\mathrm{SoC})$, $C_1 = C_1(\mathrm{SoC})$, $C_2 = C_2(\mathrm{SoC})$ by piecewise cubic Hermite polynomial interpolation (PCHIP).
3. Identify the open-circuit voltage decomposition (5) by linear regression.
4. Fix Q and R as in Sect. 3.3.
5. For another set of points $\mathrm{SoC}_k \in [0, 1]$, identify the gain parameters $\mathbf{L}_k = \mathbf{L}(\mathrm{SoC}_k)$ by solving the corresponding CARE (19) and apply PCHIP interpolation to establish a continuously differentiable function $\mathrm{SoC} \to \mathbf{L}(\mathrm{SoC})$.

During operation, a new prediction $\hat{s}(t_j)$ for $\mathrm{SoC}(t_j)$ is obtained via Eq. (9), by numerically integrating the observer dynamics (12) using a table lookup at $\hat{s}(t_{j-1})$.

4 Application: Lithium-Titanate Oxide Battery Cells

Battery cells with the LTO chemistry have been reported to withstand high currents, repeated cycling and operation in hot and cold temperatures without significant degradation [5]. This makes them prime candidates for Heavy-Duty Electric Vehicle (HDEV) applications, including harvesters, mining dumpers, race cars and trucks, among others. Recently, second-order ECMs for LTO cells were presented in [5, 17], to understand the cells' heat generation characteristics in dynamic discharge conditions. Here we utilize those models and design computationally lightweight parametric Luenberger observers for robust SoC estimation for the LTO cells.

4.1 Model Parameters

In [5, 17], Immonen and Hurri carried out data-based second-order ECM modeling for 23 Ah prismatic Toshiba SCiB™ LTO cells, and reported model parameter values identified through Discharge Pulse Power Characterization (DPPC) tests at C-rate 1. We utilize their published parameter data and PCHIP interpolation to find continuously differentiable functions $V_0(\text{SoC})$, $R_s(\text{SoC})$, $R_1(\text{SoC})$, $R_2(\text{SoC})$, $C_1(\text{SoC})$ and $C_2(\text{SoC})$ for the ECM. By regression, we further obtain $V_0^a = 0.74$ V and $V_0^b = 1.86$ V, with a nonlinear residual $V_0^{res}(\text{SoC})$.

4.2 Observer Gain Design

For the present example application, let us choose $Q = \text{diag}([0.01, 0.01, 0.5]) \in \mathbb{R}^{3\times3}$ and $R = 1$. Then the CARE (19) can be solved numerically in Matlab for any given SoC $\in [0, 1]$. By discretizing the interval [0, 1] at resolution $\Delta\text{SoC} = 0.01$, we obtain the gains $\mathbf{L}_k$ as shown in Fig. 2. Clearly the conditions of Proposition 1 are satisfied for every SoC value. Moreover, the three components of the vectors $\mathbf{L}_k$ remain relatively constant across the entire SoC interval [0, 1]. A continuously differentiable function SoC $\rightarrow \mathbf{L}(\text{SoC})$ is obtained by PCHIP interpolation.

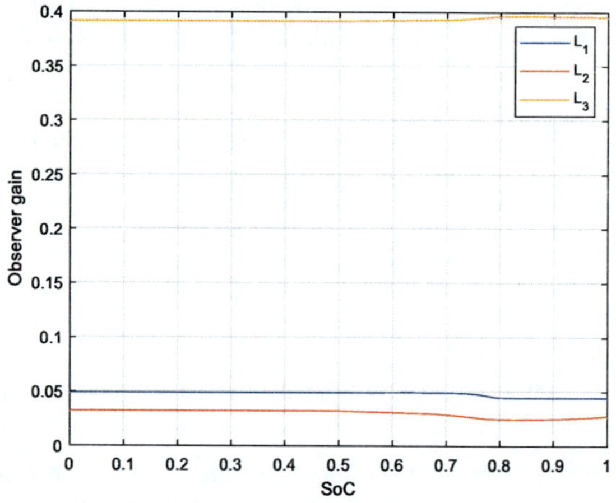

Fig. 2 Numerical design of the Luenberger observer gain $\mathbf{L} = (L_1\ L_2\ L_3)^T \in \mathbb{R}^{3\times1}$ by solution of the CARE (19)

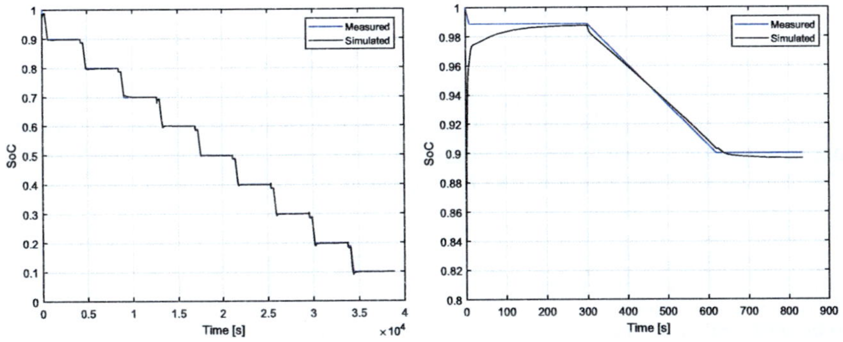

(a) Full 4C DPPC test. (b) First 830 seconds of the 4C DPPC test.

Fig. 3 A SoC estimate for the 4C DPPC test compared to measurements [5, 17], using the false initial SoC estimate $\hat{s}(0) = 0.8$. The true $\mathrm{SoC}(0) = 1$ (fully charged)

4.3 SoC Estimation in 4C DPPC Test

For out-of-sample validation, we utilize the 4C DPPC test, which has not been used in model identification [5, 17]. The test involves repeating the following four steps 10 times or until voltage is below 1.5 V: Constant-current (CC) discharge at 92 A for 10 s; Rest 290 s; CC discharge at 23 A for 320 s; Rest 60 minutes. In the laboratory measurements, the true battery SoC was calculated by Coulomb counting, starting from a fully charged cell.

Figure 3 displays the performance of the Luenberger observer (12), with the gain designed in Sect. 4.2, under the *false* assumption that $\mathrm{SoC}(0) = 0.8$. Clearly, the SoC estimate remains accurate despite any modeling errors and nonlinearities. If the initial SoC estimate is correct, i.e. $\hat{s}(0) = \mathrm{SoC}(0) = 1$, then the root mean square error (RMSE) of the SoC estimate is $4.32 \cdot 10^{-3}$, indicating good performance.

5 Conclusions

In this article, we have introduced a method for designing a Luenberger observer for computationally lightweight on-line battery SoC estimation, based on a given second-order ECM for the battery. We presented both a theoretical justification and easy-to-verify design conditions for the gain parameters. These conditions do not depend on the battery's nominal capacity, and hence the proposed observer displays some robustness with respect to battery aging. The proposed design methodology connects naturally to the LQR optimal control framework, thus reducing the need for trial and error in Luenberger observer gain tuning. As an application, we considered SoC estimation for LTO chemistry cells which are interesting for heavy-duty electric vehicle batteries, among others. The presented numerical results show insensitivity to a poor initial estimate and good accuracy over a standard dynamic discharge test that has not been used in identifying the ECM parameters.

An important topic for future work is considering (slowly) time-varying ECMs for batteries, and establishing more general sufficient conditions for Luenberger state observer estimate convergence than those presented herein. Moreover, for practical BMS implementation, the proposed methodology should be optimized with respect to run-time energy consumption.

References

1. Yu, Z., Huai, R., Xiao, L.: State-of-charge estimation for lithium-ion batteries using a kalman filter based on local linearization. Energies **8**(8), 7854–7873 (2015)
2. How, D.N., Hannan, M.H. Lipu, Ker, P.J.: State of charge estimation for lithium-ion batteries using model-based and data-driven methods: a review. IEEE Access **7**, 136 116–136 136 (2019)
3. Gyenes, B., Stevens, D., Chevrier, V., Dahn, J.: Understanding anomalous behavior in coulombic efficiency measurements on li-ion batteries. J. Electrochem. Soc. **162**(3), A278 (2014)
4. Rabah, M., Immonen, E., Shahsavari, S., Haghbayan, M.-H., Murashko, K., Immonen, P.: Capacity loss estimation for li-ion batteries based on a semi-empirical model. Anwendungen und Konzepte der Wirtschaftsinformatik, no. 14 (2021)
5. Immonen, E., Hurri, J.: Incremental thermo-electric CFD modeling of a high-energy lithium-titanate oxide battery cell in different temperatures: a comparative study. Appl. Thermal Eng. **197**, 117260 (2021)
6. Zhou, W., Zheng, Y., Pan, Z., Lu, Q.: Review on the battery model and SOC estimation method. Processes **9**(9), 1685 (2021)
7. Ali, M.U., Zafar, A., Nengroo, S.H., Hussain, S., Junaid Alvi, M., Kim, H.-J.: Towards a smarter battery management system for electric vehicle applications: a critical review of lithium-ion battery state of charge estimation. Energies **12**(3), 446 (2019)
8. Shrivastava, P., Soon, T.K., Idris, M.Y.I.B., Mekhilef, S.: Overview of model-based online state-of-charge estimation using kalman filter family for lithium-ion batteries. Renew. Sustain. Energy Rev. **113**, 109233 (2019)
9. Wang, Y., Tian, J., Sun, Z., Wang, L., Xu, R., Li, M., Chen, Z.: A comprehensive review of battery modeling and state estimation approaches for advanced battery management systems. Renew. Sustain. Energy Rev. **131**, 110015 (2020)
10. Vidal, C., Malysz, P., Kollmeyer, P., Emadi, A.: Machine learning applied to electrified vehicle battery state of charge and state of health estimation: State-of-the-art. IEEE Access **8**, 52 796–52 814 (2020)
11. Lievre, A., Pelissier, S., Sari, A., Venet, P., Hijazi, A.: Luenberger observer for soc determination of lithium-ion cells in mild hybrid vehicles, compared to a kalman filter. In: 2015 Tenth International Conference on Ecological Vehicles and Renewable Energies (EVER), pp. 1–7. IEEE (2015)
12. Meng, J., Ricco, M., Luo, G., Swierczynski, M., Stroe, D.-I., Stroe, A.-I., Teodorescu, R.: An overview of online implementable SOC estimation methods for lithium-ion batteries. In: 2017 International Conference on Optimization of Electrical and Electronic Equipment (OPTIM) & 2017 International Aegean Conference on Electrical Machines and Power Electronics (ACEMP), pp. 573–580. IEEE (2017)
13. Rahimi-Eichi, H., Chow, M.-Y.: Adaptive parameter identification and state-of-charge estimation of lithium-ion batteries. In: IECON 2012-38th Annual Conference on IEEE Industrial Electronics Society, pp. 4012–4017. IEEE (2012)
14. Tang, X., Liu, B., Gao, F., Lv, Z.: State-of-charge estimation for li-ion power batteries based on a tuning free observer. Energies **9**(9) (2016). https://www.mdpi.com/1996-1073/9/9/675

15. Manthopoulos, A., Wang, X.: A review and comparison of lithium-ion battery SOC estimation methods for electric vehicles. In: IECON: The 46th Annual Conference of the IEEE Industrial Electronics Society, vol. 2020, pp. 2385–2392. IEEE (2020)
16. Turku university of applied sciences e-rallycross car project. https://erallycross.turkuamk.fi/en/main-page/. Accessed 10 Jan. 2022
17. Immonen, E., Hurri, J.: Equivalent circuit modeling of a high-energy lto battery cell for an electric rallycross car. In: IEEE 30th International Symposium on Industrial Electronics (ISIE), vol. 2021, pp. 1–5. IEEE (2021)
18. Ferrante, A., Lepschy, A., Viaro, U.: A simple proof of the routh test. IEEE Trans. Autom. Control **44**(6), 1306–1309 (1999)
19. Rotella, F., Zambettakis, I.: On functional observers for linear time-varying systems. IEEE Trans. Autom. Control **58**(5), 1354–1360 (2012)
20. Sontag, E.D.: Mathematical Control Theory: Deterministic Finite Dimensional Systems, vol. 6. Springer Science & Business Media (2013)

Fault Detection and Diagnosis of PV Systems Using Kalman-Filter Algorithm Based on Multi-zone Polynomial Regression

Yehya Al-Rifai, Adriana Aguilera-Gonzalez, and Ionel Vechiu

Abstract Faults must be timely diagnosed, detected, and identified to enhance photovoltaic (PV) system's dependability. In this context, this paper presents a novel Fault Detection and Diagnosis (FDD) methodology based on a hybrid combination of model-based, through Kalman Filter (KF), and a statistical data-driven regression approach for online monitoring of a PV system's DC side. This statistical approach is formulated on Multi-Zone non-linear Polynomial regression (MZP) techniques of PV characteristics under Global Maximum Power Points (GMPP) at the array level. In particular, the proposed method effectively detects intermittent soft Short-Circuit (SC) even at very low irradiation. The performance of the proposed FDD methods is evaluated via MATLAB/Simulink®considering varying weather conditions.

1 Introduction

With the growing concerns about the climate change, PV technology represents an outstanding alternative carbon-free and sustainable energy generation to replace depleted conventional resources.

Despite the fact that PV systems require low maintenance, they are still vulnerable to faults that can occur without early warning due to the harsh working environment. As indicated in [1], faults on the DC side have the severest impact that dramatically deteriorates PV output performance. Thereby, this paper targets fault on the DC side, in particular intermittent soft SC at extreme weather conditions.

Undoubtedly, protection systems are paramount to prevent catastrophic failures. However, the conventional protection systems alone fail to preserve the PV required

Y. Al-Rifai (✉) · A. Aguilera-Gonzalez · I. Vechiu
University of Bordeaux, Estia Institute of Technology, F-64210 Bidart, France
e-mail: y.alrifai@estia.fr

A. Aguilera-Gonzalez
e-mail: a.aguilera-gonzalez@estia.fr

I. Vechiu
e-mail: i.vechiu@estia.fr

D. Theilliol et al. (eds.), *Recent Developments in Model-Based and Data-Driven Methods for Advanced Control and Diagnosis*, Studies in Systems, Decision and Control 467,
https://doi.org/10.1007/978-3-031-27540-1_4

performance in abnormal circumstances due to several factors as: fault at low irradiation, hidden fault by the Maximum Power Points Tracking (MPPT), or nonlinear Power-Voltage $(P-V)$ characteristics [2]. Thus, an FDD system is essential to improve PV system dependability, while reducing maintenance costs and downtime.

Consequently, FDD has become a central research field in PV systems. For example, a remote monitoring method is used in [3] to detect abnormal power dissipation based on environmental data from satellite-based observations. This multi-sensor method is undeniably less accurate than the on-site measurement, and not affordable for small-scale projects. Besides, the environmental data are subject to noise and interruption, leading to erroneous power yield estimation.

Recently, electrical-based FDD methods have become more commonly adopted, as they require only the measurement of the output electrical signatures [4]. An interesting FDD technique is performed in [5] based on Current-Voltage $(I-V)$ curve analysis to detect several faults in a PV system. However, this offline method requires a special tool for $I-V$ curve checking and not suitable for online monitoring.

Another method is proposed in [6] to detect power lowering at the array level. This method is based on the relationship between PV module temperature, current, and voltage under GMPP conditions by using Polynomial Regression (PR) at Standard fixed Temperature (ST) (25 °C). Although, this method was confined to detect Partial Shading (PS) at high irradiation. However, the $I-V$ characteristics become highly non-linear at low irradiation, where the simple cubic PR function is definitely not accurate, resulting in false and missing alarm.

To overcome these issues, this paper proposes a novel FDD approach by combining a statistical data-driven regression analysis and a model-based through KF, for robustness against high uncertainties and measurement noises. This technique is derived from [6] and developed to detect faults even at very-low irradiation based on MZP interpolation according to the current level. Further, to improve the projection function accuracy into ST, the average value of Temperature Coefficients (TCs) is divided into Multi-Zone Average value (MZA) accordingly. Aiming to detect soft faults, an adaptive threshold is proposed based on a PR in state of temperature variable. To evaluate the performance of the proposed FDD approach, virtual intermittent soft SCs fault are analyzed via MATLAB/Simulink®at various weather conditions.

The rest of this paper is organized as follows: Sect. 2 describes the PV model. Section 3 outlines the performance of the proposed estimation methods. Then, the FDD strategy with the KF is presented in Sect. 4. Next, the simulation and results are addressed in Sect. 5. Finally, conclusion is given in Sect. 6.

2 PV Array Characteristics and Modeling

PV system naturally has nonlinear time-variant $P-V$ characteristics that vary according to irradiation and temperature [7]. Thus, PV modelling plays a vital role in FDD to accurately emulate its $I-V$ operating point at various weather conditions. Several models have been addressed in the literature, but the most common

Table 1 Soltech-1STH-350-WH PV module parameters

Parameter	Value	Units
Maximum power ($\boldsymbol{P_{MPP}}$)	349.59	W
Open circuit voltage ($\boldsymbol{V_{OC}}$)	51.5	V
Short-Circuit current ($\boldsymbol{I_{SC}}$)	9.4	A
Voltage at P_{MAX} ($\boldsymbol{V_{MPP}}$)	43	V
Current at P_{MAX} ($\boldsymbol{I_{MPP}}$)	8.13	A
Temperature coefficient of V_{OC} ($\boldsymbol{\alpha}$)	−0.36	%/°C
Temperature coefficient of I_{SC} ($\boldsymbol{\beta}$)	0.09	%/°C
Configuration	9 Series—21 Parallel	

due to simplicity and efficacy is the single diode model with additional shunt resistance in parallel [8]. This five-parameters model is adopted by 'Simscape' library and employed in this paper to generate the fitting/test-data-set. For the simulation, Soltech-1STH-350-WH is used, and its electrical parameters are listed in Table 1. Indeed, a PV system strongly depends on weather conditions. The temperature influence on I and V under MPP condition is mathematically expressed in Eq. (1) as follow:

$$\begin{aligned} I_{mpp}(T) &= I_{T_{ref},mpp} + \beta(T - T_{stc}) \\ V_{mpp}(T) &= V_{T_{ref},mpp} + \alpha(T - T_{stc}) \end{aligned} \tag{1}$$

where $I_{T_{ref},mpp}$, $V_{T_{ref},mpp}$ denote the output current and voltage projected to ST, respectively. T is the instantaneous temperature, and T_{stc} is the ST. α and β represent the TCs for voltage and current, respectively. This paper employs the perturbation and observation technique to track the MPP at varying weather conditions [8].

3 Regression Estimation Methods

3.1 Estimation Based on the PR Method

The PR method presented in [6] is able to detect power lowering based on a cubic PR at ST with variable irradiation, as expressed in Eq. (2):

$$V_{ref,diag}(I) = \sum_{k=0}^{n} b_k I_{T_{ref},mpp}^{n-k} \tag{2}$$

where $V_{ref,diag}$ is the reference diagnosis voltage that is later estimated by KF, b_k represents the coefficients of the polynomial function, $I_{T_{ref},mpp}$ is the instantaneous

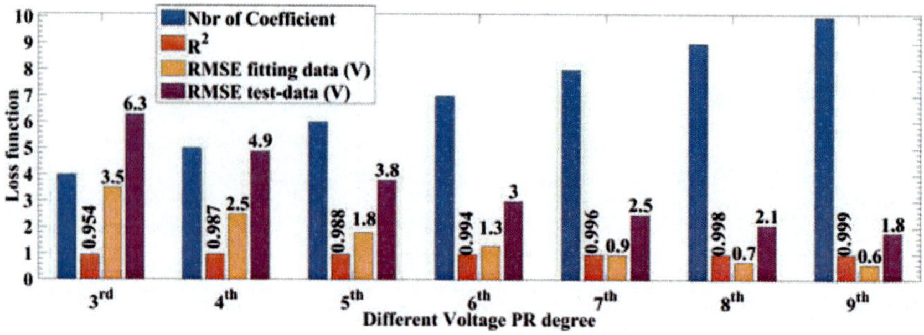

Fig. 1 Loss function of $V = f(I)$ PR method under fitting and test-data at different degrees

projected MPP current to ST by Eq. (1). Indeed, to detect faults, $V_{ref,diag}$ is compared with the projected $V_{T_{ref},mpp}$ by Eq. (1). Thereby, if a discrepancy exceeds a predefined threshold, an alarm is then triggered indicating the occurrence of PS. However, the method was tested at 600 W/m^2, and not performed against other faults. Further, the authors had not proven the choice of the third-degree function over higher/lower degree accuracy. Thus, this method could generate false alarm at low irradiation where the $I - V$ relationship is strongly nonlinear.

Therefore, this paper performs initially a loss function to evaluate the fitting accuracy of PR at different irradiation, and thereby, to extract the best fitting polynomial degree that covers the whole irradiation scope. For that, a fitting data-set is firstly generated using 1-single diode model with additional shunt resistance, at wide temperature and irradiation range ($[-5 : 50]$ °C, $[0 : 1000]$ W/m^2). Subsequently, the PR is evaluated by using RMSE (root-mean-square error), MAPE (mean absolute percentage error), and R^2 (R-squared). Then, to avoid overfitting, the different PR degrees are re-evaluated with another test-dataset. The loss function at different degrees with the fitting/test-data is analyzed in Fig. 1.

Obviously, the test data reveals that the cubic voltage PR ($V(I)$) is bad-fitted with an RMSE higher than 6.3 V. However, the Estimation Error (EE) slightly decreases while increasing PR's order (to reach less than 2 V at the 9th degree). Although, this EE will lessen the effectiveness of FDD for detecting soft faults, beside the high PR degree poses higher computational complexity.

In order to reach a better prediction accuracy with the least EE possible by acceptable PR's degree, an improved method is proposed based on voltage MZP.

3.2 MZP Regression Proposed Method

Aiming to improve the prediction accuracy of the PR, the MPP curve at ST is divided into three-zones as depicted in Fig. 2. The first zone represents the variation of MPP at Very Low irradiation (VLG) below 100 W/m^2. The second stands for the non-linear variation at Low irradiation (LG) below $400\,\mathrm{W/m^2}$. The last zone represents the MPP variation at High irradiation (HG). Since this method is irradiation sensor-less, the zone is rather split with respect to current magnitude, which is an irradiation's image.

Remarkably, comparing with the PR method of [6], the RMSEs based on MZP method have significantly dropped, as indicated by the legend box in Fig. 2. As appreciated, the cubic MZP over all zones are more reliable than the 9th PR degree. To emphasize the MZP's robustness, likewise, the loss function is employed under test-data to select the best fitting order of voltage MZP at each zone, as illustrated in Fig. 3.

Obviously, the test-data confirm that the 3rd order voltage MZP is well-fitted with trivial RMSE at high and low irradiation. Although, its RMSE at VLG is slightly high. Whereas, the 4th degree is remarkably more reliable with higher estimation accuracy over all 3-zones. Indeed, the latter outperform specially at VLG with RMSE less than (0.9 V) comparing with the 3rd degree of (1.57 V). Thereby, the 4th order is selected for all three regions. The voltage MZP function is represented in Eq. (3) as follows:

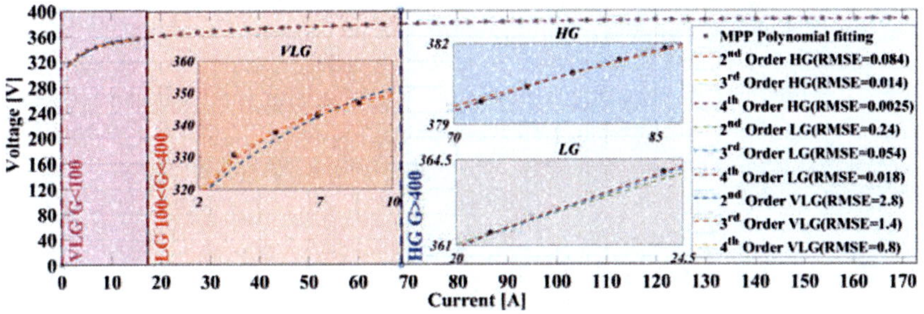

Fig. 2 MZP of V(I) at ST divided into 3-zones (High, low, and very low irradiation)

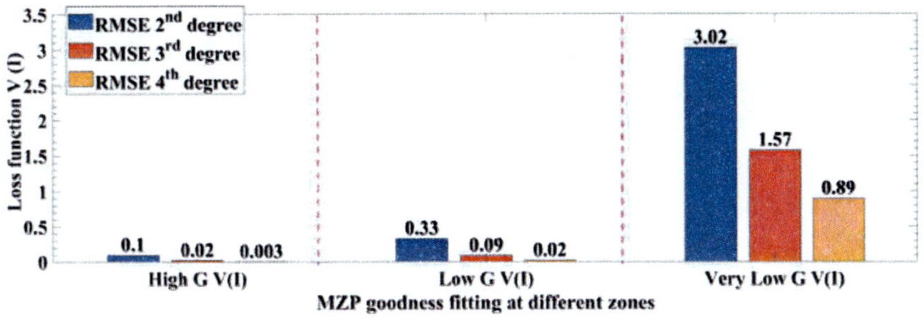

Fig. 3 MZP goodness-fitting at different degree in each zone

$$V_{ref,diag}(I) = \begin{cases} \sum_{k=0}^{4} l_k I_{T_{ref},mpp}^{4-k} & \forall I \geq 68\,A \\ \sum_{k=0}^{4} m_k I_{T_{ref},mpp}^{4-k} & 17\text{A} < I < 68\,\text{A} \\ \sum_{k=0}^{4} n_k I_{T_{ref},mpp}^{4-k} & 0 < I \leq 17\,A \\ 0, & I = 0A \end{cases} \tag{3}$$

where, l_k,m_k and n_k denote the polynomial coefficients at high, low, and very low irradiation, respectively. Briefly, the MZP broadly outperform the PR estimation method, with notably lower EE over all zones. Thus, due to its perfect prediction accuracy covering all irradiation scope, the proposed MZP allows setting the threshold closer to the reference diagnosis indicator, which will certainly increase the likelihood of detecting soft faults even at low irradiation.

Although, this technique is a two-steps methodology where the instantaneous MPPs voltage and current are initially projected to the ST by Eq. (1) based on TCs. However, (α and β) have an average value, where the projection EE rises at extreme weather conditions. Therefore, to address this issue, the values of TCs are divided into MZA according to the current level.

Figure 4 illustrates the EE of the projection function Eq. (1) using an average value of TCs in comparison with the proposed method based on MZA according to the current level. As appreciated, with the proposed method the EEs for both current and voltage projection functions have significantly reduced, even at extreme temperature, making the projection to ST curve error-less. Generally, this overall proposed methods significantly diminish the EE, leading to a higher prediction accuracy at various weather conditions, that crucially improve the FDD performance.

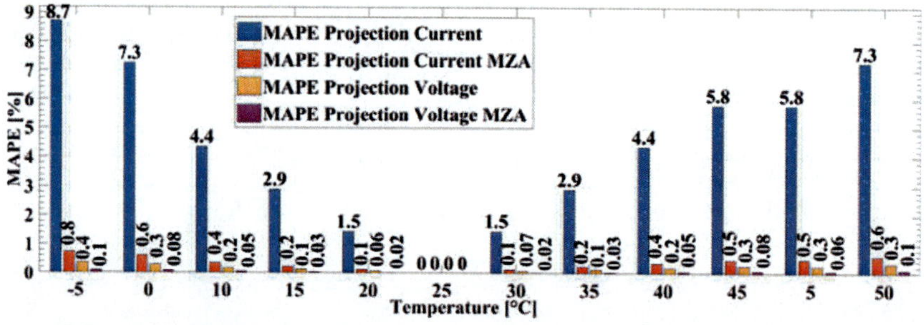

Fig. 4 EE comparison between projection function using average value and MZA value at different temperature

4 Fault Detection and Diagnosis Method

Model-based fault detection is a simpler and cheaper alternative method than physical redundancy [9]. This approach consists of two consecutive stages: residual generation and residual evaluation. The structure of the proposed FDD algorithm is shown in Fig. 5. The PV configuration represents actually 9-series and 21-parallels connections.

4.1 Residual Generation by Using KF

For the residual generation, the recursive algorithm KF is employed due to its robustness against high uncertainties and measurement noises in the stochastic framework [10, 11]. Thus, two KFs are employed in this paper to estimate firstly the projection variable $V_{T_{ref},mpp}$ and $I_{T_{ref},mpp}$ in Eq. (1), and then applied in FDD to estimate the independent variable $V_{ref,diag}$ in Eq. (3). For that, the proposed MZP equation can be formulated with the following state Eq. (4):

$$
\begin{aligned}
x(k+1) &= \Phi x(k) + \Gamma \omega(k) \\
y(k) &= C x(k) \\
z(k) &= y(k) + \upsilon(k)
\end{aligned}
\tag{4}
$$

where Φ, Γ and c are the state, noise process and output matrices, respectively. The state vector $x(k)$ represents the dependent and independent variable in Eq. (3), and $y(k)$ is the output vector. $z(k)$ denotes the measured data. For a quick divergence in faulty case, the dependent variable is estimated and not measured. ω and υ are respectively the process and measurement noises. Indeed, the KF equation fall into two phase process. The first represents the time update phase, where the KF algorithm

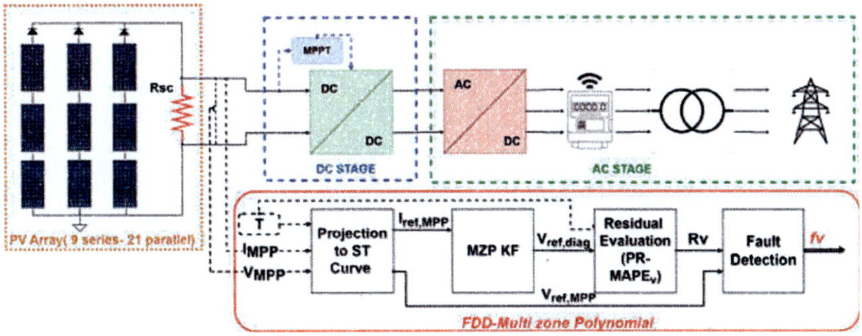

Fig. 5 General FDD scheme of the proposed diagnosis method based on MZP

predicts the state by projecting ahead the current state and error covariance to get a priori estimate for the next time steps as expressed in Eq. (5):

$$\begin{aligned} \hat{x}_k^- &= \Phi \hat{x}_{(k-1)} \\ P_k^- &= \Phi P_{(k-1)} \Phi^T + Q \end{aligned} \quad (5)$$

where, $\hat{x}_k^-$, P_k^- are the priori state and covariance estimate at iteration (k) from the previous iteration, $(k-1)$, respectively. Q is the process noise covariance. Secondly, the next phase represents the measurement update. In this phase, the priori estimates are then corrected by the Kalman gain K_k through the statistical filtered noisy measurement to obtain the posteriori states as given in Eq. (6).

$$\begin{aligned} k_k &= P^- C^T (C P_k^- C^T + R)^{-1} \\ \hat{x}_k &= \hat{x}_k^- + K_k (z_k - C \hat{x}_k^-) \\ P_k &= (\Upsilon_\eta - K_k C) P_k^- \end{aligned} \quad (6)$$

where, $\hat{x}_k$ is the update state, the covariance P_k is a positive definite symmetric matrix with relatively large initial value $P_0 = \text{diag}(4)$ that declares low credibility for initial guess. Υ_η is an identity matrix. R is the measured noise covariance. The matrices Q and R are diagonal initially predefined 10^{-4} [12].

4.2 Residual Evaluation

This stage examines the system's health status by checking the consistency between the projected measurement $V_{T_{ref},mpp}$ and the predicted value $V_{ref,diag}$. If the discrepancy exceeds a defined threshold, a fault alarm is triggered. Aiming to minimize the false and missing alarms by addressing the projection function and MZP uncertainties, this paper proposes an adaptive threshold based on a quadratic PR in state of temperature variable, as formulated in Eq. (7):

$$MAPE_V(T) = \sum_{k=1}^{2} \bar{\Lambda}_k T^{2-k} \quad (7)$$

where, T is the instantaneous-measured temperature and $\bar{\Lambda}_k$ is the polynomial coefficient based on the EE of MZP and projection function at different temperature, as calculated previously. Thus, the health status can be examined by Eq. (8):

$$(1 - MAPE_V(T)) V_{ref,diag}(k) \leq V_{T_{ref},mpp} \leq (1 + MAPE_V(T)) V_{ref,diag}(k) \quad (8)$$

Thereby, to trigger a true alarm, a Boolean fault vector $f_{(V)} \in \{0, 1\}$ is settled according to Eq. (9):

$$f_V = \begin{cases} 1 & if \ \ |V_{T_{ref},mpp}| > (1 \pm MAPE_V)V_{ref,diag} \\ 0, & otherwise \end{cases} \quad (9)$$

Therefore, if the fault signature is 1, an alarm is generated indicating the occurrence of faults. Otherwise, the PV system is considered in healthy state.

5 FDD Simulation Results

To investigate the robustness of the proposed FDD method, simulations under healthy and faulty cases are performed through MATLAB/Simulink®. The diagnosis test is performed under test-data to investigate the reliability of the projection function at variable temperature and the robustness of PR and MZP at different irradiation zone (high, low, and very low irradiation), as shown in Fig. 6.

5.1 *Healthy Scenario*

This scenario considers a normal condition of the PV system. A comparison between PR and MZP FDD methods under healthy state is illustrated in Fig. 7. As appreciated, the proposed MZP threshold in state of temperature variable is inline and much closer to the reference value than the PR threshold based on a constant MAPE. Thus, the latter will undoubtedly conceal soft anomalies. Further, due to the high estimation reliability along with the threshold robustness, the reference value $V_{T_{ref},mpp}$ is always within the boundaries (upper and lower MZP threshold) defined in Eq. (8) using MZP

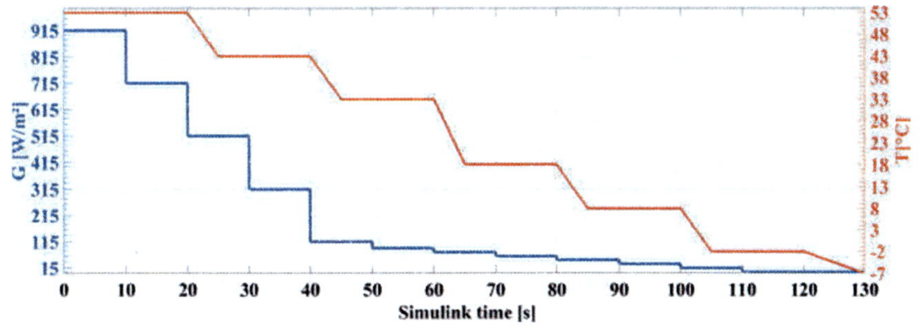

Fig. 6 Temperature and irradiation profiles during simulation in diagnosis test

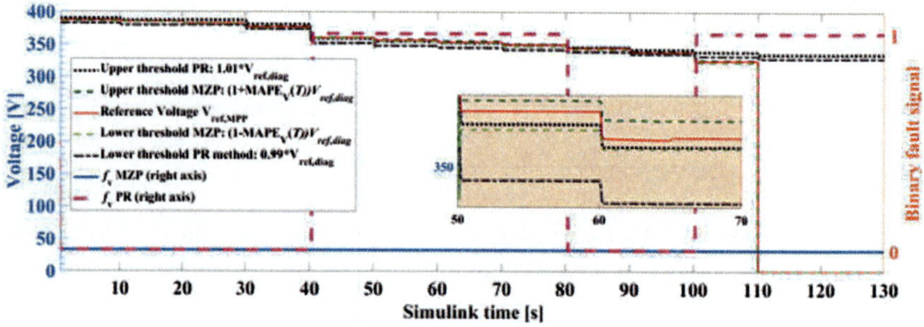

Fig. 7 Healthy scenario: comparison between PR and MZP FDD methods

method. Accordingly, the fault vector f_v MZP remains 0 during the simulation, confirming a healthy state.

In contrast, due to the bad fitting of the cubic PR function along with the low threshold reliability, the discrepancies are relatively large. Consequently, the threshold gets violated in almost all zones using PR methods. In particular, at 0 W/m^2, the threshold sharply deviates from the reference value, since the PR's intercept is different from 0. Thereby, false alarms are generated. In general, the MZP FDD method is more accurate and robust.

5.2 *Faulty Scenario: Intermittent Soft Short Circuit*

Further, to examine the performance of the two FDD methods under abnormal behavior, artificial intermittent soft SCs are induced at each zone, as depicted in Fig. 8. The time duration of the intermittent fault is considered 2 s, with a soft magnitude of $R_{sc} = 50\Omega$. The SC current (marked in solid red line) magnitude varies between ([6.7 : 7.5] A) during the diagnosis test, that represents 4.2% at HG and more than 50% at LG of the total MPP current generated. Remarkably, due to its high accuracy the threshold gets violated each time the SCs faults are activated into the PV plant using MZP FDD method, despite its soft magnitude at HG, as illustrated in Fig. 8. Thereby, true alarms are generated accordingly, indicating the presence of abnormal behavior. On the other side, due to its low estimation reliability along with the threshold uncertainties, the PR method is not able to reveal the soft anomalies in the PV plant at HG. In addition, at LG and VLG The PR fault signals has the same false alarm profile as under normal condition due to the bad fitting accuracy. Exceptionally, a true alarm is generated at $t = 85$ s where the hard SC current become relatively large, representing 100% of total MPP current. Generally, it is possible to conclude that the overall proposed methods are more reliable and robust, that able to detect intermittent very soft SC faults even at VLG.

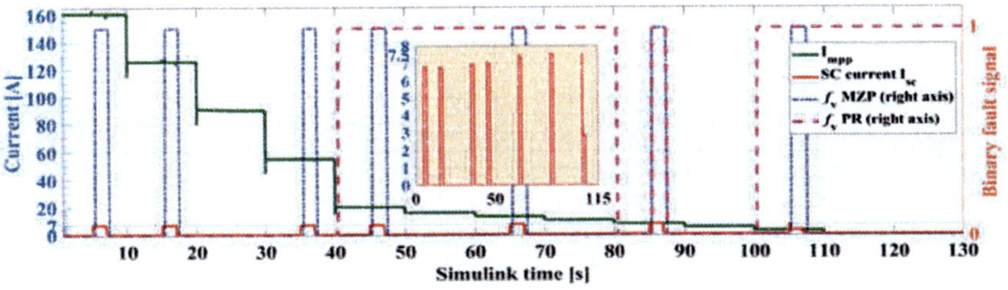

(a) .

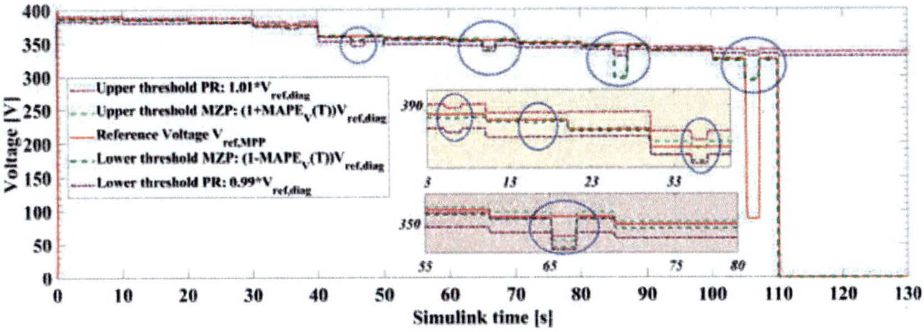

(b) Fault signature.

Fig. 8 Faulty scenario: Comparison between PR and MZP FDD method under soft SCs **a** SC current magnitude and fault signals, **b** voltage diagnosis indicator

6 Conclusion

This paper proposed a novel FDD method based on a KF formulated on MZP approach to detect faults in PV systems. The regression method uses the relationship between voltage, current, and temperature under GMPP with irradiation sensorless. In addition, this paper proposes a MZA TCs, according to the current level, to improve the accuracy of the projection functions. Further, an adaptive threshold based on a quadratic PR in state of temperature variable is formulated to precisely examine the health status of the PV plant. The result reveals the effectiveness of the proposed FDD method to detect intermittent soft SC faults even at very-low irradiation while minimizing missing and false alarms.

References

1. Gunduz, H., Jayaweera, D.: Reliability assessment of a power system with cyber-physical interactive operation of photovoltaic systems. Int. J. Electrical Power Energy Syst. **101**, 371–384 (2018)
2. Singh, B., Goyal, S., Kumar, P.: Solar PV cell materials and technologies: analyzing the recent developments. Mater. Today: Proc. **43**, 2843–2849 (2021)

3. Drews, A., Keizer, A., Beyer, H., Lorenz, E., Betcke, J.: Monitoring and remote failure detection of grid-connected PV systems based on satellite observations. Solar Energy **81**(4), 548–564 (2007)
4. Li, B., Delpha, C., Migan, A., Diallo, D.: Fault diagnosis of photovoltaic panels using full I–V characteristics and machine learning techniques. Energy Convers. Manag. **248** (2021)
5. Li, B., Migan-Dubois, A., Delpha, C., Diallo, D.: Evaluation and improvement of IEC 60891 correction methods for I-V curves of defective photovoltaic panels. Solar Energy **216**, 225–237 (2021, 3)
6. Kang, B., Kim, S., Bae, S., Park, J.: Diagnosis of output power lowering in a PV array by using the kalman-filter algorithm. IEEE Trans. Energy Convers. **27**, 885–894 (2012)
7. Barth, N., Jovanovic, R., Ahzi, S., Khaleel, M.: PV panel single and double diode models: optimization of the parameters and temperature dependence. Solar Energy Mater. Solar Cells **148**, 87–98 (2016)
8. Jendoubi, A., Tlili, F., Bacha, F.: Sliding mode control for a grid connected PV-system using interpolation polynomial MPPT approach. Math. Comput. Simul. **167**, 202–218 (2020)
9. Blanke, M., Kinnaert, M., Lunze, J., Staroswiecki, M.: Diagnosis and Fault-Tolerant Control. Springer, Berlin, Heidelberg (2016)
10. Isermann, R.: Fault-diagnosis systems: an introduction from fault detection to fault tolerance. Springer (2006)
11. Kalman, R.: A new approach to linear filtering and prediction problems. J. Basic Eng. **82**, 35–45 (1960)
12. Lynch, C., OMahony, M., Scully, T.: Simplified method to derive the kalman filter covariance matrices to predict wind speeds from a NWP model. Energy Procedia **62**, 676-685 (2014)

Parity-Space and Multiple-Model Based Approaches to Measurement Fault Estimation

Ivo Punčochář and Ondřej Straka

Abstract The paper compares two approaches to measurement fault estimation for linear discrete-time stochastic systems. The first fault estimator utilizes the parity-space approach. It assumes that at most a limited number of components of the measurement fault can be non-zero. A fault is detected using the χ^2 test applied to the parity-space based residuals and then the indices of non-zero components and fault itself are estimated. The second fault estimator is based on the multiple-model approach. The space of measurement faults is quantized to construct a set of models and a variable structure interacting multiple-model estimator is employed to estimate the state and the measurement fault. Both approaches are compared in a numerical example.

1 Introduction

Fault diagnosis is an established research field [5, 8] with many applications [3, 6]. The approaches to fault diagnosis differ in the level of the information that is inferred from the available data. The least information is provided by simple fault detection (a binary decision regarding fault existence), followed by fault isolation (determination of which fault has occurred) and fault estimation (estimation of fault signal time profile) provides the most information about a fault. The approaches to fault diagnosis can roughly be divided into parity-space based, observer based, and parameter estimation based. Although the basic fault detectors based on parity-space and observer approaches do not inherently provide a fault estimate, there are several extensions in this direction, see e.g. [4, 10–12]. On the other hand, fault detectors based on parameter estimation inherently provide a fault estimate [5].

I. Punčochář (✉) · O. Straka
University of West Bohemia, Univerzitní 8, 301 00 Pilsen, Czech Republic
e-mail: ivop@kky.zcu.cz

O. Straka
e-mail: straka30@kky.zcu.cz

D. Theilliol et al. (eds.), *Recent Developments in Model-Based and Data-Driven Methods for Advanced Control and Diagnosis*, Studies in Systems, Decision and Control 467,
https://doi.org/10.1007/978-3-031-27540-1_5

The aim of this paper is to compare two selected fault estimators. The first one is based on the parity-space approach and represents an extension of an idea presented in [9] to simultaneous faults in more that one component of measurement fault. The second one is based on multiple-model estimator and utilizes a variable structure interacting multiple-model (VSIMM) [7] to reduce computational demands related to a large number of models.

The paper is organized as follows. Section 2 deals with the problem formulation. The parity space based fault estimator is described in Sect. 3. Section 4 deals with the fault estimator that is based on the VSIMM estimator. Both fault estimators are compared through a numerical example in Sect. 5. Section 6 concludes the paper.

2 Problem Formulation

Let us assume that a monitored system can be described at each time instant $k \in \mathcal{T} = \{0, 1, \ldots\}$ by the following linear time-invariant discrete-time stochastic model[1]

$$\mathbf{x}_{k+1} = \mathbf{A}\mathbf{x}_k + \mathbf{G}\mathbf{w}_k, \tag{1}$$

$$\mathbf{y}_k = \mathbf{C}\mathbf{x}_k + \mathbf{F}\mathbf{f}_k + \mathbf{H}\mathbf{v}_k, \tag{2}$$

where $\mathbf{x}_k \in \mathbb{R}^{n_x}$ is a state, $\mathbf{w}_k \in \mathbb{R}^{n_w}$ is a state noise, $\mathbf{y}_k \in \mathbb{R}^{n_y}$ is a measurement, $\mathbf{f}_k \in \mathcal{F} \subseteq \mathbb{R}^{n_f}$ is a measurement fault, and $\mathbf{v}_k \in \mathbb{R}^{n_v}$ is a measurement noise. The state noise and the measurement noise are assumed to be mutually independent white noises with the standard Gaussian distributions. The initial condition $\mathbf{x}_0$ is assumed to be independent of both noise processes and it has the Gaussian distribution with the mean $\hat{\mathbf{x}}_{0|-1}$ and covariance matrix $\mathbf{P}_{0|-1}$. The matrices $\mathbf{A} \in \mathbb{R}^{n_x \times n_x}$, $\mathbf{G} \in \mathbb{R}^{n_x \times n_w}$, $\mathbf{C} \in \mathbb{R}^{n_y \times n_x}$, $\mathbf{F} \in \mathbb{R}^{n_y \times n_f}$, $\mathbf{H} \in \mathbb{R}^{n_y \times n_v}$ are assumed to be known. The following three assumptions are made

A1: $n_y > \text{rank}(\mathbf{C}) \geq 1$,
A2: $\text{rank}(\mathbf{H}) = n_y$,
A3: $\text{rank}(\mathbf{F}) = n_f$.

The assumption A1 means that the matrix $\mathbf{C}$ differs from the zero matrix and it is possible to design a residual generator using parity-space approach even if only the measurement equation is considered. The assumption A2 ensures that no linear combination of the components of the measurement $\mathbf{y}_k$ is free of the measurement noise $\mathbf{v}_k$. The assumption A3 means that the model of the fault is not overparametrized. We do not have particular assumption regarding the measurement faults $\mathbf{f}_k$ except that fault-free behaviour is represented by zero value of the measurement fault and non-zero value signifies a measurement fault.

The aim is to design a fault estimator of the form

[1] Note that a time-varying model and a known input to the monitored system can also be considered at the expense of more complicated notation.

$$\hat{\mathbf{f}}_k = \boldsymbol{\sigma}_k\left(\mathbf{y}_0^k\right), \tag{3}$$

where $\hat{\mathbf{f}}_k \in \mathbb{R}^{n_f}$ is an estimate of the measurement fault $\mathbf{f}_k$, $\boldsymbol{\sigma}_k : \mathbb{R}^{(k+1)n_y} \mapsto \mathbb{R}^{n_f}$ is a function that describes the fault estimator,[2] and $\mathbf{y}_0^k = [\mathbf{y}_0^{\mathrm{T}}, \mathbf{y}_1^{\mathrm{T}}, \ldots, \mathbf{y}_k^{\mathrm{T}}]^{\mathrm{T}}$ is a sequence of measurements obtained up to time instant k.

3 Measurement Fault Estimator Based on Parity-Space Approach

The first fault estimator uses the parity-space approach. It is an extension of an idea presented in [9] to estimating measurement faults with more than one non-zero component. A linear residual generator that uses only the current measurement $\mathbf{y}_k$ to generate a residuum $\mathbf{r}_k$ is given as

$$\mathbf{r}_k = \mathbf{W}\mathbf{y}_k, \tag{4}$$

where $\mathbf{r}_k \in \mathbb{R}^{n_r}$ is the residuum and $\mathbf{W} \in \mathbb{R}^{n_r \times n_y}$ is a full row rank matrix with $n_r = n_y - \mathrm{rank}(\mathbf{C})$ that satisfies the following two conditions

$$\mathbf{W}\mathbf{C} = \mathbf{0}_{n_r \times n_x}, \tag{5}$$

$$\mathbf{W}\mathbf{H}\mathbf{H}^{\mathrm{T}}\mathbf{W}^{\mathrm{T}} = \mathbf{I}_{n_r \times n_r}. \tag{6}$$

Since only the mean value of the residuum $\mathbf{r}_k$ depends on the fault $\mathbf{f}_k$, a decision generator can make a binary decision using the following rule

$$d_k = \begin{cases} 0 & \text{if } S_k \leq S_{\mathrm{th}} \\ 1 & \text{if } S_k > S_{\mathrm{th}} \end{cases}, \tag{7}$$

where $d_k \in \{0, 1\}$ is a decision about fault existence ($d_k = 1$ indicates a fault), $S_k = \mathbf{r}_k^{\mathrm{T}}\mathbf{r}_k \in [0, \infty)$ is the χ^2 test statistic, and $S_{\mathrm{th}} \in [0, \infty)$ is a selected threshold.[3] If $\mathbf{f}_k = \mathbf{0}_{n_f \times 1}$, the statistic S_k has the central χ^2 distribution with n_r degrees of freedom. For a non-zero fault $\mathbf{f}_k$, the statistic S_k has the non-central χ^2 distribution with n_r degrees of freedom and the parameter of non-centrality $\lambda_k = \mathbf{f}_k^{\mathrm{T}}\mathbf{F}^{\mathrm{T}}\mathbf{W}^{\mathrm{T}}\mathbf{W}\mathbf{F}\mathbf{f}_k$. The threshold S_{th} can be computed as

$$S_{\mathrm{th}} = F^{-1}\left(1 - P_{\mathrm{fa}}, n_r\right), \tag{8}$$

[2] Note that a particular fault estimator might not necessarily use the whole sequence of past measurements to generate the fault estimate.

[3] Although the condition (6) can be omitted, a more computationally demanding expression for the statistic $S_k = \mathbf{r}_k^{\mathrm{T}}(\mathbf{W}\mathbf{H}\mathbf{H}^{\mathrm{T}}\mathbf{W}^{\mathrm{T}})^{-1}\mathbf{r}_k$ must be evaluated on-line.

where $F^{-1} : [0, 1] \times \mathbb{N} \mapsto [0, \infty]$ is the inverse cumulative distribution function of the central χ^2 distribution with n_r degrees of freedom and $P_{\text{fa}} \in [0, 1]$ is a desired probability of the false alarm.

If the decision $d_k = 0$ is generated, the estimate of the measurement fault is simply set to zero, i.e., $\hat{\mathbf{f}}_k = \mathbf{0}_{n_f \times 1}$. If decision $d_k = 1$ is generated, a fault estimate $\hat{\mathbf{f}}_k$ is computed with the assumption that at most $n^* \in \mathcal{I} = \{1, \ldots, n_f\}$ components of the measurement fault $\mathbf{f}_k$ are non-zero. A reasonable value n^* is proposed in the following discussion and the index k is omitted for notational convenience. First, it should be noted that the subspace of undetectable faults is given as

$$\mathcal{F}_{\text{us}} = \left\{\mathbf{f} \in \mathbb{R}^{n_f} : \mathbf{Mf} = \mathbf{0}\right\}, \tag{9}$$

where $\mathbf{M} = \mathbf{WF} \in \mathbb{R}^{n_r \times n_f}$ and $\text{rank}(\mathbf{M}) = m \leq \min(n_r, n_f)$. If the number of non-zero components of $\mathbf{f}$ is greater than m, there exist non-zero undetectable faults. However, the converse is not true and there can be non-zero undetectable faults even if the number of non-zero components of $\mathbf{f}$ is less than or equal to m.[4] Therefore, the remainder of the analysis considers $n^* \leq m$. For a particular $n \in \mathcal{I}_{\text{red}} = \{1, \ldots, m\}$, let us denote the i-th n-tuple of indices of non-zero components of $\mathbf{f}$ as

$$\mathbf{s}_i = [s_{i,1}, s_{i,2}, \ldots, s_{i,n}]^{\text{T}}, \tag{10}$$

where $s_{i,j} \in \mathcal{I}$, $s_{i,j} \neq s_{i,j'}$ for $j \neq j'$, $j, j' \in \{1, \ldots, n\}$, and $i \in \mathcal{I}_{\text{comb}} = \{1, \ldots, N_n\}$. The number of different n-tuples is denoted N_n and it is given by the combination number

$$N_n = \binom{n_f}{n} = \frac{n_f!}{n!(n_f - n)!}. \tag{11}$$

The measurement fault for the i-th n-tuple can be written as

$$\mathbf{f} = \sum_{j=1}^{n} \mathbf{e}_{s_{i,j}} \alpha_{i,j}, \tag{12}$$

where $\mathbf{e}_{s_{i,j}}$ is the $s_{i,j}$-th column of the identity matrix $\mathbf{I}_{n_f \times n_f}$ and $\alpha_{i,j}$ is the signed magnitude of the fault in the $s_{i,j}$-th component of $\mathbf{f}$. The the mean value of the residuum can be written as

$$\text{E}\{\mathbf{r}\} = \mathbf{Mf} = \sum_{j=1}^{n} \mathbf{Me}_{s_{i,j}} \alpha_{i,j} = \mathbf{M}_i \boldsymbol{\alpha}_i, \tag{13}$$

[4] Note that there are no undetectable faults if $\mathbf{M}$ has the full column rank, i.e., $m = n_f$.

where $\mathbf{M}_i = [[\mathbf{M}]_{:,s_{i,1}}, [\mathbf{M}]_{:,s_{i,2}} \cdots, [\mathbf{M}]_{:,s_{i,n}}] \in \mathbb{R}^{n_r \times n}$ is a matrix[5] consisting of selected columns of $\mathbf{M}$ and $\boldsymbol{\alpha}_i = [\alpha_{i,1}, \alpha_{i,2}, \cdots, \alpha_{i,n}]^{\mathrm{T}}$. If the fault given by the i-th n-tuple has truly occurred, then the mean value of the residuum lies in the subspace $\mathcal{R}_i$ that is spanned by the columns of $\mathbf{M}_i$. Thus, we are interested in relationships between subspaces $\mathcal{R}_i$ for $i = 1, \ldots, N_n$.

Since the condition $\mathcal{R}_i \cap \mathcal{R}_j = \{\mathbf{0}\}$ is too strong to be satisfied except for $n = 1$, it seems practically reasonable to consider the i−th and j−th n-tuple to be marginally distinguishable if

$$\dim(\mathcal{R}_i \cap \mathcal{R}_j) < \min(\dim(\mathcal{R}_i), \dim(\mathcal{R}_j)). \tag{14}$$

This condition admits the existence of the mean value of residuum that can result from different multiple faults. The greatest n, for which the matrices $\mathbf{M}_i$ has the full column rank and the above condition is satisfied for all pairs $i, j \in \mathcal{I}_{\text{comb}}$, is denoted as n^*.

For a selected n^*, we can estimate the index j of the n^*-tuple of non-zero elements as the one that minimizes the distance of the residual vector $\mathbf{r}$ to the subspaces $\mathcal{R}_j$ for $j = \{1, \ldots, N_{n^*}\}$

$$\hat{j} = \underset{j \in \{1,\ldots,N_{n^*}\}}{\arg\min}\ \mathbf{r}^{\mathrm{T}} \left(\mathbf{I} - \mathbf{M}_j \left(\mathbf{M}_j^{\mathrm{T}} \mathbf{M}_j\right)^{-1} \mathbf{M}_j^{\mathrm{T}}\right) \mathbf{r}. \tag{15}$$

If a tie occurs in the minimization, it can be interpreted as that some n^*-tuples cannot be distinguished. Once $\hat{j}$ is determined, the estimate of the fault $\mathbf{f}$ can be computed as

$$\hat{\mathbf{f}} = \mathbf{M}_{\hat{j}} \hat{\boldsymbol{\alpha}}, \tag{16}$$

where $\hat{\boldsymbol{\alpha}}_k$ is a least square estimate

$$\hat{\boldsymbol{\alpha}} = \left(\mathbf{M}_{\hat{j}}^{\mathrm{T}} \mathbf{M}_{\hat{j}}\right)^{-1} \mathbf{M}_{\hat{j}}^{\mathrm{T}} \mathbf{r}. \tag{17}$$

This fault estimator requires an additional assumption in the form of the maximum number of simultaneous faults. It should improve the quality of the fault estimates if the assumption is satisfied. When faults occur in more measurements than assumed, there are no guarantees regarding the quality of the fault estimate.

[5] Notation $[\mathbf{M}]_{:,i}$ denotes the i-th column of the matrix $\mathbf{M}$.

4 Measurement Fault Estimator Based on Multiple-Model Approach

The multiple-model approach to fault estimation requires a set of models that represent fault-free and faulty behaviours of the system. A discrete set $\bar{\mathcal{F}} \subset \mathcal{F}$ of possible values of the measurement faults can be set up as the Cartesian product of discrete sets for individual dimensions of the measurement fault

$$\bar{\mathcal{F}} = \bar{\mathcal{F}}_1 \times \bar{\mathcal{F}}_2 \times \ldots \times \bar{\mathcal{F}}_{n_f}, \tag{18}$$

where $\bar{\mathcal{F}}_i = \left\{ \bar{f}_i^{(1)}, \bar{f}_i^{(2)}, \ldots, \bar{f}_i^{(M_i)} \right\}$ for each $i \in \{1, 2, \ldots, n_f\}$ and $\bar{f}_i^{(l_i)} \in \mathbb{R}$ denotes the discrete value of the i-th component of the fault vector. It is assumed that $0 \in \bar{\mathcal{F}}_i$ and $\bar{f}_i^{(\ell_i)} < \bar{f}_i^{(\ell_i+1)}$ for each $\ell_i \in \{1, \ldots, M_i - 1\}$. The discrete set $\bar{\mathcal{F}} = \{\bar{\mathbf{f}}^{(1)}, \bar{\mathbf{f}}^{(2)}, \ldots, \bar{\mathbf{f}}^{(M)}\}$ with $M = \prod_{i=1}^{n_f} M_i$ creates a regular grid of points in the n_f-dimensional space. The selection of the grid is driven by the following two factors. The set $\bar{\mathcal{F}}$ should cover the part of the space $\mathcal{F}$ where the faults are expected to lie and the spacing of the regular grid should be chosen to provide a desired accuracy of the fault estimates. Thus, the measurement model is

$$\mathbf{y}_k = \mathbf{C}\mathbf{x}_k + \mathbf{F}\mathbf{f}(\mu_k) + \mathbf{H}\mathbf{v}_k, \tag{19}$$

where $\mu_k \in \mathcal{M} = \{1, \ldots, M\}$ is a model index and $\mathbf{f} : \{1, \ldots, M\} \mapsto \bar{\mathcal{F}}$ is a bijective function defined as

$$\mathbf{f}(j) = \bar{\mathbf{f}}^j. \tag{20}$$

We also complement the original dynamical model with a dynamical model that describes time behaviour of the measurement fault. For each dimension of the measurement fault, we consider an auxiliary Markov chain

$$P(\nu_{i,k+1} = \bar{f}_i^{(\ell_i)} | \nu_{i,k} = \bar{f}_i^{(\ell_i')}) = \begin{cases} p & \text{if } \ell_i = \ell_i' - 1, \\ p & \text{if } \ell_i = \ell_i' + 1, \\ 1 - 2p & \text{if } \ell_i = \ell_i', \\ 0 & \text{otherwise.} \end{cases} \tag{21}$$

If we assume that the switching is independent among the dimensions, the transition probabilities of a complete Markov chain are given as

$$P(\nu_{1,k+1} = \bar{f}_1^{(\ell_1)}, \ldots, \nu_{n_f,k+1} = \bar{f}_{n_f}^{(\ell_{n_f})} | \nu_{1,k+1} = \bar{f}_1^{(\ell'_1)}, \ldots, \nu_{n_f,k+1} = \bar{f}_{n_f}^{(\ell'_{n_f})}) = \prod_{i=1}^{n_f} P(\nu_{i,k+1} = \bar{f}_i^{(\ell_i)} | \nu_{i,k} = \bar{f}_i^{(\ell'_i)}). \tag{22}$$

This transition probability matrix could be far from sparse, which is a key factor for lowering computational demands of the multiple-model estimator. Therefore, the transition probabilities that corresponds to a change in more then one component of the fault vector are forced to zero and the remaining non-zero probabilities are renormalized. Finally, we define the transition probabilities of the Markov chain for $\mu_k \in \{1, \ldots, M\}$ using the bijective function $\mathbf{f}$ as

$$P(\mu_{k+1}|\mu_k) = P(\boldsymbol{\nu}_{k+1} = \mathbf{f}(\mu_{k+1})|\boldsymbol{\nu}_k = \mathbf{f}(\mu_k)). \tag{23}$$

The initial probability $P(\mu_0)$ of this Markov chain is usually selected such that the fault-free case gets the probability one. A multiple-model estimator can be used to compute an estimate the model index μ_k based on all available measurements $\mathbf{y}_0^k$. However, the standard multiple-model estimators like GPB or IMM cannot be used due to a high number of models. This issue can be avoided to a certain degree by using the VSIMM estimator [7] that is an extension of the standard interacting multiple-model (IMM) estimator [2]. It considers the original set $\mathcal{M}$ and its subset $\mathcal{M}_{\mathrm{a},k}$ that includes models assumed to be active at the time instant k. At each time instant k, the VSIMM performs the following three steps:

- Filtering step—The standard filtering step of the IMM algorithm is used to compute pdf $p(\mathbf{x}_k, \mu_k|\mathbf{y}_0^k, \mathcal{M}_{\mathrm{a},k})$ using the current measurement $\mathbf{y}_k$ and the predictive pdf $p(\mathbf{x}_k, \mu_k|\mathbf{y}_{k-1}, \mathcal{M}_{\mathrm{a},k-1}, \mathcal{M}_{\mathrm{a},k})$.
- Adaptation step—The filtering model probabilities $P(\mu_k|\mathbf{y}_0^k, \mathcal{M}_{\mathrm{a},k})$ and the transition probabilities are used to design set $\mathcal{M}_{\mathrm{a},k+1}$ that is an extension or reduction of the set $\mathcal{M}_{\mathrm{a},k}$. The parameters that govern the adaptation are the minimum number of active models $N_{\min}$, a lower thresholds P_{L} and an upper threshold P_{U} to which the filtering model probabilities are compared.
- Predictive step—The standard predictive step of the IMM algorithm is used to compute the predictive pdf $p(\mathbf{x}_{k+1}, \mu_{k+1}|\mathbf{y}_k, \mathcal{M}_{\mathrm{a},k}, \mathcal{M}_{\mathrm{a},k+1})$ using the filtering pdf $p(\mathbf{x}_k, \mu_k|\mathbf{y}_0^k, \mathcal{M}_{\mathrm{a},k})$ and dynamical models.

Since the VSIMM is implemented without any modification, the details are omitted here and an interested reader is referred to [7].

5 Numerical Comparison of Fault Estimators

The presented fault estimators are compared by means of a numerical example. Instead of making the comparison on equal ground, the goal is to demonstrate the main feature of both estimators. Although the intended application is detection of measurement faults in s satellite navigation system, a model of a lower dimensionality is used to ease the presentation. The discrete-time version of the continuous white noise acceleration model [1] for a scalar position is

$$\mathbf{x}_{k+1} = \begin{bmatrix} 1 & T_s \\ 0 & 1 \end{bmatrix} \mathbf{x}_k + \sqrt{q} \begin{bmatrix} \sqrt{\frac{T_s^3}{3}} & 0 \\ \frac{\sqrt{3T_s}}{2} & \frac{\sqrt{T_s}}{2} \end{bmatrix} \mathbf{w}_k, \tag{24}$$

where $T_s = 1$ is the sampling period, $q = 0.1$ is the intensity of the continuous-time process noise, and $n_x = n_w = 2$. It is assumed that four measurements of the position are available and the measurement model is

$$\mathbf{y}_k = \begin{bmatrix} 1 & 2 & 3 & 4 \\ 0 & 0 & 0 & 0 \end{bmatrix}^{\mathrm{T}} \mathbf{x}_k + \mathbf{f}_k + 0.01\mathbf{v}_k, \tag{25}$$

where $\mathbf{F} = \mathbf{I}_{n_y \times n_y}$ and $n_y = n_f = n_v = 4$. The initial state has the mean value $\mathbf{x}_{0|-1} = [5, 0.02]^{\mathrm{T}}$ and covariance matrix $\mathbf{P}_{0|-1} = \mathrm{diag}(10, 0.01)$. The simulated time profile of the considered measurement fault is given in the left graph of Fig. 1. The first half of the simulation, i.e., the time interval $\{0, \ldots, 250\}$, contains the measurement faults $\mathbf{f}_k$ that have various combinations of non-zero values in the first three components including the case where all three components are non-zero simultaneously ($\{120, \ldots, 149\}$). Note that it violates the assumption of the parity-space based fault estimator. The second half of the simulation, i.e., the time interval $\{251, \ldots, 500\}$, contains the measurement faults $\mathbf{f}_k$ that are gradually brought up to the fault vector $\mathbf{f}_k \in \mathcal{F}_{\mathrm{us}} = \mathrm{span}\{[1, 2, 3, 4]^{\mathrm{T}}\}$ that is undetectable by the parity-space approach. The gradual rise is simulated to satisfy the assumption encoded by $P(\mu_{k+1}|\mu_k)$ on the time behaviour of the measurement fault. Note that the sudden disappearance of the measurement fault at the end of the simulation violates the assumption of multiple-model based fault estimator.

The parity-space based fault estimator is designed as described in Sect. 3. The desired probability of false alarm is set to $P_{\mathrm{fa}} = 10^{-3}$. The criterion of marginal distinguishability reveals that the assumed maximum number of non-zero components of a measurement fault is $n^* = 2$ and all matrices $\mathbf{M}_i$ have the full column rank. Since $n_f = 4$, we have $N_{n^*} = 6$. The multiple-model based fault estimator is designed as described in Sect. 4 and parameters are chosen as follows. The grid of discrete measurement faults is chosen as $\bar{\mathcal{F}}_i = \{-4, -3, -2, -1, 0, 1, 2, 3, 4\}$ for $i = 1, 2, 3, 4$ and the total number of models is 6561. The probability for the auxiliary Markov chain is selected $p = 0.2$.The adaptation parameters of the VSIMM are chosen as $N_{\min} = 9$, $P_{\mathrm{L}} = 0.1$, and $P_{\mathrm{U}} = 0.9$. The number of models that were kept active by the multiple-model fault estimator is given in the right graph of Fig. 1.

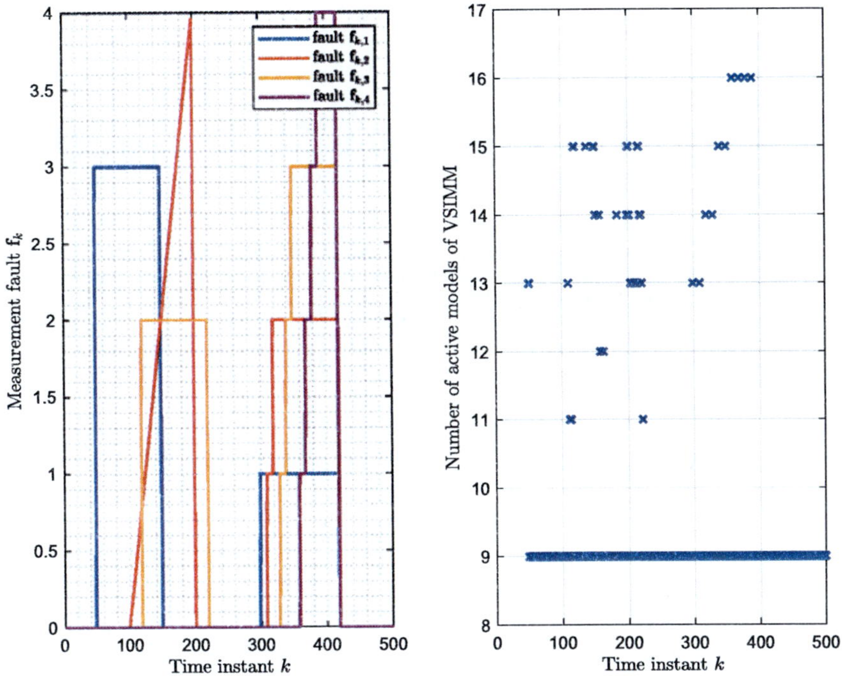

Fig. 1 The time profile of measurement faults and the number of models used by the VSIMM

Figure 2 illustrates the estimates of measurement fault computed by the parity-space based fault estimator (left graph) and by the multiple-model based fault estimator (right graph). It can be seen that both fault estimators provide a decent quality of estimate in the time intervals where their respective assumptions are satisfied. Although the estimates of the multiple-model base fault estimator are naturally impacted by the coarseness of selected grid $\bar{\mathcal{F}}$, this estimator is able to estimate the measurement fault that is even undetectable by the fault estimator based on the parity-space approach.

The main features of both fault estimators can be summarized as follows. The parity-space based fault estimator is theoretically and computationally simple. The only parameters that must be selected are the required probability of the false alarm P_{fa} and the number of the maximum non-zero components of the measurement fault n^*. However, this number cannot be chosen arbitrarily high due to structural constraints imposed by the model and projection operation performed within the fault estimator. Moreover, this fault estimator is suitable for estimating the additive faults only. The multiple-model based fault estimator is more complex and its high computational demands are partially mitigated by the use of the VSIMM. Besides the parameters of the VSIMM estimator itself $N_{\min}$, P_{L}, and P_{H}, also a dynamical model

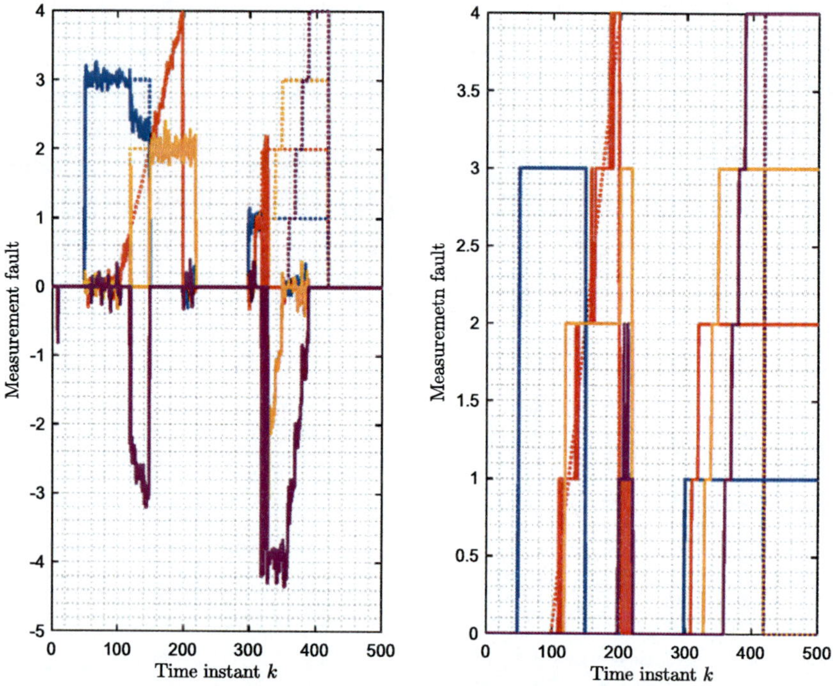

Fig. 2 The measurement faults (dotted lines) and their estimates (solid lines) generated by the parity-space based (left graph) and multiple-model based (right graph) fault estimators

of the fault needs to be chosen. Due to measurement fault model, this estimator is able to estimate faults that parity-space based fault estimator cannot and is suitable for multiplicative faults as well.

6 Conclusion

The paper presented two fault estimators that are based parity-space and multiple-model approaches. The main advantage of the parity-space base fault estimator is its simplicity and low computational demands. The multiple-model based fault estimator can provide better estimation capabilities but its computational demands are higher and an accurate dynamical model of the measurement fault is needed.

Acknowledgements The work was supported by the Czech Science Foundation under grant 22-11101S.

References

1. Bar-Shalom, Y., Li, X.R., Kirubarajan, T.: Estimation with Applications to Tracking and Navigation. Wiley, New York, NY, USA (2001)
2. Blom, H.A.P., Bar-Shalom, Y.: The interacting multiple model algorithm for systems with Markovian switching coefficients. IEEE Trans. Autom. Control **33**(8), 780–783 (1988)
3. Chiang, L.H., Russell, E.L., Braatz, R.D.: Fault Detection and Diagnosis in Industrial Systems, 1st edn. Springer, London, UK (2001)
4. Gao, Z., Ding, S.X., Ma, Y.: Robust fault estimation approach and its application in vehicle lateral dynamic systems. Optim. Control Appl. Methods **28**(3), 143–456 (2007)
5. Isermann, R.: Fault-Diagnosis Systems: An Introduction from Fault Detection to Fault Tolerance. Springer, Berlin, Germany (2006)
6. Isermann, R.: Fault-Diagnosis Applications. Springer, Heidelberg, Germany (2011)
7. Li, X.R.: Multitarget/Multisensor Tracking: Applications and Advances, chap. Engineer's Guide to Variable-Structure Multiple-Model Estimation for Tracking, Artech House (2000)
8. Patton, R.J., Frank, P.M., Clark, R.N.: Issues of Fault Diagnosis for Dynamic Systems, 1st edn. Springer, London, UK (2000)
9. Punčochář, I., Straka, O., Šimandl, M.: Confidence regions for multi-sensor state estimation under faulty measurements. In: Proceedings of 2013 Conference on Control and Fault-Tolerant Systems, pp. 202–207. Nice, France (2013)
10. Sun, X., Patton, R.J.: Robust actuator multiplicative fault estimation with unknown input decoupling for a wind turbine system. In: 2013 Conference on Control and Fault-Tolerant Systems (2013)
11. Wang, Z., Rodrigues, M., Theilliol, D., Shen, Y.: Sensor fault estimation filter design for discrete-time linear time-varying systems. Acta Automatica Sinica **40**(10), 2364–2369 (2014)
12. Zhong, M., Ding, S.X., Han, Q.L., Ding, Q.: Parity space-based fault estimation for linear discrete time-varying systems. IEEE Trans. Autom. Control **55**(7), 1726–1731 (2010)

Diagnosis and Prognosis

Online Condition Monitoring of a Vacuum Process Based on Adaptive Notch Filters

Mohammad F. Yakhni, S. Cauet, A. Sakout, H. Assoum, and M. El-Gohary

Abstract Vacuum systems play an essential role in industrial installations. This system is subject to various operating conditions, making it susceptible to failures that can cause real damage in the working environment. Online condition monitoring is therefore necessary to quickly detect any malfunction and avoid its consequences. It increases the life of system components while reducing maintenance costs and downtime. Various techniques can perform this task. In this paper, we have relied on the transient motor current signature analysis. The system studied is a vacuum system located in the Municipal Technical Center of Poitier, France. Matlab/Simulink program was used to build a digital twin of this system and create several types of faults at different operating speeds. Four Adaptive Notch Filter techniques were developed, based different structures, which are: Chambers' all-pass structure, Regalia's all-pass solution, Cho, Choi & Lee's all-pass method, and M'Sirdi's structure, in order to identify the system faults. M'Sirdi's structure, which is the most recent technology, was discussed in detail. The comparison between the results of the four methods was presented, where the simulation results proved their effectiveness in achieving the desired goal, with the superiority of the M'Sirdi structure.

M. F. Yakhni (✉) · S. Cauet
Université de Poitiers, 86000 Poitiers, France
e-mail: mohammad.yakhni@univ-lr.fr

S. Cauet
e-mail: sebastien.cauet@univ-poitiers.fr

M. F. Yakhni · A. Sakout
Université de La Rochelle, 17000 La Rochelle, France
e-mail: asakout@univ-lr.fr

M. F. Yakhni · H. Assoum · M. El-Gohary
Beirut Arab University, Beirut, Lebanon
e-mail: h.assoum@bau.edu.lb

M. El-Gohary
e-mail: m.gohary@bau.edu.lb

M. El-Gohary
Alexandria University, Alexandria, Egypt

D. Theilliol et al. (eds.), *Recent Developments in Model-Based and Data-Driven Methods for Advanced Control and Diagnosis*, Studies in Systems, Decision and Control 467,
https://doi.org/10.1007/978-3-031-27540-1_6

1 Introduction

There is an increasing demand to enhance the availability and reliability of industrial processes. Any unexpected malfunction can lead to expensive downtime, destruction of equipment, and even danger to human life. Autonomous online health monitoring systems with integrated fault detection algorithms provide early alerts on mechanical and electrical failures to prevent system malfunction. The types of faults differ according to the system concerned [21]. In this paper, we deal with a unit used for suctioning sawdust in the Municipal Technical Centre (MTC) of Poitiers. These units mainly consist of an AC motor, a fan, shafts, bearings, ducts, etc. [22]. In most applications, the fan is not directly linked to the electric motor, primarily to serve two purposes. The first one is to protect the driver, as the fan is subject to various operating and environmental disturbances. Meanwhile, the second objective is to regulate the rotational speed and torque of the fan by adjusting the gear or pulley diameter [10]. So, in brief, we can state that the suction system can be separated into two major parts. The first is the electric motor, which we will consider, in this paper, as an induction motor (IM) as it is the most widely used in the industrial fields, including suction systems. The second section is the pure mechanical unit, which comprises the rest of the components. Thus, we will deal with these systems as Fan/Motor systems. The IM is affected by various faults, that can be divided into two types, electrical and mechanical. The most common electrical defercts are: the stator faults [7] and the voltage unbalance [8], While for mechanical ones, the most common are: the presence of eccentricity [24], the bearings damage [17] or the broken rotor bar [1]. Regarding the fan side, it may be affected by various faults, like any rotating mechanism. The most significant ones are: fan unbalance, belt defects, shaft misalignment, and bearing problems [11].

Vibration and acoustic emission analysis is still the most used approach, but both are expensive technologies in terms of complexity and sensors. An efficient method that may accomplish this task is motor current signature analysis (MCSA) since all of these defects affect the electric motor either directly or indirectly. Since most applications require the driver to operate at variable speeds, we go for the transient MCSA, since the current varies with load and speed [21]. Since the system under study is in a company and not in an academic laboratory, it is impractical to manipulate it by creating faults and seeing its results. Therefore, a digital twin, created by [22], was used to simulate the actual system.

Since the concept was first coined by John Vickers and Michael Grieves [9], many authors have attempted to define the term Digital Twin, beginning with the aerospace industry [12], focusing on structural mechanics, materials science, and long-term performance prediction of air and spacecraft [18]. With the growth of Industry 4.0, the focus has shifted to manufacturing [12]. It has been adopted in many types of previous research and has shown satisfactory performance in achieving its goals [6, 20].

To process the current signals, four different Adaptive Notch Filters (ANFs) structure were used. ANF approach has been used in many researches to estimate the frequencies of sinusoidal signals. Various structure were adopted to develop the filter. Chaochao et al. [3] describes a technique for calculating the power system frequency using the second-order ANF. The suggested method accurately calculates the power system frequency, according to simulation findings. The suggested methods make use of Newton's method and stochastic optimization, are simple for using, and converge quickly on a global scale. A novel improved ANF with phase shift was presented by [4]. And used to reduce synchronous vibration for the rotor of active magnetic bearings. Simulation and experimentation show the suggested ANF's efficacy and adaptability in eliminating synchronous regulated current over a wide operating speed range. A variable step-size ANF utilizing a mixed gradient technique is provided by [23] for frequency estimation in order to enhance the performance of ANF. Coriolis mass flow meter application and simulation results both attest to the efficacy of the suggested approach. A parallel ANF technique was created by [19] to get a precise estimate of the frequency needed to generate a sinusoidal reference signal. The performance of the noise reduction system is enhanced by the proposed algorithm's accurate estimation.

Punchalard [14] developed a limited IIR ANF for frequency estimation of a single real tone imbedded in Gaussian noise using a weighted least squares technique. The suggested approach performs better than several earlier methods in terms of both convergence rate and steady-state mean square error. Results from computer simulations are presented to support the claim.

This document is structured into four main sections. Section 2 describes the MTC vacuum system. It then discusses the development of the digital twin. The four ANF methods are presented in Sect. 3. The simulation results are presented in Sect. 4 and discussed. The last section is the conclusion of this article with a portal for future work.

2 Vacuum Model

The MTC of Poitiers, France, is a public organization that contributes with its different resources to the realization of projects for the town. The carpentry is part of its workshops. Among the devices in this section, what concerns us is the sawdust suction device. We are interested primarily in monitoring the operating condition of the fan/motor unit. The system to be controlled is represented in Fig. 1. Figure 2 presents a detailed view of this arrangement. In this article, the mathematical system that the one developed by [22] was used. The concept of the free body diagram and Newton's second law were used to build it. It was divided into three masses: Pulley 1, Pulley 2 and the Fan, where their three equations of motions are presented below in Eqs. (1) to (3), respectively:

Fig. 1 The CTM fan/motor unit **a** fan side **b** transmission and IM side

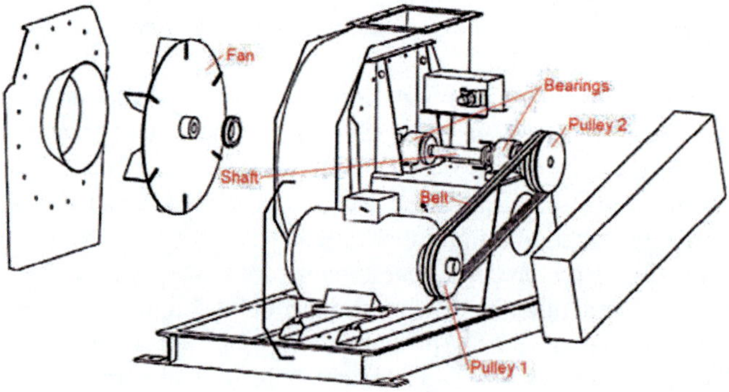

Fig. 2 The real system with labels

$$J_1\ddot{\theta}_1 = T_e - 2K_b(R_{p1}\theta_1 - R_{p2}\theta_2) \tag{1}$$

$$J_2\ddot{\theta}_2 = 2K_b(R_{p1}\theta_1 - R_{p2}\theta_2) - K_s(\theta_2 - \theta_F) - B_r(\dot{\theta}_2 - \dot{\theta}_F) \tag{2}$$

$$J_F\ddot{\theta}_F = K_s(\theta_2 - \theta_F) + B_r(\dot{\theta}_2 - \dot{\theta}_F) + T_a \tag{3}$$

Let θ_1, θ_2, and θ_F be the rotational angle of pulley 1, pulley 2, and the fan respectively, whereas J_1, J_2, and J_F be the polar moments of inertia of the three weights respectively. T_e is the resulting electromagnetic torque of the IM, and T_a is the load torque, which will be discussed in the next paragraph. The shaft has a torsional stiffness K_s, while three belts were modeled as flexible components with equivalent linear stiffnesses K_b. The overall friction coefficient of the two bearings is B_r. The defect is added to the model as a pair T_d as a sine wave, and all defect values will be added to T_a. These two parameters are expressed as Eqs. (4) and (5), Where A and f_d are the amplitude and frequency of the defect, that varies with time (t) respectively, while

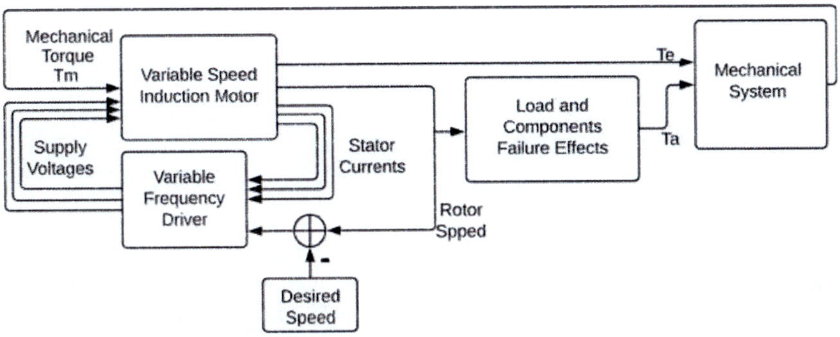

Fig. 3 Simulink model diagram

Q, w and ΔP are the flow rate, the fan speed (rad/s) and the pressure difference on the fan.

$$T_d = A \times cos(2\pi f_d t) \tag{4}$$

$$T_a = (Q \times \Delta P/w) + T_d \tag{5}$$

The value of T_a varies according to two cases: the normal case and in case of the presence of one or more defects in the system. In the first one, T_a is the aerodynamic torque produced by the fan motion. When a defect appears, its value has added to the equation of this torque. In this paper, we discuss the most common faults in the system, each with a particular frequency proportional to the speed of rotation [21].

Matlab/Simulink software was used to create the model. Figure 3 shows the over all model that consists of a three-phases variable speed IM, driven with a variable frequency drive. The desired speed is set by the user. Feedback from the mechanical system (T_m), which is the sum of the torques on the fan ($J_F\ddot{\theta}_F$), was utilized to control the motor performance.

3 Adaptive Notch Filtering (ANF) Methods

Constant speed operation is not possible in most real-world applications, so it is necessary to apply an analysis method that accommodates all working environments, not just stationary signals. To this end, the technique adopted in this research achieves this goal and can be used for online condition monitoring without the need to store the data and analyze it at a later stage [21]. One phase of the three-phase current is processed, and noise is added to this signal to make it as close as possible to reality. The first step is to filter the current using a variable bandwidth filter because the amplitude of the target frequency is much smaller than the fundamental frequency, so we need to isolate the region where the frequency of the defect appears to track it.

Each defect has a particular signature in current, where the frequency is modulated to appear in the current as expressed by Eq. (6), with f_e is fundamental frequency of the supply power source.

$$f_{modulated} = f_d \pm f_e \tag{6}$$

There is a relationship between these frequencies and the rotation speed [22]. For this, variable frequency estimation techniques were used, where four Adaptive Notch Filter (ANF) structures to track transient frequencies were developed, which are: Chambers' all-pass structure [2], Regalia's all-pass solution [15], Cho, Choi & Lee's all-pass method [5] and M'Sirdi's structure [11]. The ANF structure is adjusted to track these specific frequencies. ANFs are well suited for predicting the frequencies of the sinusoidal components as shown in the ANF transfer function (Eq. (7)), where the i_{th} sinusoidal component is filtered by the second order ANF. T_e is the sampling period.

$$H_i(z) = \frac{1 - a_i z^{-1} + z^{-2}}{1 - r a_i z^{-1} + r^2 z^{-2}} \tag{7}$$

with:

$$a_i = -2cos(2\pi f_i T_e) \tag{8}$$

The transfer function of the second-order notch filter is independent of amplitudes or phases. The bandwidth of the notch filter is determined by the parameter $0 < r < 1$, while the frequency of the notch filter is determined by the value a_i. Since a perturbed sinusoidal signal has p sinusoidal components, let us consider the transfer function of the notch filter with p cells of the second-order adaptive filter in cascade, which is given by Eq. 9.

$$\prod_{i=1}^{p} H_i(z) \tag{9}$$

We assume that the bandwidth parameter r is the same for the notches without sacrificing generality. Therefore, frequency independence determines the independence of the characteristics of each second-order cell. After prefiltering the signal through each filter, each cell can be modified independently of the others [11, 13]. Correlations between frequencies can be exploited to reduce the computational burden of the comb filter. If all frequency components are independent, so we can write Eq. (10) as follows:

$$\tilde{y}_k^j = \prod_{\substack{i=1 \\ i \neq j}}^{p} H_i(q) y_k \tag{10}$$

If we consider that the center-frequency cascade filters f_i of $H_i(z)$ (for $i = 1...p$ and $i \neq j$) have converged, the filter $H_i(z)$ will remove one of the jth sinusoidal components [11, 13]. Then the error ε_k of prediction is obtained by removing the jth component, we can thus write:

$$\tilde{y}_k^j = \frac{1 - a_i z^{-1} + z^{-2}}{1 - r a_i z^{-1} + r^2 z^{-2}} y_k \tag{11}$$

$$\varepsilon_k = H_i(q)\tilde{y}_k^j = \frac{1 - a_j q^{-1} + q^{-2}}{1 - r a_j q^{-1} + q^2 z^{-2}} \tilde{y}_k^j \tag{12}$$

Therefore, a single global minimum for the mean square error (MSE) criterion will be created, with each notch removing a frequency component. For the estimation process, we derive the estimation gradient using the Output Error Prediction Method (OPEM) approach according to Eqs. (13) [16]:

$$\psi_{k-1}^j = \frac{d\varepsilon_k}{da_j} = \frac{(1-r)(1-rq^{-2})}{1 - r a_j q^{-1} + r^2 q^{-2}} \tilde{y}_{k-1}^j \tag{13}$$

In terms of computational complexity, the RML (Recursive Maximum Likelihood) algorithm results mainly from the calculation of the gradient and the filtering (Eqs. (11), (12) and (13)). By switching to an approximate RML method (ARML) [11, 13], a first approximation can be constructed to reduce the computational complexity. Instead of using the Eq. (13), we can use Eq. (14), shown below:

$$\psi_{k-1}^j = \frac{d\varepsilon_k}{da_j} = \frac{S_{k-1}^j - r\varepsilon_{k-1}}{1 - r a_j q^{-1} + r^2 q^{-2}} \tilde{y}_k^j \tag{14}$$

with S_k^j is calculated as shown below:

$$forj = (p-1), \quad ..., 1 \ do$$

$$y_k^j = \prod_{l=j+1}^{p} H_l(q) y_k \tag{15}$$

$$S_k^j = y_k^j - y_k^{j-1} \tag{16}$$

The RML algorithm can be summarized as follows:

$$\begin{cases} for \ j = (p-1), \ ..., 1 \ do \\ \tilde{a}_k^j = \tilde{a}_{k-1}^j + F_k^j \psi_{k-1}^j \varepsilon_k \\ F_k^j = F_{k-1}^j / (\lambda + \psi_{k-1}^j F_{k-1}^j \psi_{k-1}^j) \end{cases} \tag{17}$$

where F_k^j is the matching gain, y_k is the input signal to the cascade filter, and r_k is the forgetting factor. The bandwidth or what we call the depolarization factor r_k must be positive. In the application, this parameter varies exponentially in time from 0 or r_0 to 1 or r_f, according to the expression (18):

$$r_k = r_d r_{k-1} + (1 - r_d) - r_f \tag{18}$$

These developed techniques have a shortcoming in that, in the absence of malfunction, the output of the ANF is random. Therefore, it must be adjusted. To do this, a fake defect whose frequency is close to the frequency of the actual one is added to the system. By tuning the magnitude of this defect, we can create a threshold that will determine the severity of the real malfunction. The estimated frequency is biased towards the frequency with the highest magnitude and constrained between two values, which are the value of the actual and fake defect frequencies.

4 Results and Discussions

This unit consists mainly of Pulley 1 linked directly to the motor. The power is delivered to Pulley 2 through three belts, each with a length (L) equal to 2300 mm of type SPB Vbelt. The fan is coupled to Pulley 2 by a drive shaft with a diameter of 45 mm and a length of 80 mm. The diameter of pulleys 1 and 2 (R_{p1} and R_{p2}) is 160 and 225 mm respectively. The two bearings supporting the fan shaft are 22210 EK type. The fan is made of 60A steel, like the other elements, and has eight blades. The other components have little impact on the basic function of the system, so we ignored their impact on the modeling process.

To test the effectiveness of the developed system, we simulated for 11 s under the following conditions: the defect occurred at t = 5 s, while the rotor speed varies from 2950, which is the nominal speed, to 2500 rpm at $t = 6.$s Seven types of defects were added to the system, which are: fan imbalance, bearing inner ring and balls defects, belts malfunction, broken rotor bar, notches harmonics, and eccentricity or load effects. All parameters used to creates these faults are as that used by [22]. Note that f_r is the rotational frequency of the rotor, and N_b is the fan blades number. The initialisation parameters of each ANF structure are as follow, for Chambers' structure: $\alpha = 0.9385$, $\psi_0 = 1$, $\mu = 0.0000127$, $\gamma = 0.99$ and $\beta = cos(w_0)$, with w_0 is the initialization frequency, while the parameters of Reglia's are: $\psi_0 = 1$, $\mu = 0.0000125$, $\gamma = 0.99$ and $\beta = 0.02$. For Cho's method: $\alpha = 0.95$, $\gamma = 0.5$ and $\lambda = 0.9989$, $a_0 = -cos(w_0 \times T_s)$, where $T_s = 0.0001$ s is the sampling time and $k_1 = 1$. Finally the M'Sirdi's ANF has the parameters estimates $\hat{a}_{ik}$ is initially set to zero ($\hat{a}_{i0} = 0$). The ANF bandwidth are set at $r_0 = 0.1$, at the beginning and exponentially increases to $r_f = 0.8$ witha factor of $r_d = 0.99$, as Eq. 18, and $\lambda_0 = 0.99$.

In this paper, We will show the performance of ANFs with the most popular defect in this type of systems, wich is the fan imbalance, that varies with the rotation frequency as follow: $N_b \times f_r$, and what applies to it applies to the other kinds of malfunctions. Figure 4 shows the desired operating speed with the actual rotor frequency (f_r). The performance of the four developed techniques in case of this type of malfunction is shown in Fig. 5.

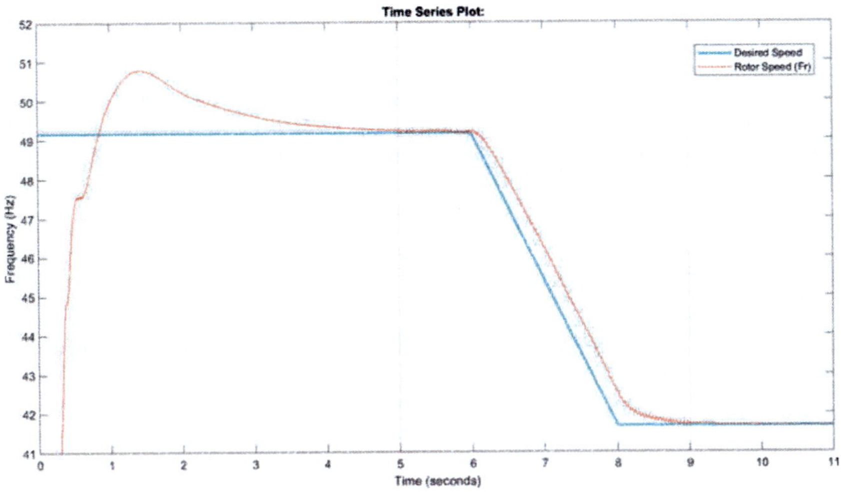

Fig. 4 The Desired and the actual rotor speed

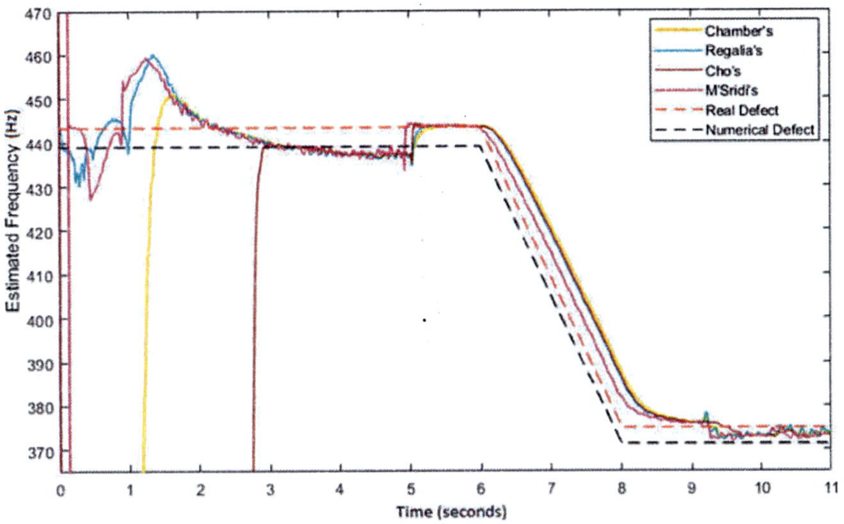

Fig. 5 Estimated frequencies for fan imbalance in case of four different ANF methods

The simulation results proved the effectiveness of the proposed approach in achieving the desired objective. All four techniques were able to track the frequency of faults and show their severity, by moving towards or away from the frequency of the real defect. It should be emphasized that the technology developed by M'Sridi delivers the best results in terms of response time and stability.

5 Conclusion

Electrical Signals can be used to detect vacuum system faults. Each defect in the system has particular frequencies. These frequencies are proportional to the motor rotational speed, so the TMCSA was used as a condition monitoring technique for this system. Four variable frequency estimation techniques based on ANF structures have been used to achieve this goal. A numerical defect is added to the system in order to control the output of the ANF and determine the severity of the malfunction by adjusting its amplitude. At last, as a proposition of future works, the created approach can be utilized with other kinds of industrial systems and diverse types of defects. Other approaches can moreover be adopted in processing the current signals. As well another theme of research, is the usage of the artificial intelligence, like Artificial Neural Networks, Fuzzy and Neuro-Fuzzy logic, to make the alert that determine the severity of the defects.

References

1. Asad, B., Vaimann, T., Belahcen, A., Kallaste, A., Rassolkin, A., Heidari, H.: The low voltage start-up test of induction motor for the detection of broken bars. In: 2020 International Conference on Electrical Machines (ICEM), vol. 1, pp. 1481–1487. IEEE (2020)
2. Chambers, J.A., Constantinides, A.G.: Frequency tracking using constrained adaptive notch filters synthesised from all pass sections. In: IEE Proceedings F (Radar and Signal Processing), vol. 137, pp. 475–481. IET (1990)
3. Chaochao, J., Yixin, S., Huajun, Z., Shilin, L.: Power system frequency estimation based on adaptive notch filter. In: 2016 International Conference on Industrial Informatics-Computing Technology, Intelligent Technology, Industrial Information Integration (ICIICII), pp. 191–194. IEEE (2016)
4. Chen, Q., Liu, G., Han, B.: Unbalance vibration suppression for AMBs system using adaptive notch filter. Mech. Syst. Signal Process. **93**, 136–150 (2017)
5. Cho, N.I., Choi, C.-H., Lee, S.U.: Adaptive line enhancement by using an IIR lattice notch filter. IEEE Trans. Acoust. Speech Signal Process. **37**(4), 585–589 (1989)
6. Chowdhury, S., Yedavalli, R.K.: Dynamics of belt-pulley-shaft systems. Mech. Mach. Theory **98**, 199–215 (2016)
7. Glowacz, A., Glowacz, W., Glowacz, Z., Kozik, J.: Early fault diagnosis of bearing and stator faults of the single-phase induction motor using acoustic signals. Measurement **113**, 1–9 (2018)
8. Gonzalez-Cordoba, J.L., Osornio-Rios, R.A., Granados-Lieberman, D., de Romero-Troncoso, R.J., Valtierra-Rodriguez, M.: Thermal-impact-based protection of induction motors under voltage unbalance conditions. IEEE Trans. Energy Convers. **33**(4), 1748–1756 (2018)
9. Grieves, M.: Manufacturing Excellence through Virtual Factory Replication (2015)

10. Kaya, D., Kılıç, F., Hüseyin Öztürk, H.: Energy Management and Energy Efficiency in Industry: Practical Examples. Springer Nature (2021)
11. M'Sirdi, N.K., Monneau, A., Naamane, A.: Adaptive notch filters for prediction of narrow band signals. In: 2018 7th International Conference on Systems and Control (ICSC), pp. 403–408. IEEE (2018)
12. Negri, E., Fumagalli, L., Macchi, M.: A review of the roles of digital twin in cps-based production systems. Procedia Manuf. **11**, 939–948 (2017)
13. Pei, S.-C., Tseng, C.-C.: Real time cascade adaptive notch filter scheme for sinusoidal parameter estimation. Signal Process. **39**(1–2), 117–130 (1994)
14. Punchalard, R.: Frequency estimation based on WLS-constrained adaptive notch filter. In: 2020 17th International Conference on Electrical Engineering/Electronics, Computer, Telecommunications and Information Technology (ECTI-CON), pp. 353–356. IEEE (2020)
15. Regalia, P.A.: An improved lattice-based adaptive IIR notch filter. IEEE Trans. Signal Process. **39**(9), 2124–2128 (1991)
16. Stearns, S.: Error surfaces of recursive adaptive filters. IEEE Trans. Circuits Syst. **28**(6), 603–606 (1981)
17. Toma, R.N., Prosvirin, A.E., Kim, J.-M.: Bearing fault diagnosis of induction motors using a genetic algorithm and machine learning classifiers. Sensors **20**(7), 1884 (2020)
18. Tuegel, E.J., Ingraffea, A.R., Eason, T.G., Spottswood, S.M.: Reengineering aircraft structural life prediction using a digital twin. Int. J. Aerosp. Eng. **2011** (2011)
19. Wang, H., Sun, H., Sun, Y., Ming, W., Yang, J.: A narrowband active noise control system with a frequency estimation algorithm based on parallel adaptive notch filter. Signal Process. **154**, 108–119 (2019)
20. Yakhni, M.F., Ali, M.N., El-Gohary, M.A.: Magnetorheological damper voltage control using artificial neural network for optimum vehicle ride comfort. J. Mech. Eng. Sci. **15**(1), 7648–7661 (2021)
21. Yakhni, M.F., Cauet, S., Sakout, A., Assoum, H., Etien, E., Rambault, L., El-Gohary, M.: Variable speed induction motors' fault detection based on transient motor current signatures analysis: a review. Mech. Syst. Signal Process. **184**, 109737 (2023)
22. Yakhni, M.F., Hosni, H., Cauet, S., Sakout, A., Etien, E., Rambault, L., Assoum, H., El-Gohary, M.: Design of a digital twin for an industrial vacuum process: a predictive maintenance approach. Machines **10**(8), 686 (2022)
23. Yang, H., Tu, Y., Li, M.: A variable step-size adaptive notch filter for frequency estimation using combined gradient algorithm. J. Phys.: Conf. Ser. **1187**, 032085 (2019). IOP Publishing
24. Yassa, N., Rachek, M., Houassine, H.: Motor current signature analysis for the air gap eccentricity detection in the squirrel cage induction machines. Energy Procedia **162**, 251–262 (2019)

A Study of OBF-ARMAX Performance for Modelling of a Mechanical System Excited by a Low Frequency Signal for Condition Monitoring

Oscar Bautista Gonzalez and Daniel Rönnow

Abstract A digital twin of a mechanical system (a pair of axial rolls in a ring mill used in a steel plant) with poles close to the unit circle and the real axis in the discrete pole-zero map was built. The system was excited by a signal concentrated in the low-frequency band. For this particular case, it is shown that the ad-hoc combination of ARMAX and orthonormal basis filter model structures outperform model structures based on either ARMAX or orthonormal basis functions when estimating the poles of the basis by analyzing the data in the frequency domain. The followed modelling methodology of the system is described in detail to help replicate the work for similar systems in the steel industry. Real production data from a steel plant were used in contrast to previous studies, where the combination of ARX and ARMAX with orthonormal basis filter model structures was evaluated using simulated data instead of real data. We believe that the resultant model can be used when having systems with poles close to the unit circle and real axis and poor excited input signal concentrated in the low frequency band. The resultant model can be used for condition monitoring and failure detection.

1 Introduction

The fourth industrial revolution significantly improves the industry's efficiency through its digitalization using technologies based on cloud computing, digital twins, and IoT. Digital twins are virtual representations of objects, systems, or processes [1]. Digital twins bring the opportunity of monitoring systems status, real-time communication between the virtual representation and the plant, and failure prediction, which needs a model of the system's behaviour [2]. In the steel industry, digital twins

O. B. Gonzalez (✉) · D. Rönnow
Department of Electrical Engineering, Mathematics and Science, University of Gävle, 80176 Gävle, Sweden
e-mail: Oscar.Bautista.Gonzalez@hig.se

D. Rönnow
e-mail: daniel.ronnow@hig.se

D. Theilliol et al. (eds.), *Recent Developments in Model-Based and Data-Driven Methods for Advanced Control and Diagnosis*, Studies in Systems, Decision and Control 467,
https://doi.org/10.1007/978-3-031-27540-1_7

are needed at each stage of the product life cycle, but digital twins of more machines and production processes are still needed [3].

The field of system identification provides a toolkit of methods to describe the behavior of dynamical systems. ARX, ARMAX and ANN are common model structures of such methods where arithmetic complexity plays an essential role in conditioning monitoring of real systems. However, other methods are available, like the ones based on orthonormal basis filters (OBF) [4–7]. The last group presents advantages compared to the first ones regarding the statistical properties of the estimators by having orthonormal inputs when solving the least squares normal equations, models linear in the parameters, and the possibility to encode prior information about the system in the model structure before estimation [8]. A combination of both methods (ARMAX and OBF) in an ad-hoc manner can be found in the literature for the case of simulated data. For example, both methods were combined for identifying systems running in closed-loop where OBF-ARX outperforms ARMAX for control purposes [9]. OBF-ARMAX has been used for identifying closed-loop systems for both stable and unstable systems by exciting the system with an extra signal [10]. A high-order ARX system was identified using an OBF model structure, achieving a lower number of parameters than the ARX case [11]. ARMAX and Laguerre functions have been used for modelling the implementation of an efficient Model predictive control (MPC) controller. Laguerre was selected to provide an efficient number of parameters [12].

In the author's previous work, a digital twin of a radial-axial ring rolling (RARR) machine was designed to monitor the machine's status over time by modelling its five subsystems. It was found that ARMAX SISO models gave poor model results for some subsystems [6]. In this work, one of the RARR's subsystems, i.e., the axial canonical rolls, which was modelled unsuccessfully in previous work through a SISO ARMAX model structure, is successfully modelled with a SISO OBF-ARMAX. To our knowledge, it is the first time that OBF-ARMAX has been used with experimental data, and for modelling systems with complex conjugate poles close to the unit circle and the positive real axis in the discrete pole-zero map by using signals of in the low frequency band with poor excitation. This particular system was investigated and validated by modelling real data from a steel plant during production and numerical experiments. The resultant model can be used for condition monitoring of the system and failure detection.

2 System

The analyzed RARR machine from Banning (1971) is used in an operating steel plant. A mechanical scheme of the RARR system can be found in [6]. The RARR can be divided into five different subsystems: (1) the horizontal subsystem with the mandrel, (2) the canonical rolls, (3) the vertical tower with the canonical rolls attached to it, (4) and (5) which are the lower and upper arms, repectively [6]. The canonical rolls' subsystem plays an essential role in the forming process of the ring and is studied in this work. The system operates in different batches (cycles), which can be divided into three stages. (1) The initialization corresponds to the rapid movement of the tools until a pre-defined position close to the ring. (2) A slow movement of the tools until touching the ring is defined as preparation. (3) The rolling process is when the tools are in contact with the ring, and the forming process takes place. The modelled data correspond to the forming process. The control signal of the axial rolls is selected as the input and, the position signal is selected as the output signal. A IBA-DAQ system is used to collect data at 0.01 s sampling time, which is given by the system design.

3 Methodology

A standard procedure in time series analysis for dynamical systems inspired this work's methodology [5]. The data from the RARR's canonical rolls during production in a steel plant of one ring were selected as training data for parameter estimation and data of the following ring as test data for model validation. Simulations were used to investigate further the selected models with signals and systems based on the knowledge of the real system obtained in the data analysis. Six steps were followed and iterated:

1. Data collection: Data from production time in the steel plant were collected. The data during the forming process was used for the following steps.
2. Data analysis: The collected data was analyzed in the time and frequency domain. The input and output signal were plotted in the time domain for several cycles of the process. Relatively similar relations between input and output were observed. No new frequency components in the output were observed compared to the input. Hence, a linear time-invariant (LTI) system was assumed after the analysis. The system's poles were estimated by analyzing data in the frequency domain. These are the poles which are used as prior information in the next step.
3. Model structure selection: After establishing that the system was an LTI system, ARX and ARMAX model structures were considered as candidates. OBF was considered after analyzing the modelling results provided by ARX and ARMAX. Finally, OBF-ARX and OBF-ARMAX were studied. A pair of complex conjugate poles was selected for building the orthonormal basis filters in the case of OBF, OBF-ARX, and OBF-ARMAX as described in the previous step.
4. Data cleaning: Small noise and no outliers were observed in the signals in the time domain. The mean value of signals was removed before parameter estimation.
5. Parameter estimation: The least-square criterion was used for estimating the parameters of each selected model in step 3. The least-square method was used for estimating the parameters of the ARX, OBF-ARX, and OBF-ARX models by using the function *arx* in MATLAB. An iterative method was used for the estimation of ARMAX and OBF-ARMAX's parameters with the function *armax* [13]. Training data were used for estimation.

6. Model evaluation:

   a. Model validation: Cross-validation with the test data, pole-zero map and residual analysis were considered for validating the models. Cross-validation is based on

$$fit = 1 - \frac{||\mathbf{y} - \hat{\mathbf{y}}||}{||\mathbf{y} - \overline{\mathbf{y}}||}, \quad (1)$$

   where $\mathbf{y}$ is a vector containing the measured output, $\hat{\mathbf{y}}$ is a vector of the model's output, and $\overline{\mathbf{y}}$ is the mean value of $\mathbf{y}$. The pole-zero map was studied to check the overfitting and stability of the model. The error signals were analyzed in the time domain.

   b. Simulations: The same models, which were selected in step 3, were further investigated by formulating a numerical example with input signals concentrated in the low frequency band, and a system with complex conjugate poles close to the unit circle and the real axis. The modelling of such a formulated system was done as with the RARR's canonical rolls, with the difference that the output signal was obtained by simulation.

The above modelling procedure was considered finished with relatively good results in the model validation, i.e., similar accuracy levels given by (6) were observed for the training and test data, no overlapping between zeros and poles in the pole-zero map, poles inside the unit circle, and residuals with small amplitudes were considered as desired validation results. In the other case, the procedure was re-started, going from step 6a back to step 3. In the case of OBF, OBF-ARX and OBF-ARMAX, a new number of basis functions or a new pair of complex conjugates was selected in addition. The pole selection was repeated with prior information about the location of the poles.

4 Models

A LTI set of models for SISO systems with $y(t)$ as the scalar output, $u(t)$ as the scalar input and $e(t)$ as the white-noise processes of finite variance is defined as

$$y(t) = G(q, \boldsymbol{\theta})u(t) + H(q, \boldsymbol{\theta})e(t), \quad (2)$$

where $G(q, \boldsymbol{\theta})$ and $H(q, \boldsymbol{\theta})$ are scalar transfer functions with rational functions of the shift operator q as entries, and $\boldsymbol{\theta}$ is the parameter vector which is an element of a subset of $\mathbb{R}^d$ [4]. The scalar transfer functions can be factorized as

$$G(q, \boldsymbol{\theta}) = \frac{B(q)}{A(q)}, H(q, \boldsymbol{\theta}) = \frac{C(q)}{A(q)}, \quad (3)$$

where

$$\begin{aligned} A(q) &= 1 + a_1 q^{-1} + a_2 q^{-2} + \ldots + a_{na} q^{-na}, \\ B(q) &= b_1 q^{-1} + b_2 q^{-2} + \ldots + b_{nb} q^{-nb}, \\ C(q) &= 1 + c_1 q^{-1} + c_2 q^{-2} + \ldots + c_{nc} q^{-nc}. \end{aligned} \tag{4}$$

The model structure given by (2)–(4) is identifiable. $\boldsymbol{\theta}$ includes all the coefficients of the polynomials in (4). The one-step ahead predictor is

$$\hat{y}(t|\boldsymbol{\theta}) = H^{-1}(q, \boldsymbol{\theta}) G(q, \boldsymbol{\theta}) u(t) - (1 - H(q, \boldsymbol{\theta}) y(t) \tag{5}$$

as indicated in [4, 5].

The identification of the parameters in $\boldsymbol{\theta}$ was performed using the given training dataset $D = \{((u(1), y(1)), ..., (..., u(N), y(N))\}$. The least square criterion is used for estimating $\boldsymbol{\theta}$ by

$$\hat{\boldsymbol{\theta}} = \arg\min_{\boldsymbol{\theta}} \frac{1}{N} \sum_{t=1}^{N} \left(y(t) - \hat{y}(t|\boldsymbol{\theta})\right)^2. \tag{6}$$

More information on the parametrization of the model structures of ARX, ARMAX and OBF models can be found in [6].

In the case of the OBF-ARX ad-hoc framework, (2) contains

$$G(q, \boldsymbol{\theta}) = \frac{G_{OBF}(q)}{A(q)}, \; H(q, \boldsymbol{\theta}) = \frac{1}{A(q)}, \tag{7}$$

where

$$G_{OBF}(q, \boldsymbol{\theta}) = \sum_{k=0}^{p-1} \theta_k \mathcal{B}_k(q), \; H(q, \boldsymbol{\theta}) = 1. \tag{8}$$

More information on how to construct $\mathcal{B}_k(q)$ can be found in [8] where prior information can be used in the selection of the poles $\{\xi_0, \xi_1, ..., \xi_{p-1}\}$ [8, 9]. The one-step ahead predictor can be written as (5) where

$$\boldsymbol{\theta} = \left(a_1 \ldots a_{na} \; \theta_0 \ldots \theta_{p-1}\right)^{\mathsf{T}} \tag{9}$$

is the $((na + p) \times 1)$ parameter vector and $\boldsymbol{\varphi}(t)$ is the $((na + p) \times 1)$ regressor vector where

$$\varphi(t, \theta) = \left(-y(t-1) \ldots - y\,(t - n_a) \;\; \mathcal{B}_0(q) u(t) \ldots \mathcal{B}_{p-1}(q) u\,(t)\right)^T. \tag{10}$$

The optimization problem in (6) gives a closed-form solution when considering (9) and (10).

For the OBF-ARMAX ad-hoc framework, (2) contains the same transfer functions as indicated in (7), but $H(q, \boldsymbol{\theta}) = C(q)/A(q)$ where $G_{OBF}(q)$ is the same as in (8) [10].

The one-step ahead predictor can be written as (5) where

$$\boldsymbol{\theta} = \left(a_1 \ldots a_{na}\ \theta_0 \ldots \theta_{p-1}\ c_1 \ldots c_{nc}\right)^{\mathsf{T}} \tag{11}$$

is the $((na + p + nc) \times 1)$ parameter vector and $\boldsymbol{\varphi}(t)$ is the $((na + p + nc) \times 1)$ regressor vector where

$$\begin{aligned}\varphi(t, \boldsymbol{\theta}) = (-y(t-1) \ldots - y\,(t-n_a) \quad \mathcal{B}_0(q)u(t) \ldots \\ \mathcal{B}_{p-1}(q)u(t) \quad \varepsilon(t-1, \boldsymbol{\theta}) \quad \ldots \quad \varepsilon\,(t-n_c, \boldsymbol{\theta}))^T,\end{aligned} \tag{12}$$

having $\varepsilon(t-k, \boldsymbol{\theta}) = y(t-k) - \hat{y}(t-k|\boldsymbol{\theta})$ [4, 5]. The optimization problem in (6) is solved by an iterative method.

5 Results

This section is divided into the modelling of the RARR's canonical rolls using real production data and a numerical example for investigating how OBF-ARMAX compares to the other cases for similar systems and signals. This work compares how well the different model structures fit the data. A comparison in terms of computational cost between ARX, ARMAX and OBF of similar systems can be found in [6], where the reader can find guidelines to analyze the computational complexity of OBF-ARX and OBF-ARMAX.

5.1 *Modelling of the RARR's Canonical Rolls*

The modelling results of the canonical rolls are presented in Table 1. In the case of the ARX model with 40 parameters, $\boldsymbol{\theta}$ and $\boldsymbol{\varphi}(t)$ become (40×1) vectors with $na = nb = 20$. The ARMAX model contains $\boldsymbol{\theta}$ and $\boldsymbol{\varphi}(t)$ as (12×1) vectors with $na = nb = nc = 4$. The same vectors become (34×1) vectors for the OBF model. The specific expressions of $\boldsymbol{\theta}$ and $\boldsymbol{\varphi}(t)$ in the mentioned cases can be found in [6]. OBF-ARX with 12 parameters has $\boldsymbol{\theta}$ in (9) and $\boldsymbol{\phi}(t)$ in (10) as (12×1) vectors where $na = 2$ and $p = 10$. Finally, (11) and (12) become (24×1) vectors with $na = nc = 2$ and $p = 20$. The parenthesis in Table 1 for OBF, OBF-ARX and OBF-ARMAX in the last column corresponds to the parameter ξ_k and ξ_i in (14) of [6], which are equal to each other in our case. The poles must be defined in terms of magnitude and phase.

Table 1 Modelling results for the RARR's canonical rolls and the numerical example. First column indicates if the row is about the canonical rolls or the numerical example. The second column shows the selected model structure, the third and fourth are the values of (6) for each model for the training and validation data, respectively, and the fourth one indicates the number of parameters. "uns." indicates unstability and "(2)" the required two parameters that define the pair of complex conjugate poles of the basis functions

Type	Model	Training	Validation	#p
RARR's canonical rolls	ARX	0.225	0.224	40
	ARMAX (uns.)	0.990	0.987	12
	OBF	0.610	0.562	34(2)
	OBF-ARX	0.364	0.310	12(2)
	OBF-ARMAX	0.983	0.967	24(2)
Numerical experiment	ARX	0.018	0.015	2
	ARX(uns.)	0.979	0.973	3
	ARMAX	0.017	0.015	3
	ARMAX (uns.)	0.979	0.962	5
	OBF	0.998	0.843	18(2)
	OBF-ARX	0.994	0.861	21(2)
	OBF-ARMAX	0.966	0.926	22(2)

The ARX models gave poorer results than ARMAX. But, ARMAX gave an unstable model, which is not reflected in the fit values of the table due to the short time of the process and its slow dynamics. A stable model and better results were given by using OBF to create the transfer function between the input and output signals. OBF-ARX didn't give any improvement in comparison to previous cases. Finally, the use of OBF-ARMAX where $G_{OBF}(q)$ is encoded with complex conjugate poles close to the real positive axis, and the unit circle in the pole-zero map gave the best result. The number of parameters is smaller than for ARX and OBF. Two models' outputs and their corresponding error signals are shown in Fig. 1. Figure 1a shows a small similarity between the OBF's output and the real output, whereas OBF-ARMAX's output is much closer to the real output. Figure 1b presents information about the error signals. OBF's residuals have periodical components of 1 and 2 s and some high frequency components. OBF-ARMAX's residuals are in the high frequency range, but much smaller in amplitude than in the OBF case. The high frequency components of the OBF's error signal are caused by the $G_{obf}(q)$ structure. The functions $\mathcal{B}_k(q)$ are built by placing a pair of complex conjugates poles at the opposite side from (a mirror operation in the imaginary axis) the dominating pole, which is close to one in the unit circle. High frequency components are amplified as a result of these poles. The parameters $\boldsymbol{\theta}$ of these basis functions are not well identified because of the small amplitudes of the excitation signal at high frequencies.

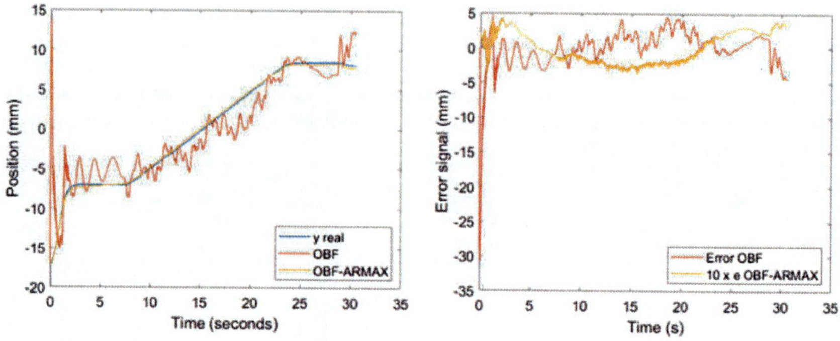

(a) (b)

Fig. 1 Modelling of the RARR's canonical rolls. **a** Measured real (y real) data and modelled output signals (y OBF and y OBF-ARMAX) for the investigated models. **b** Respective error signals (e OBF and e OBF-ARMAX)

5.2 Numerical Example

The system formulation to generate the data of the numerical example follows (2) and (3), (4) becomes $A(q) = 1 - 3.11q^{-1} + 3.695q^{-2} - 2.061q^{-3} + 0.4755q^{-4}$, $B(q) = 0.1q^{-1}$ and $C(q) = 1 - q^{-1} + 0.2q^{-2}$. The signal $u(t)$ is a filtered white noise for excitation in the low frequency band as the real input signal, and $e(t)$ is a white noise signal with variance equal to 1. $A(q)$ was selected according to the frequency response of the real system, which is studied in the data analysis according to step 2 in sect. 3, and $B(q)$ and $C(q)$ are chosen based on the experience of modelling the other subsystems of the machine, reported in [6].

The modelling results of this numerical example are shown in Table 1. In the case of the ARX model with 2 and 3 parameters, $\boldsymbol{\theta}$ and $\boldsymbol{\varphi}(t)$ become (2×1) and (3×1) vectors, respectively. The ARMAX model with 3 and 5 parameters contain the $\boldsymbol{\theta}$ and $\boldsymbol{\varphi}(t)$ as (3×1) and (5×1) vectors, respectively. The same vectors become a (18×1) and (5×1) vectors, respectively, for the OBF model. The specific expressions of $\boldsymbol{\theta}$ and $\boldsymbol{\varphi}(t)$ for the mentioned cases can be found in [6]. OBF-ARX with 21 parameters has $\boldsymbol{\theta}$ in (9) and $\boldsymbol{\phi}(t)$ in (10) as (21×1) vectors where $na = 3$ and $p = 18$. Finally, (11) and (12) become (22×1) vectors with $na = nc = 2$ and $p = 18$. The parenthesis in Table 1 for OBF, OBF-ARX and OBF-ARMAX in the last column corresponds to the poles of the scalar transfer functions $\mathcal{B}_k(q)$ in (8) which are complex conjugate poles and must be defined in terms of magnitude and angular frequency.

According to the numerical example results in Table 1, ARX and ARMAX give poor fit values when the number of parameters is small and excellent results when the number of parameters is high. But the models become unstable for the last cases. The stability problem is not reflected in the metrics due to computing the system's output for short time intervals and the slow dynamics of the system. OBF substantially improves model error but with a significant overfitting problem. This overfitting

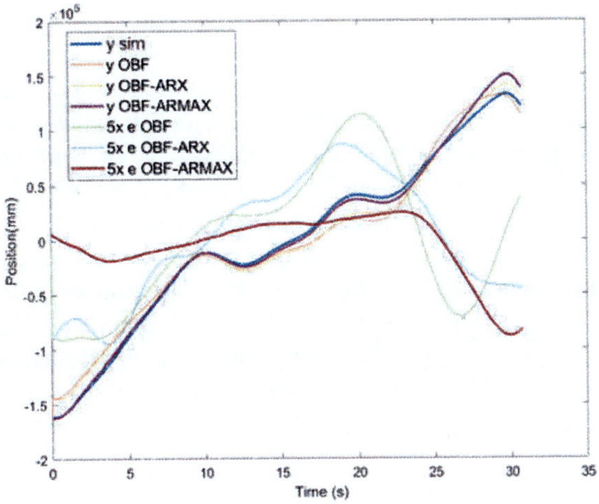

Fig. 2 Numerical example. Simulated (y sim) and modelled output signals for the investigated models. The respective error signals are shown (e)

problem was reduced when applying OBF-ARX and OBF-ARMAX, where the latter outperformed all other models. OBF, OBF-ARX, and OBF-ARMAX were formulated by setting complex conjugate poles close to the unit circle and the real positive axis in the pole-zero map when defining $G_{OBF}(q)$. Figure 2 shows the simulated output and the model's output for the different ad-hoc model structures and the error signals corresponding to the difference between the simulated and each models' output. The residuals corresponding to the OBF model have a periodical component corresponding to 10 s, and 5 s for the OBF-ARX case. The difference between the simulated and the OBF-ARMAX's output is the smallest, with no apparent periodical components and which offset is bigger in the initial and final stages of the process. The measured output signal shows no dynamics corresponding to having complex conjugate poles because the poles are close to the real axis in the discrete pole-zero map, and its magnitude is small relative to the gain of the system.

Comparing the results from the RARR's canonical rolls and the numerical example in Table 1, the results for ARX and ARMAX gave poor modelling results and models with stability problems for both. The modelling results of OBF in the numerical example was better than in the real case in terms of the fit metric and the number of parameters. In the case of OBF-ARX, the numerical example gave better performance than the real case study, and the OBF-ARMAX case gave similar results for both with almost the same number of parameters. OBF-ARMAX outperforms other models in terms of its fit value.

6 Conclusions

A conventional model selection methodology was used and ARX, ARMAX, OBF, OBF-ARX, and OBF-ARMAX were selected for modelling the RARR's canonical rolls. More flexible methods, such as ANN, were discarded after the data analysis step. We conclude that OBF-ARMAX outperforms ARX, ARMAX, OBF and OBF-ARX for these type of systems, i.e., systems with a pair of complex conjugate poles close to the unit circle and real axis, and an exciting signal concentrated in the low-frequency band. Simulations were used to corroborate the results obtained for the real production data. The problem of estimating the parameters of the ARMAX model structure could be complemented by adding some inequality constraints that take into account information of the pole positions obtained in the frequency analysis. Choosing a model structure based on OBF-ARMAX brings the possibility of encoding prior information of the pole positions into the model. It also gives good identification properties due to the orthonormal input signals. OBF-ARMAX is more demanding in terms of computations than the other methods. We suggest that the obtained OBF-ARMAX can be used for monitoring the system and predicting anomalies in a way similar to other machines in the industry, such as in [1, 14].

References

1. Roy, R.B., Mishra, D., Pal, S.K., Chakravarty, T., Panda, S., Chandra, M.G., Pal, A., Misra, P., Chakravarty, D., Misra, S.: Digital twin: current scenario and a case study on a manufacturing process. Int. J. Adv. Manuf. Technol. **107**, 3691–3714 (2020)
2. He, B., Bai, K.: Digital twin-based sustainable intelligent manufacturing: a review. Adv. Manuf. **9**, 1–21 (2021)
3. Xiang, F., Zhi, Z., Jiang, G.: Digital twins technolgy and its data fusion in iron and steel product life cycle. In: 2018 IEEE 15th International Conference on Networking, Sensing and Control (ICNSC), pp. 1–5 (2018)
4. Ljung, L.: System Identification: Theory for the User. Pearson, 2nd edn (1998)
5. Söderstrom, T., Stoica, P.: System Identification. Prentice-Hall, Inc. (1988)
6. Bautista Gonzalez, O., Rönnow, D.: Time series modelling of a radial-axial ring rolling system. Int. J. Model. Identif. Control.
7. Akouemo, H.N., Povinelli, R.J.: Data improving in time series using ARX and ANN models. IEEE Trans. Power Systems. **32**, 3352–3359 (2017)
8. Ninness, B., Gustafsson, F.: A unifying construction of orthonormal bases for system identification. IEEE Trans. Autom. Control. **42**, 515–521 (1997)
9. Tufa, L.D., Ramasamy, M.: Closed-loop identification of systems with uncertain time delays using ARX-OBF structure. J. Process. Control. **21**, 1148–1154 (2011)
10. Tufa, L.D., Ramasamy, M.: Closed-loop system identification using OBF-ARMAX model. J. Appl. Sciences. **10**, 3175–3182 (2010)
11. Muddu, M., Narang, A., Patwardhan, S.C.: Reparametrized ARX models for predictive control of staged and packed bed distillation columns. Control. Eng. Pract. **18**, 114–130 (2010)
12. Nia, S.H.H., Lundh, M.: Robust MPC design using orthonormal basis function for the processes with ARMAX model. In: Proceeding of the 2014 IEEE Emerging Technology and Factory Automation (ETFA), pp. 1–8 (2014)
13. Ljung, L.: System Identification Toolbox: User's Guide. MathWorks Incorporated (1995)
14. Mattsson, P., Zachariah, D., Björsell, N.: Flexible models for smart maintenance. In: 2019 IEEE International Conference on Industrial Technology (ICIT), pp. 1772–1777 (2019)

Pre-localization of Two Leaks in a Water Pipeline Using Hydraulic and Spatial Constraints

Lizeth Torres and Cristina Verde

Abstract This article presents a method to estimate frequency distributions of the parameters associated with the presence of two leaks in a main water pipeline: the locations and the emitter coefficients. Since the method is offline, it can be used to locate simultaneous or sequential leaks; the only requirement is that both leaks must be present when the data for diagnosis are acquired. The method requires as input information pressure and flow rate data to be measured at the pipeline boundaries. In a first stage, hydraulic quasi-steady state relations are computed by using the input information. These relations are used to calculate hydraulic and spatial constraints. In a second stage, Monte Carlo simulations are used together with the previously defined restrictions to obtain frequency distributions of the parameters associated with both leaks. To test the feasibility and potentiality of the method, it was implemented and tested in MATLAB with synthetic data.

1 Introduction

The location of leaks in pipelines has been approached in various ways depending on the type of leak that is searched. Leaks can be roughly classified as: **(i)** single, **(ii)** sequential and **(iii)** simultaneous. The localization of the third class of leaks is one of the most challenging problems since it is not possible to estimate the *exact locations* of two or more simultaneous leaks by using flow rates and pressures taken at the pipeline boundaries when the flow is in steady state [1]. Therefore, the unique possibility to estimate the *exact locations* of simultaneous leaks with hydraulic data is to force the flow to be in unsteady or steady-oscillatory state as done by Navarro et al. [2].

L. Torres (✉) · C. Verde
Instituto de Ingeniería UNAM, Coyoacán, Mexico City, Mexico
e-mail: ftorreso@iingen.unam.mx

C. Verde
e-mail: verde@unam.mx

D. Theilliol et al. (eds.), *Recent Developments in Model-Based and Data-Driven Methods for Advanced Control and Diagnosis*, Studies in Systems, Decision and Control 467,
https://doi.org/10.1007/978-3-031-27540-1_8

1.1 Related Work

There are few works that present a solution to the problem of locating of two leaks (simultaneous or sequential leaks). Some of these works have been taken up to propose the method presented in this work; for this reason, they are briefly following described. In [3], Verde et al. presented an offline algorithm to identify the parameters associated with two leaks: two positions and two emitter coefficients. This algorithm is based on a family of nonlinear lumped models with the same steady state behavior in leak condition and expressed in terms of an equivalent position. Each member of the family is parameterized with different positions and emitter coefficients from those of another member of the family. To determine which of the members is the correct model, i.e., the model with the closest position values and coefficients to the real ones, the transient response of each member is compared with the transient response of the real pipe with two leaks. The parameters of the member that generates a transient response more similar to the real flow response of the pipeline are the parameters that are taken as the best estimations. The key to the success of this algorithm is the generation of a transient to evaluate the responses of the models because it allows the identifiability of the parameters. For improvement purposes, in [4] Verde et al. proposed an algorithm similar to that proposed in [3], but with a significant change: the number of parameters to be identified were reduced to three by using some hydraulic constraints, which were derived from equations of motion and continuity. Navarro et al. submitted the most recent solution in [2], where an approach based on the numerical differentiation of the boundary measurements was proposed. Although the procedure for inverting the nonlinear map was not presented, the authors claim that the leak parameters can be recovered from the boundary measurements and their time derivatives. A requirement for using this approach is that the derivatives of the boundary measurements must be non-zero, preferably oscillatory, as well as nearly noise-free.

1.2 The Essence of the Proposed Method

Taking inspiration from the mentioned works, and in order to take a step forward towards a realistic and implementable solution, this article presents a leak pre-localization method based on a 2-stage offline algorithm. The main strategy of this algorithm is the use of Monte Carlo simulations to estimate the frequency distributions of the parameters that are associated with the presence of two leaks: **(i)** positions and **(ii)** emitter coefficients; see Fig. 1. This task is done by replacing each hydraulic variable that is not exactly known with a range of values (a probability distribution). The algorithm then computes the results over and over again, each iteration using a different set of random values generated from probability distributions, which change at each iteration. In a first stage, the parameters of a fictitious leak are calculated. These fictitious parameters are associated to the parameters of two real leaks that are actually present in the pipeline. These fictitious parameters will

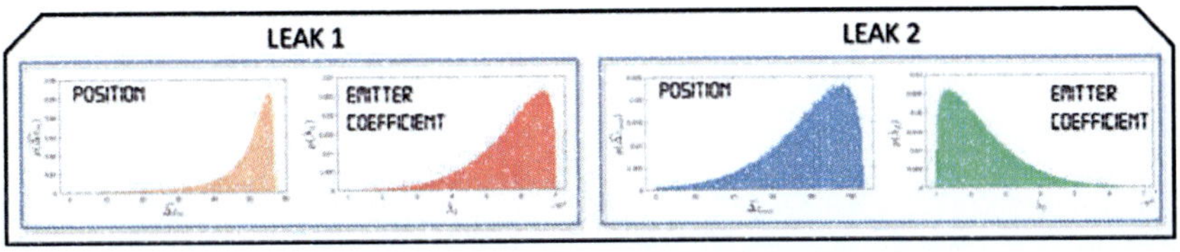

Fig. 1 The goal of the method is the estimation of the probability distributions of the parameters associated with each leak. Notice that each leak has associated a position and an emitter coefficient

help, in the next stage, to define spatial constraints. In a second stage, frequency distributions of the parameters associated with both leaks are obtained by means of Monte Carlo simulations and the spatial constraints. To obtain the parameters of the fictitious leak, the rigid water column (RWC) theory is used. Models under this theory can be obtained by assuming that the walls of the pipelines are rigid and the flowing fluid is incompressible [5–8].

1.3 Disadvantages and Advantages

(i) The proposed method is based on an offline algorithm. An advantage of this fact is that the leak time is not relevant for the method and, therefore, it can be applied for both sequential and simultaneous leaks. **(ii)** Since the method is based on algebraic equations, the sampling time is not important. **(iii)** This method assumes the existence of two leaks, reason by which it is not a complete diagnosis approach, but solely a localization approach, i.e. it does not detect if there are one, two or more leaks in the pipeline. Detecting the amount of leaks before their location is still a pending issue. **(iv)** This method does not provide a unique and *exact solution*, but frequency distributions of the parameters associated with the leaks. **(v)** The main advantage of the proposed method is that it does not require forcing the behavior of the flow in an unsteady state, which is not possible in most real pipelines operating in the world. The price to pay, however, is to get only likely locations. Section 2 presents the physics-based mathematical models, as well as auxiliary equations used for the design and implementation of the algorithm. Section 3 explains with details the main algorithm of the method. Section 4 presents some simulation results that show the method feasibility. Section 5 concludes this work.

2 Rigid Water Column Model

The proposed method involves useful mathematical relationships, which can be derived from the so-called rigid water column model (RWC model). The assumptions and physics behind the RCW model can be consulted in [9, 10]. By assuming that the flow rate is unidirectional and the pipeline has not extractions and/or devices (such as valves or elbows), the RWC model for a pipeline section (denoted by Δz) is expressed by the following equation:

$$\dot{Q} = \theta \frac{\Delta H}{\Delta z} - \alpha Q^{\gamma}, \tag{1}$$

where Q is the flow rate flowing through the pipeline section $\Delta z = z_L - z_0$. z_0 denotes the coordinate of the upstream boundary of the pipeline and z_L the downstream boundary. $\Delta H = H_{in} - H_{out}$ is the pressure head loss along Δz, where H_{in} and H_{out} are the upstream and downstream hydraulic head,[1] respectively. $\theta = gA_r$, where g is the gravity acceleration and A_r is the cross-sectional area. α and $1 \leq \gamma \leq 2$ are parameters related with the friction losses, and they can be associated to physical parameters of the pipeline and fluid by means of the formula employed to describe such losses (e.g. the Darcy-Weisbach equation, the Hazen-Willian equation or the Gaukler-Strikler formula).

2.1 *One-Leak Moment Equations*

The two following equations express the flow motion along a pipeline with a leak, i.e., describe the flow trough two different sections of the pipeline:

$$\begin{aligned} \dot{\bar{Q}}_{in} &= \frac{\theta}{\Delta z_{in}} \left(\bar{H}_{in} - \bar{H}_{\ell}\right) - \alpha \bar{Q}_{in}^{\gamma}, \\ \dot{\bar{Q}}_{out} &= \frac{\theta}{\Delta z_{out}} \left(\bar{H}_{\ell} - \bar{H}_{out}\right) - \alpha \bar{Q}_{out}^{\gamma}. \end{aligned} \tag{2}$$

where $\bar{Q}_{in}$ and $\bar{Q}_{out}$ are the flow rates before and after the leak, respectively, $\Delta z_{in} = z_\ell - z_0$ and $\Delta z_{out} = L - z_\ell$ are the sections to the left and the right of the leak, respectively, $z_0 = 0$ is the origin coordinate, $z_L = L$ is the downstream-edge coordinate, z_ℓ is the leak coordinate (leak position), $\bar{H}_\ell$ is the hydraulic head at the leak position, H_{in} and H_{out} are the hydraulic heads at the upstream and downstream boundaries, respectively. Check Fig. 2 for a graphic illustration of this description. The total flow rate of the leak can be calculated in two ways: with the equation $\bar{Q}_\ell = C_\ell A_\ell \sqrt{2g\bar{H}_\ell}$, where C_ℓ is the discharge coefficient and A_ℓ is the leak area, or with

$$\bar{Q}_\ell = \bar{Q}_{in} - \bar{Q}_{out}. \tag{3}$$

Be aware that $\bar{\ }$ over a variable means that the variable is associated to the one-leak model.

---

[1] The hydraulic head is defined as $H = \dfrac{P}{g\rho} + z$, where P is pressure, ρ is density, g is the gravity acceleration and z is the elevation at the piezometer bottom.

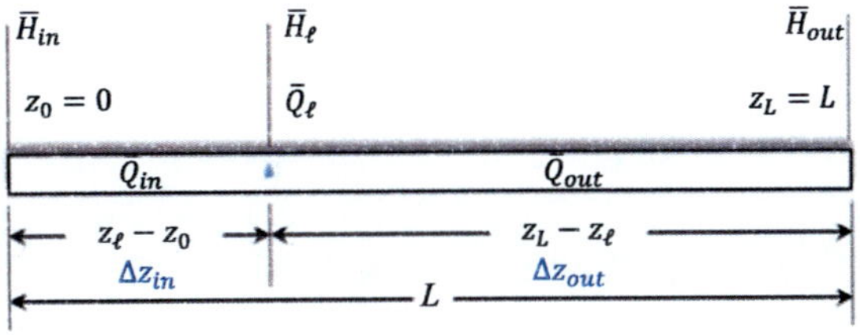

Fig. 2 Pipeline with a single leak

2.2 *Two-Leaks Moment Equations*

The three equations that the describe the flow motion in a pipeline with two leaks are the following:

$$\begin{aligned} \dot{Q}_{in} &= \frac{\theta}{\Delta z_{in}} \left(H_{in} - H_{\ell_1}\right) - \alpha \left(Q_{in}\right)^{\gamma}, \\ \dot{Q}_{12} &= \frac{\theta}{\Delta z_{12}} \left(H_{\ell_1} - H_{\ell_2}\right) - \alpha \left(Q_{12}\right)^{\gamma}, \\ \dot{Q}_{out} &= \frac{\theta}{\Delta z_{out}} \left(H_{\ell_2} - H_{out}\right) - \alpha \left(Q_{out}\right)^{\gamma}, \end{aligned} \tag{4}$$

where Q_{in}, Q_{12} and Q_{out} are flow rates trough sections $\Delta z_{in} = z_0 - z_{\ell_1}$, $\Delta z_{12} = z_{\ell_2} - z_{\ell_1}$ and $\Delta z_{out} = L - z_{\ell_2}$, respectively. z_0, z_{ℓ_1}, z_{ℓ_2} and L are the coordinates of the upstream edge, the position of the first leak, the position of the second leak and the pipe downstream edge, respectively. H_{in} and H_{out} are the hydraulic heads at the inlet an outlet ends, whereas H_{ℓ_1} and H_{ℓ_2} are the hydraulic heads at the leak coordinates. The flow rates of both leaks can be described by the following equations:

$$Q_{\ell_1} = \lambda_{\ell_1}\sqrt{2gH_{\ell_1}}, \quad Q_{\ell_2} = \lambda_{\ell_2}\sqrt{2gH_{\ell_2}}, \tag{5}$$

where $\lambda_{\ell_1} = C_{\ell_1}A_{\ell_1}$ and $\lambda_{\ell_2} = C_{\ell_2}A_{\ell_2}$ are the emitter coefficients. C_{ℓ_1} and C_{ℓ_2} are the discharge coefficients, respectively, and A_{ℓ_1} and A_{ℓ_2} are their areas. A visual description of the variables involved in the moment equations is given in Fig. 3.

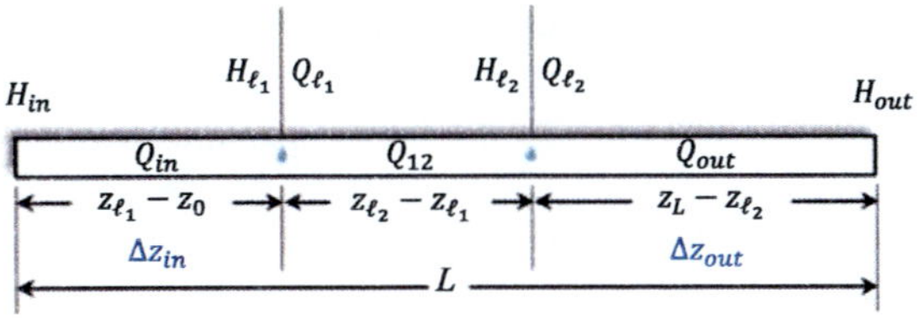

Fig. 3 Pipeline with two leaks

3 Hydraulic Relations

3.1 *Parameters of a Fictitious (Equivalent) Leak*

An equivalent leak is a fictitious single leak with parameters (a position and an emitter coefficient) calculated with information acquired at the boundaries of a pipe that actually has two real leaks. The equivalent leak parameters are very useful because they can be used to define hydraulic constraints, which in turn can be used to form probability distributions of the two actual leak parameters. The parameters of a equivalent leak can obtained by assuming that the flow in a pipeline with two leaks is in steady state. If this assumption holds, then

$$Q_{in} = \bar{Q}_{in} \text{ and } Q_{out} = \bar{Q}_{out},$$

which means that the flow rates at the boundaries of a pipeline with a single fictitious leak are equivalent to the boundary flow rates of a pipeline with two leaks. There are four parameters associated to the fictitious leak: equivalent position, equivalent flow rate, equivalent pressure and equivalent emitter coefficient. The constitutive laws of these parameters are following described. It is worth mentioning that these constitutive laws are hereinafter called *equivalence relations*, since they relate the parameters of a fictitious leak with those of two real leaks.

3.1.1 Equivalent Position

In [11] was shown that the information of the equivalent position helps to delimit the section where each leak is located without implanting additional sensors to those already located on the boundaries. The constitutive law of the equivalent position, which can be obtained from equations (2) in steady state, is given by

$$z_\ell = \frac{\theta\left(\bar{H}_{in} - \bar{H}_{out}\right) - L\alpha\left(\bar{Q}_{out}\right)^\gamma}{\alpha\left(\left(\bar{Q}_{in}\right)^\gamma - \left(\bar{Q}_{out}\right)^\gamma\right)}; \quad \text{where } z_{\ell_1} < z_\ell < z_{\ell_2}. \tag{6}$$

3.1.2 Equivalent Flow Rate

The equivalent flow rate is nothing more than the sum of the two leakage discharges: Q_{ℓ_1} and Q_{ℓ_2}. The equivalent flow rate, Q_ℓ, can be estimated using the leak discharges if the positions of the leaks are known. On the contrary, it can be estimated with

$$Q_\ell = Q_{\ell_1} + Q_{\ell_2} = \bar{Q}_{in} - \bar{Q}_{out}. \tag{7}$$

3.1.3 Equivalent Hydraulic Head

The equivalent hydraulic head results from calculating the pressure at the assumed coordinate (position) of the equivalent leak by using (2) in steady state, that is

$$H_\ell = \bar{H}_{in} - \frac{\alpha z_\ell}{\theta} \bar{Q}_{in}^\gamma; \quad H_\ell = \bar{H}_{out} + \frac{\alpha(L - z_\ell)}{\theta} \bar{Q}_{out}^\gamma. \tag{8}$$

3.1.4 Equivalent Emitter Coefficient

There exist two manners to calculate the equivalent emitter coefficient. The first one is by using the following relation:

$$\lambda_\ell = \frac{Q_\ell}{\sqrt{H_\ell}} \tag{9}$$

The second one is by adding both leak coefficients:

$$\lambda_\ell = C_{\ell_1} A_{\ell_1} \sqrt{2g} + C_{\ell_2} A_{\ell_2} \sqrt{2g} = \lambda_{\ell_1} + \lambda_{\ell_2}. \tag{10}$$

3.2 *Estimation of the Two-Leaks Parameters*

The following equations are used to estimate the parameters of both leaks. The hydraulic heads, H_{ℓ_1} and H_{ℓ_2}, can be calculated with (4) in steady state, such that (5) can be rewritten as follows:

$$\begin{aligned} Q_{\ell_1} &= \lambda_{\ell_1} \sqrt{2g \left(H_{in} - \frac{\alpha(z_0 - z_{\ell_1})}{\theta} (Q_{in})^\gamma \right)}, \\ Q_{\ell_2} &= \lambda_{\ell_2} \sqrt{2g \left(H_{out} + \frac{\alpha(L - z_{\ell_2})}{\theta} (Q_{out})^\gamma \right)}. \end{aligned} \tag{11}$$

where λ_{ℓ_1} and λ_{ℓ_2} can be calculated from (10). The leak position can be calculated from sections Δz_{in} and Δz_{out} as follows: $z_{\ell_1} = \Delta z_{in}$ and $z_{\ell_2} = L - \Delta z_{out}$.

3.3 *Constraints*

The algorithm uses the following constraints to remove (filter) estimates that do not make physical sense:

- The position of the first leak is positive and it is not located at the upstream edge.

$$\Delta z_{in} > 0 \tag{12}$$

- The position of the first leak is before the fictitious equivalent leak.

$$\Delta z_{in} < z_{\ell} \tag{13}$$

- The position of the second leak must be before the downstream edge.

$$(L - \Delta z_{out}) < L \tag{14}$$

- The position of the second leak must be after the fictitious equivalent leak.

$$(L - \Delta z_{out}) > z_{\ell} \tag{15}$$

- The emitter coefficients of both leaks should be limited by an upper bound and a lower bound.

$$\lambda_{\ell_1}^{min} < \lambda_{\ell_1} < \lambda_{\ell_1}^{max}; \ \lambda_{\ell_2}^{min} < \lambda_{\ell_2} < \lambda_{\ell_2}^{max}, \tag{16}$$

  where the maximum values of λ_{ℓ_1} and λ_{ℓ_2} can be calculated as follows

$$\lambda_{\ell_1}^{max} = \lambda_{\ell_2}^{max} = \frac{Q_{\ell}}{\sqrt{H_{\ell}}}.$$

4 Method Description

The method is based on the following Monte Carlo-based algorithm, which estimate the frequency distributions of the parameters associated with the two leaks, i.e., λ_{ℓ_1}, λ_{ℓ_2}, Δz_{in} and Δz_{out}.

Algorithm

1. Load physical and algorithm parameters: $\alpha, \gamma, \theta, N, \lambda_{\ell_1}^{min}, \lambda_{\ell_2}^{min}, \lambda_{\ell_1}^{max}, \lambda_{\ell_2}^{max}$.
2. Compute z_{ℓ}, Q_{ℓ}, H_{ℓ} and λ_{ℓ} with (6), (7), (8) and (9), respectively.
3. Start a loop to generate random values for $\lambda_{\ell_1}^i$ and Δz_{in}^i from uniform distributions. i is the index used to denote each random value (a particle). This loop ends when $i = N$, where N is a predefined value.
for $i = initial\ value : step : final\ value\ (N)$
(a) Generate random sample sets for $\lambda_{\ell_1}^i$ from a bounded prior (uniform) probability distribution:

$$\lambda_{\ell_1}^i \sim U[\lambda_{\ell_1}^{min}, \lambda_{\ell_1}^{max}]$$

(b) Generate random sample sets for Δz_{in}^i from a bounded prior (uniform) probability distribution:

$$\Delta z_{in}^i \sim U[0, z_\ell]$$

(c) Calculate: (1) $\lambda_{\ell_2}^i$ from λ_ℓ and $\lambda_{\ell_1}^i$, (2) $Q_{\ell_1}^i$ from $\lambda_{\ell_1}^i$ and Δz_{in}^i, (3) $Q_{\ell_2}^i$ from Q_ℓ and $Q_{\ell_1}^i$, (4) Δz_{out}^i from $Q_{\ell_2}^i$ and $\lambda_{\ell_2}^i$. It is important to follow the order, since it is a staggered estimation. The calculated values together with the values generated in this step compose a i set.
(d) Verify that the set of values i comply with restrictions (12)–(16). If they do not meet them, the set of values is discarded, if they do meet them, the set of values are saved in vectors as follows:

$$\mathbf{\Delta Z}_{in}[j] \leftarrow \Delta z_{in}^j, \quad \mathbf{\Delta Z}_{out}[j] \leftarrow \Delta z_{out}^j, \quad \mathbf{\Lambda}_{\ell_1}[j] \leftarrow \lambda_{\ell_1}^j, \quad \mathbf{\Lambda}_{\ell_2}[j] \leftarrow \lambda_{\ell_2}^j;$$

where j is the index of each element that meets the restrictions.
End
4. From the vectors, generate frequency distributions for the parameters of the two leaks.
5. Calculate the mean and standard deviation of each distribution.

5 Simulation Tests

The results presented in this section were obtained from simulations executed in Matlab. For this purpose a simulator was implemented by using the momentum and continuity equations that describe the flow in a pipeline. These equations were approximated by using finite differences and were numerically solved by using the Runge-Kutta method. The parameters used for the simulator were: $A_r = 0.0045365$ (m^2), $g = 9.81$ (m/s^2), $L = 163.72$ (m) $\alpha = 19.43$, $\gamma = 1.8794$. The boundary conditions that were imposed for the numerical solutions were $H_{in} = 20.5$ (m) and $H_{out} = 5.8$ (m). The solver time step was fixed at $\Delta t_s = 0.01$ (s). The algorithm was launched using the followin parameters: $\lambda_{\ell_1}^{min} = \lambda_{\ell_2}^{min} = 0.1\lambda_\ell$ and $N = 1 \times 10^5$. In Table 1, three cases of two leaks in a pipeline are presented.

Table 1 Leak cases

Case number	Parameter values
Case 1	$\Delta z_{in} = 20$ (m), $\Delta z_{out} = 100$ (m), $\lambda_1 = 6 \times 10^{-4}$, $\lambda_2 = 3.5 \times 10^{-4}$
Case 2	$\Delta z_{in} = 50$ (m), $\Delta z_{out} = 90$ (m), $\lambda_1 = 5 \times 10^{-4}$, $\lambda_2 = 3 \times 10^{-4}$
Case 3	$\Delta z_{in} = 60$ (m), $\Delta z_{out} = 30$ (m), $\lambda_1 = 4 \times 10^{-4}$, $\lambda_2 = 4 \times 10^{-4}$

Table 2 Results of the pre-localization

Case number	Pre-localization results
Case 1	$\Delta z_{in}^{\mu} = 20.47$, $\Delta z_{in}^{\sigma} = 9.33$, $\Delta z_{out}^{\mu} = 86$, $\Delta z_{out}^{\sigma} = 36.16$
Case 2	$\Delta z_{in}^{\mu} = 37.96$, $\Delta z_{in}^{\sigma} = 15.30$, $\Delta z_{out}^{\mu} = 62.16$, $\Delta z_{out}^{\sigma} = 30.56$
Case 3	$\Delta z_{in}^{\mu} = 65.90$,$\Delta z_{in}^{\sigma} = 21.96$, $\Delta z_{out}^{\mu} = 40.59$, $\Delta z_{out}^{\sigma} = 21.31$

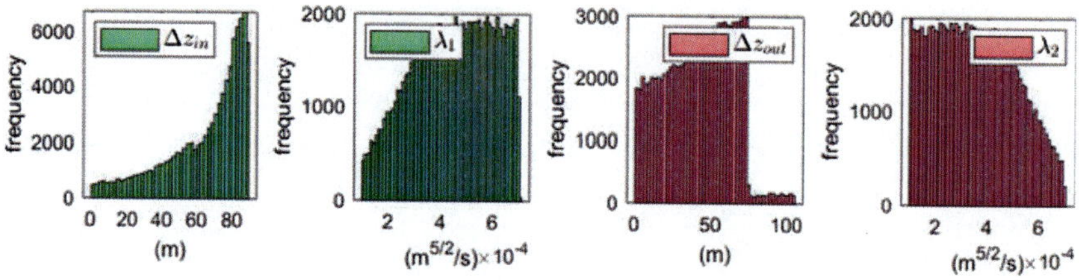

Fig. 4 Frequency distributions of the leaks' parameters for Case 3

The results of the prelocalization are summarized in Table 2. The values with the super-index μ are the means of the frequency distribution of Δz_{in} and Δz_{out}. The values with the super-index σ are the standard deviations of the frequency distribution of Δz_{in} and Δz_{out}. The frequency distributions of the results for Case 3 are illustrated in Fig. 4.

6 Conclusions

A method for pre-localizing two leaks in a pipelines was presented. This method does not require oscillating the flow or causing a transient. Some simulation results of three leaks' cases were presented. The algorithm is perfectible, so in future works, some improvements to the method will be presented, such as the incorporation of changes in the operating points, iterations with cost functions to reduce the standard deviation of the distributions. In addition, experimental results will also be presented to show the advantages and disadvantages of the method in the real world.

References

1. Verde, C., Visairo, N.: In: Proceedings of the American Control Conference, Boston, USA, vol. 5, pp. 4378–4383 (2004)
2. Navarro-Díaz, A., Delgado-Aguiñaga, J.A., Begovich, O., Besançon, G.: Sensors **21**(23), 8035 (2021)
3. Verde, C., Visairo, N., Gentil, S.: Adv. Water Res. **30**(8), 1711 (2007)
4. Verde, C., Molina, L., Torres, L.: J. Loss Prev. Process Ind. **29**, 177 (2014)

5. Wood, D., Funk, J., Boulos, P.: In: Proceedings of the 6th International Conferences on Pressure Surges, pp. 131–142 (1990)
6. Ivanov, K.P., Bournaski, E.G.: Comput. Methods Appl. Mech. Eng. **130**(1–2), 47 (1996)
7. Axworthy, D.H.: Water Distribution Network Modelling: from Steady State to Waterhammer. University of Toronto (1998)
8. Kaltenbacher, S., Steffelbauer, D.B., Cattani, M., Horn, M., Fuchs-Hanusch, D., Römer, K.: Proc. Comput. Control Water Ind. 2–134 (2017)
9. Cabrera, E., Garcia-Serra, J., Iglesias, P.L.: In: Improving Efficiency and Reliability in Water Distribution Systems. Springer, pp. 3–32 (1995)
10. Nault, J., Karney, B.: J. Hydraul. Eng. **142**(9), 04016025 (2016)
11. Verde, C.: Control Eng. Pract. **9**(6), 673 (2001)

Diagnosis and Failure Prognosis of Intermittent Faults: A Bond Graph Approach

Wolfgang Borutzky

Abstract Diagnosis of incipient parametric faults has been widely addressed and various model-based as well as data-based methods have been reported in the literature. In contrast to incipient faults, it is unknown if, when and for how long intermittent faults will reappear which makes their diagnosis and prognosis difficult. This paper addresses intermittent faults with a magnitude that increases over time. Such a sequence of intermittent faults may reach a failure alarm threshold and may eventually lead to a component or even a system failure. The proposed approach uses a diagnostic Bond Graph (DBG) model for an online detection, isolation and estimation of intermittent faults. Pulses of increasing height and varying width not equally spaced on the time evolution of residuals of Analytical Redundancy Relations (ARRs) obtained from an offline developed DBG indicate a degradation trend of a parameter or a variable. Certain values of this trend are concurrently extrapolated in a repeated failure prognosis resulting in a sequence of Remaining Useful Life (RUL) estimates. The proposed BG-based approach to a failure prognosis in the presence of intermittent faults can be combined with a previously presented BG-approach to incipient fault prognosis and is applicable to mode switching models. For illustration, it is applied in a case study to a small electronic circuit in an offline simulation. The case study also considers intermittent faults of constant magnitude.

Keywords Intermittent faults · Increasing fault magnitude · Diagnostic bond graph-based online fault diagnosis · Concurrent repeated failure prognosis · Remaining Useful Life (RUL) estimates

1 Introduction

Failure prognostic is a constitutive part of Prognostic and Health Management (PHM) and is of imperative importance beyond fault diagnosis for all safety critical engineering systems and processes, for supervision, automation and condition based

W. Borutzky (✉)
Bonn-Rhein-Sieg University of Applied Sciences, Sankt Augustin, Germany
e-mail: wolfgang.borutzky@h-brs.de

D. Theilliol et al. (eds.), *Recent Developments in Model-Based and Data-Driven Methods for Advanced Control and Diagnosis*, Studies in Systems, Decision and Control 467,
https://doi.org/10.1007/978-3-031-27540-1_9

maintenance (CBM) of industrial processes. Fault detection and isolation (FDI) as a prerequisite for failure prognostic has been a subject of research with regard to various applications and has become a quite mature discipline in the course of the last decades. Data-driven as well as physics model-based approaches to FDI are widely used in industry and in academia. So far, FDI has mostly addressed abrupt as well as incipient faults. In recent years, research has extended to FDI of intermittent faults.

As to failure prognosis, more recently, combinations of data-driven and model-based approaches have received increasing attention. In comparison to fault diagnosis, the combined use of model-based and data-driven methods for failure prognosis is a still rather young and is still a developing research subject with contributions from various fields. A 2016 survey on prognostics may be found in [2]. Combined bond graph model-based, data-driven approaches have been presented in [3].

Failure prognosis, so far, mostly considers incipient parametric faults so that a degradation trend can be extracted from measurements in a moving time window and can be projected into the future in order to obtain estimates of the remaining useful life (RUL). In contrast, the occurrence of intermittent faults, their duration and their magnitude are unpredictable. Accordingly, while diagnosis of intermittent faults has been subject of ongoing research, the effect of intermittent faults on the RUL of system is still a rather new topic.

With time, the number of intermittent faults may increase and may develop into a permanent fault, or their magnitude may increase until an admissible alarm threshold is reached and action is required in order to avoid a failure. The BG approach to a prognosis of intermittent faults presented in the following assumes that the magnitude of intermittent faults increases over time and is a modification of the BG model-based, data driven approach to the prognosis of incipient parametric faults in [1]. This assumption has also been adopted in a recent case study [4].

2 Bond Graph Approach to Intermittent Fault Prognosis

The proposed BG model-based approach assumes that intermittent faults of increasing magnitude happen and consists of two parts. The first part based on the evaluation of ARRs addresses the detection, isolation and estimation of intermittent faults. The second part computes a sequence of RUL estimates.

2.1 *Bond Graph Model-Based Fault Diagnosis*

The first part of the proposed BG model-based approach to failure prognosis in the presence of intermittent faults, i.e. FDI and fault estimation as a prerequisite for failure prognosis, is performed concurrently to the continuous monitoring of a system on an offline developed DBG. The latter one is obtained from a direct BG model by inverting the causality of detectors representing sensors. The reason for this causality inversion is that sensors deliver measurements, i.e. known inputs into the DBG which serves as generator of constraints between measurements and known input command signals. These constraints are called Analytical Redundancy Relations (ARRs). They may be nonlinear. In the case of BG models with switches, ARRs are even mode-dependent. Switch states may switch off and on parts of the set of ARRs. Their numerical evaluation denoted as residuals should result in values close to zero in case no faults happen. The time evolution of a residual enables to detect intermittent faults.

The dependencies of ARRs from element parameters can be captured in a structural fault signature matrix (FSM) which indicates whether a fault can be detected and, moreover, can be isolated. In the case of a unique fault signature, a potentially faulty parameter can be isolated just by inspection of the FSM. In case several elements do have the same fault signature, numerical parameter estimation can identify the parameters that deviate from their nominal values. Once a faulty parameter has been isolated, the magnitude of a fault can be estimated and its unknown faulty time behaviour, i.e. a degradation trend, can be reconstructed. As an example, in the case study in Sect. 3, the power supply of a small electronic circuit is not stable in the sense that it does not deliver a constant voltage E as expected. Instead, the voltage value is affected by intermittent faults with a magnitude that increases with time. The computation of an ARR residual is used to reconstruct the unknown time behaviour $\tilde{E}(t)$ of the faulty voltage supply. If the time evolution of a residual reveals a sequence of intermittent faults of increasing magnitude, the latter ones can be used to compute a sequence of RUL estimates.

2.2 *Repeated Prediction of RUL Estimates*

The second part of the proposed approach uses the ARR residuals obtained in the first part for a repeated prediction of RUL estimates. Once online monitoring of the health of a system has resulted in a residual different from zero, the monitoring has to proceed until the residual value returns to zero before one can conclude that an intermittent fault happened. If, instead, the residual retains the nonzero value for quite some time, one may conclude that an abrupt fault happened. In case the residual returns to zero at some later time instant, it is clear that an intermittent fault of unforeseeable length has happened. Moreover, even when the residual under consideration returns to zero, one cannot be sure that another intermittent fault will take place at any time instant later. Therefore, to be able to identify some degradation trend and to predict a RUL estimate, two consecutive intermittent faults must be detected and the latter one must be of higher magnitude. A first estimate of a RUL may be obtained from a reconstructed fault signal by determining the intersection of an interpolation line through the points (t_1, v_1), (t_2, v_2) with an alarm threshold line. The time point t_i is the time instant where a considered residual returns to zero, i.e where the length of the ith intermittent fault ends.

If subsequent intermittent faults happen while time moves on, the process can be repeated. The result is a sequence of RUL estimates that converges to zero with ongoing time. If the monotonic increase of intermittent fault magnitudes stops at some time instant, then RUL prediction is no longer possible but can resume when intermittent fault magnitudes increase again. As long as there is no intersection of an interpolating line with an alarm threshold no action is required and a system may continue its operation. However, the time interval between two consecutive intermittent faults may become shorter, their number may increase and may result in a permanent fault.

The approach to a repeated prediction of RUL estimates starting from ARR residuals obtained from a diagnostic BG is illustrated in Sect. 3 by means of an offline simulation. However, the diagnosis of intermittent faults as well as the subsequent repeated computation of RUL estimates can be performed online concurrently to the monitoring of the dynamic behaviour of a system.

3 Offline Simulation Case Study

The approach to a detection of intermittent faults and a prediction of RUL estimates is illustrated by means of the simple circuit in Fig. 1.

First, it is assumed that the element parameter do not become faulty over time but keep their nominal value. The power supply, however, is not stable. Intermittently, the constant voltage E drops to a lower value which becomes smaller over time until a value is reached at which the circuit does not function properly any more. The time evolution of the intermittent voltage supply $\tilde{E}(t)$ is depicted in Fig. 2a. As can be seen, the faults do not occur periodically and their length is not constant.

In this study, the real circuit is replaced by a BG model. To obtain pseudo measurements, the time evolution of the voltages numerically computed by means of the open source simulation program `Scilab` are overlayed with noise. These 'measurements' are then filtered before they are input into a DBG for the computation of ARRs. The offline simulation uses the parameter in Table 1.

The noisy voltages are smoothed by the `Scilab` function `lsq_splin()`, which performs a weighted least squares cubic spline fitting and provides a smoothed signal together with its time derivative. Fig. 2b shows the time evolutions of the smoothed capacitor voltages $\tilde{u}_1$, $\tilde{u}_2$.

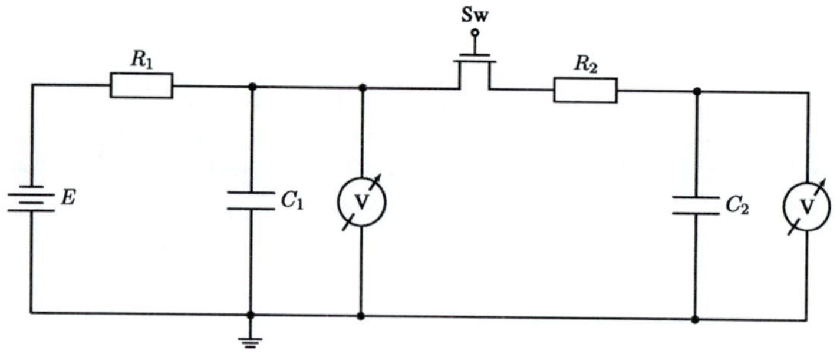

Fig. 1 Simple network with a switch and two voltage sensors

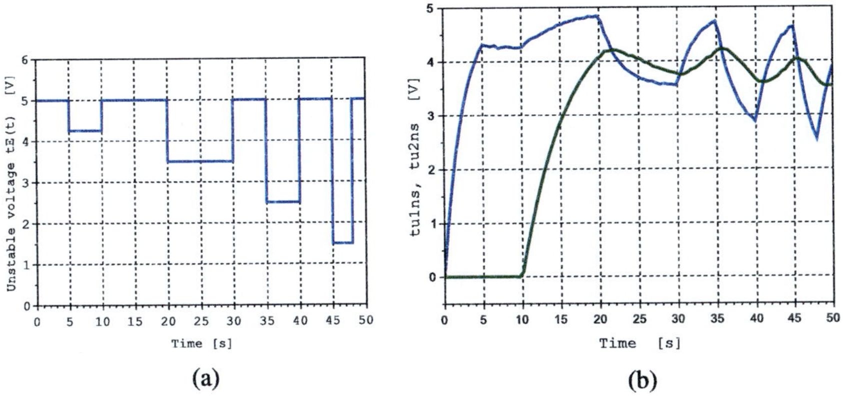

Fig. 2 **a** Voltage supply $\tilde{E}(t)$ **b** smoothed capacitor voltages $\tilde{u}_1$, $\tilde{u}_2$

Table 1 Parameters of the switched circuit in Fig. 1

Parameter	Value	Units	Meaning
E	5	V	Voltage supply
E_{thr}	1	V	Fault threshold
R_1	500	Ω	Resistor R : R_1
R_2	5	$k\Omega$	Resistor R : R_2
R_{on}	0.1	Ω	ON resistance of switch of Sw
C_1	5000	μF	Capacitor C : C_1
C_2	1000	μF	Capacitor C : C_2
$C_{2_{\mathrm{crit}}}$	400	μF	Fault threshold
t_{sw}	10	s	Switching point of Sw
t_0	5	s	Start of the decay of $\tilde{C}_2(t)$
λ	0.0308065	s^{-1}	Rate of the decay

3.1 BG Based Detection and Estimation of Intermittent Faults

The circuit schematic in Fig. 1 is transformed into the DBG in Fig. 3, in which the voltages with a tilde denote measurements from the faulty system. It is assumed that the voltage sensors themselves presented by the detectors De : $\tilde{u}_1$ and De : $\tilde{u}_2$ are faultless.

From the DBG in Fig. 3 the following two mode-dependent ARRs can be deduced.

Fig. 3 Diagnostic bond graph of the network in Fig. 1

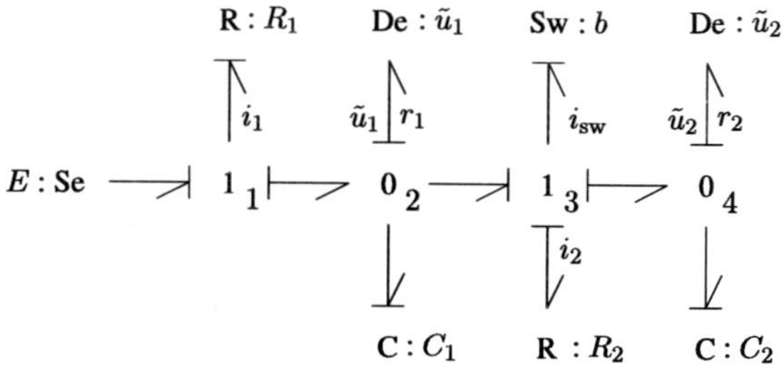

$$0_1 : \mathrm{ARR}_1 : \quad r_1 = i_1 - i_{sw} - C_1\dot{\tilde{u}}_1 = \frac{E - \tilde{u}_1}{R_1} - \frac{b^2}{R_{on} + b \cdot R_2}(\tilde{u}_1 - \tilde{u}_2) - C_1\dot{\tilde{u}}_1$$
$$= \frac{E - \tilde{u}_1}{R_1} - \frac{b}{R_2}(\tilde{u}_1 - \tilde{u}_2) - C_1\dot{\tilde{u}}_1 \qquad (1)$$
$$0_2 : \mathrm{ARR}_2 : \quad r_2 = i_{sw} - C_2\dot{\tilde{u}}_2 = \frac{b}{R_2}(\tilde{u}_1 - \tilde{u}_2) - C_2\dot{\tilde{u}}_2\,, \quad b(t) \in \{0, 1\} \qquad (2)$$

The FSM (Table 2) is a book keeping of which element parameters affect which residual. The last but one column of the FSM indicates whether a faulty parameter can be detected given the two sensors. An entry in the last column records whether a parameter is isolatable.

As can be seen, non of the element parameters is isolatable, except C : C_2. Residual r_1 depends on elements R : R_1, C : C_1 and on R : R_2 in case the switch is closed. As a result, by inspection of the FSM it cannot be decided which parameter is faulty and causes an abnormal behaviour of the circuit. By means of parameter estimation over a time window it can be shown that the element parameters, in fact, do not deviate from their nominal values. To that end, the vector of parameter dependent residuals can be used to define a cost function that is minimised

Table 2 Structural FSM for the switched network in Fig. 1

Element	Parameter	r_1	r_2	D_b	I_b
Voltage source	E	1	0	1	0
De : $\tilde{u}_1$	$\tilde{u}_1$	1	1	1	0
De : $\tilde{u}_2$	$\tilde{u}_2$	1	1	1	0
Switch	b	b	b	b	0
R : R_1	R_1	1	0	1	0
R : R_2	R_2	b	b	b	b
C : C_1	C_1	1	0	1	0
C : C_2	C_2	0	1	1	1

Given that the element parameters keep their nominal values, then numerical computation of (1) yields for the residual $\text{res}_1 = R_1 \cdot r_1$ a time evolution that clearly enables to detect the intermittent faults (Fig. 4).

Moreover, residual res_1 can be used to estimate the magnitude of the intermittent faults and to reconstruct the unknown faulty voltage supply $\tilde{E}$. If the constant voltage E in (1) is replaced by the faulty unknown one $\tilde{E}$, then ARR_1 reads:

$$0 = \frac{\tilde{E} - \tilde{u}_1}{R_1} - \frac{b}{R_2}(\tilde{u}_1 - \tilde{u}_2) - C_1\dot{\tilde{u}}_1 \tag{3}$$

Solving for $\tilde{E}$ and observing (1) gives

$$r\tilde{E} = E - R_1 r_1 = E - \text{res}_1 = \tilde{E} \tag{4}$$

$$\Delta E = \text{res}_1 \tag{5}$$

By comparison of Fig. 4a and b one can see that the residual values equal the magnitude of the intermittent faults.

3.2 RUL Prediction for Intermittent Faults

From Fig. 4 one can see that the intermittent faults return to zero at time instances $t = 10, 30, 40, 48\,s$. Connecting the values of the reconstructed time evolution of the faulty voltage supply $rE(t)$ at two consecutive time points yields lines that intersect with an alarm threshold defined to be 1 V as depicted in Fig. 5.

The time points of the intersections with the alarm threshold yield the decrease of the RUL over time shown in Fig. 6b. The offline simulation run up to $t = 50$ s takes into account that online measurement cannot look into the future. Measurements only available up to the current time instant can be used for computing ARR residuals

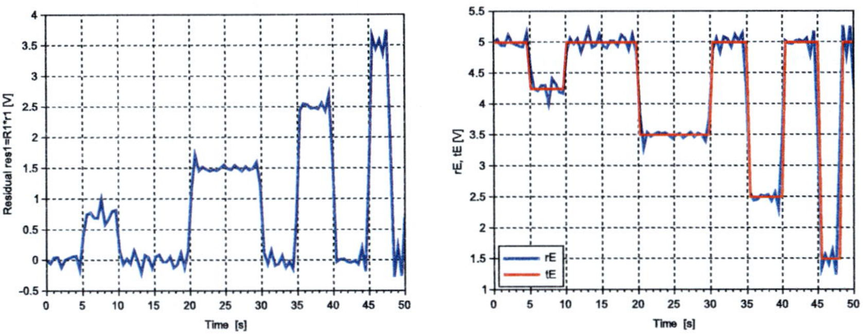

(a) (b)

Fig. 4 **a** Residual $\texttt{res}_1 = R_1 \cdot r_1$ **b** Reconstructed input voltage rtE and faulty input voltage tE

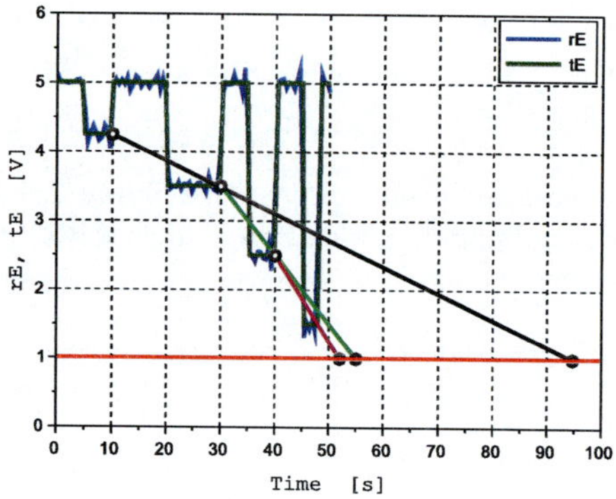

Fig. 5 Prediction of the degradation trend of the intermittent faults

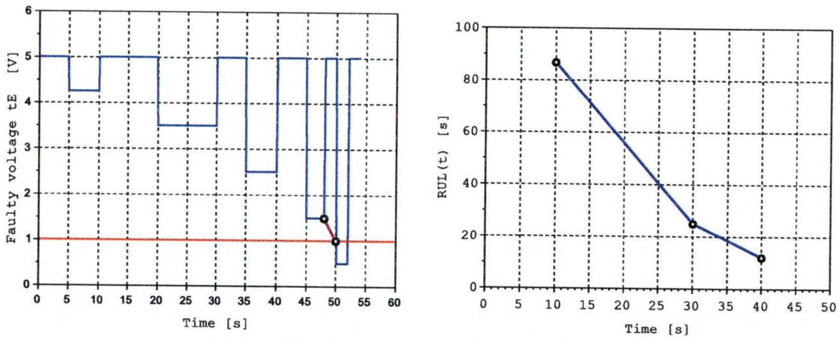

(a) (b)

Fig. 6 **a** Determination of a last RUL value different from zero **b** Decrease of the RUL over time

and for reconstruction of a time behaviour subject to intermittent faults. Although the latest intermittent fault return to zero at $t = 48$ s, we cannot be sure at that time point that another intermittent fault will follow.

Given that continued measurement yields another intermittent fault resulting in a value of the reconstructed faulty voltage $r\tilde{E}$ that is close to the alarm threshold value E_{thr} up to a value $\epsilon \geq, 0$ i.e. $|r\tilde{E}(t) - E_{\text{thr}}| \leq \epsilon$ or even below E_{thr}. As an alarm threshold is chosen well with a distance to the failure threshold, the faulty supply voltage can still cross the alarm threshold due to an intermittent fault without causing harm to the system. Let t_4 be the last time instant where $r\tilde{E}(t_4)$ still has a distance to the alarm threshold and where $r\tilde{E}$ returns to the constant value E and let t_f be the time point where $r\tilde{E}(t_f)$ touches or crosses the threshold for the first time. The last RUL value different from zero then reads $RUL(t_4) = t_f - t_4$ and $RUL(t_f) = 0$. Time instant t_4 is known as End-of-Prediction (EoP). Figure 6a shows the case where the true faulty voltage crosses the alarm threshold due to the last intermittent fault. The last RUL value different from zero is $RUL(t_4 = 48\,\text{s}) = 2\,\text{s}$ and $t_f = 50\,\text{s}$.

3.3 A Switch with Intermittent Faults

As intermittent faults are common in electronic interconnection systems, a second scenario is considered, in which it is assumed that the pass transistor in the circuit of Fig. 1 has connection problems. Instead of connecting the two capacitors permanently as of $t_{sw} = 10$ s, the switch modelling the transistor closes in an unpredictable manner for only short time spans as depicted in Fig. 7. All other components of the circuit, including the constant voltage supply are assumed to be faultless.

As can be seen from Fig. 7, the rise of capacitor voltage $\tilde{u}_1$ is very little affected by the faulty behaviour of the switch. At $t_{sw} = 10$ s, the switch closes and connects the capacitors. As of that time instant, capacitor $\mathrm{C} : C_2$ is also charged which results in a slight decline of $\tilde{u}_1$. However, since the switch immediately opens again after 2 s and remains open for 2 s, voltage $\tilde{u}_1$ increases again.At $t = 14$ s, the switch closes again and remains closed for 8 s which is enough for voltage $\tilde{u}_1$ to reach the steady-state value of 5 V.

In contrast, capacitor voltage $\tilde{u}_2$ keeps the value it has reached whenever the switch opens for a time span. As a result, $\tilde{u}_2$ reaches a steady-state value that is 0.22 V below 5 V.

Evaluating (2) and plotting the result yields the time evolution of residual $r_2(t)$ displayed in Fig. 8a. Figure 8a clearly indicates that residual r_2 is different from zero whenever the switch is open.

In case the faulty switching behaviour $\tilde{b}(t)$ were known, residual r_2 would vanish.

$$0 = \frac{\tilde{b}}{R_2}(\tilde{u}_1 - \tilde{u}_2) - C_2\dot{\tilde{u}}_2 \tag{6}$$

Substraction of (6) from (2) yields

$$r_2 = \frac{b}{R_2}(\tilde{u}_1 - \tilde{u}_2) - \frac{\tilde{b}}{R_2}(\tilde{u}_1 - \tilde{u}_2) \;=\; \frac{b - \tilde{b}}{R_2}(\tilde{u}_1 - \tilde{u}_2) \tag{7}$$

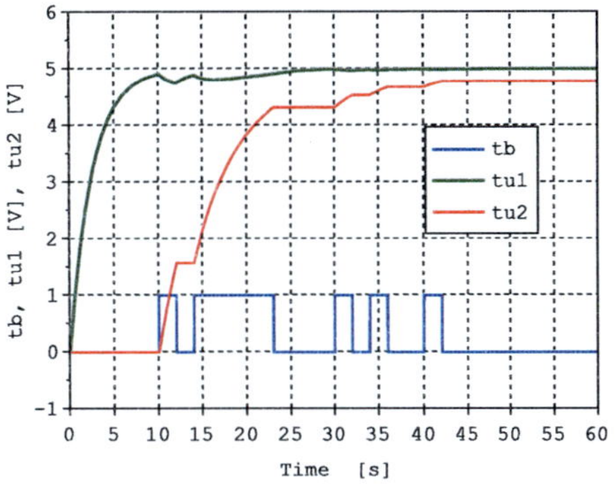

Fig. 7 Faulty switching behaviour $\tilde{b}(t)$ of the transistor and effect of the faulty switching of the transistor on the capacitor voltages u_1, u_2

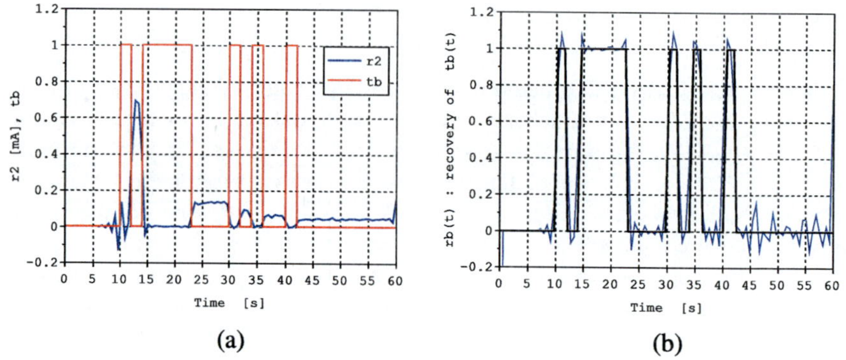

Fig. 8 **a** Time evolutions of residual $r_2(t)$ and $\tilde{b}(t)$ **b** Recovery of the time evolutions of the faulty switching function $\tilde{b}(t)$

In a similar way, subtracting (3) from (1), one obtains for residual r_1

$$r_1 = \frac{\tilde{b} - b}{R_2}(\tilde{u}_1 - \tilde{u}_2) \tag{8}$$

That is, $r_2 = -r_1$.

Solving (7) gives an equation for recovering the unknown faulty switching behaviour $\tilde{b}(t)$.

$$rb = b - \frac{R_2 \cdot r_2}{\tilde{u}_1 - \tilde{u}_2} \tag{9}$$

Figure 8b indicates that $rb(t)$ well recovers the true unknown faulty time evolution $\tilde{b}(t)$.

As a result, from measurements available up to a current time point one can compute the time evolution of an ARR residual that enables to detect intermittent faults and to recover an unknown faulty time behaviour of a system variable.

In the considered scenario, the faulty switching of the transistor does not cause a failure of the circuit. However, the transient behaviour of voltage $\tilde{u}_2$ may be too slow and may not meet requirements of the system to which the circuit is connected.

3.4 A Sensor with Intermittent Faults

Residuals derived from a DBG can also be used to detect a sensor with intermittent faults and to reconstruct the faultless signal. Let's assume that the circuit in Fig. 1 doesn't have any parametric faults. If the measurements of the capacitor voltages are correct then the residuals of ARRs (1), (2) are zero or close to zero. In contrast, if sensor De : $\tilde{u}_1'$ provides a faulty reading $\tilde{u}_1'(t)$ subject to intermittent faults due to connection problems then the residuals of both ARRs are different from zero.

Assume for simplicity that the second sensor De : $\tilde{u}_2$ provides faultless readings. ARR (2) then yields for the reconstruction of the faultless signal $\tilde{u}_1(t)$

$$R_2 \cdot r_2 = b\,(\tilde{u}_1' - \tilde{u}_1) \tag{10}$$

$$r\tilde{u}_1 = \tilde{u}_1' - R_2 \cdot r_2 = \tilde{u}_1' - \mathrm{res}_2 \quad \text{for} \quad t \geq t_{\mathrm{sw}} \tag{11}$$

Conclusion

The paper proposes a BG model-based approach to diagnosis and failure prognosis of intermittent faults with a magnitude that increases with time. Faults are detected by evaluating ARRs derived from an offline developed DBG. In an offline simulation case study of a switched electronic circuit, the constant voltage supply is subject to intermittent faults. Intermittent faults of constant magnitude on the switch as well as faulty sensor readings due to intermittent faults are also analysed. Simultaneous intermittent and incipient parameter faults are addressed elsewhere. due to the lack of space. Parameter uncertainties have been disregarded but can be well taken into account in the DBG. As intermittent faults are common in electronic interconnection systems, the approach is applied to further electronic systems in which contact problems due to corrosion, moisture, or vibration may happen in an unpredictable manner, for instance, in electromechanical relays.

References

1. Borutzky, W.: Bond Graph Modelling for Control, Fault Diagnosis and Failure Prognosis. Springer International Publishing, Switzerland (2020)
2. Elattar, H.M., Elminir, H.K., Riad, A.M.: Prognostics: a literature review. Complex Intell. Syst. **2016**(2), 125–154 (2016). https://doi.org/10.1007/s40747-016-0019-3, Open access
3. Jha, M., Dauphin-Tanguy, G., Ould Bouamama, B.: Particle filter based integrated health monitoring in bond graph framework. In: Borutzky, W. (ed.) Bond Graphs for Modelling, Control and Fault Diagnosis of Engineering Systems, pp. 233–270. Springer International Publishing, Switzerland (2017)
4. Yu, M., Lu, H., Wang, H., Xiao, C., Lan, D.: Compound fault diagnosis and sequential prognosis for electric scooter with uncertainties. Acuators **9**(128) (2020). https://doi.org/10.3390/act9040128, open access

Remaining Useful Life Estimation Based on Wavelet Decomposition: Application to Bearings in Reusable Liquid Propellant Rocket Engines

Federica Galli, Vincent Sircoulomb, Ghaleb Hoblos, Philippe Weber, and Marco Galeotta

Abstract The increasing trend towards the development of reusable rocket engines pushes towards the development of Prognosis and Health Monitoring (PHM) approaches useful to monitor the health of the system and plan its maintenance. In this perspective, we present a hybrid method for Remaining Useful Life (RUL) estimation of rocket engines turbo-pump bearings based on Maximum Overlap Discrete Wavelet Packet Transform (MODWPT) and polynomial approximation. The proposed method calculates the bearing RUL in six main steps: data acquisition, wavelet decomposition, feature extraction, degradation detection, Health Indicator (HI) computation and RUL estimation. The obtained results showed that this technique can be successfully used to separate and isolate vibration trends connected to progressive degradation. The main contribution of this work consists in proposing a monotonically increasing HI which maintains a physical meaning allowing to estimate the degradation level. The proposed HI has proven to be able to effectively quantify the level of degradation and predict it for a certain group of degradation evolution profiles.

F. Galli (✉) · V. Sircoulomb · G. Hoblos
IRSEEM, UNIROUEN, ESIGELEC, Normandie University, 76000 Rouen, France
e-mail: federica.galli@groupe-esigelec.org

V. Sircoulomb
e-mail: vincent.sircoulomb@esigelec.fr

G. Hoblos
e-mail: ghaleb.hoblos@esigelec.fr

P. Weber
CRAN, 54506 Nancy, France
e-mail: philippe.weber@univ-lorraine.fr

M. Galeotta
CNES, 75012 Paris, France
e-mail: marco.galeotta@cnes.fr

D. Theilliol et al. (eds.), *Recent Developments in Model-Based and Data-Driven Methods for Advanced Control and Diagnosis*, Studies in Systems, Decision and Control 467,
https://doi.org/10.1007/978-3-031-27540-1_10

1 Introduction

During the last decade, reusable launchers have become a reality, revealing a wide range of advantages going from cheaper costs to shorter turnaround times. A precise monitoring of a system's health state is at the base of a correct reusability. For this reason, a strong effort is put nowadays in the development of Health Monitoring Systems (HMS) and PHM approaches. As far as launchers are concerned, many parts can be reused going from the first stage (with all its associated equipment) to the fairing. In particular, the first stage engines are the most expensive and complex. Reusing a liquid propellant rocket engine (LPRE) introduces a series of additional challenges such as re-ignition, thrust modulation and maintenance performance. Given the high degree of system complexity, the development of a HMS system for a rocket engine has been approached from different angles. For example, Kawatsu et al. [1] conducted a study on a system-level focusing on fault detection and diagnosis using a model-based quantitative assessment that considers system-level interactions in a rocket engine system. Banerjee et al. [2] developed an anomaly detection and prognosis technique based on particle filtering and machine learning. Other authors focused just on a single engine component. Figure 1 shows the architecture of the Vulcain 2 rocket engine in which the most critical engine components can be recognised: turbo-pumps, combustion chamber, secondary combustion chambers.

Kanso et al. [5] and Chelouati et al. [6] studied the Remaining Useful Life (RUL) of a reusable LPRE combustion chamber. Hotte et al. [7] investigated the degradation of the walls of a reusable combustion chamber with an experimental study. Concerning turbo-pumps, they can be further divided into sub-components as shown in Fig. 2.

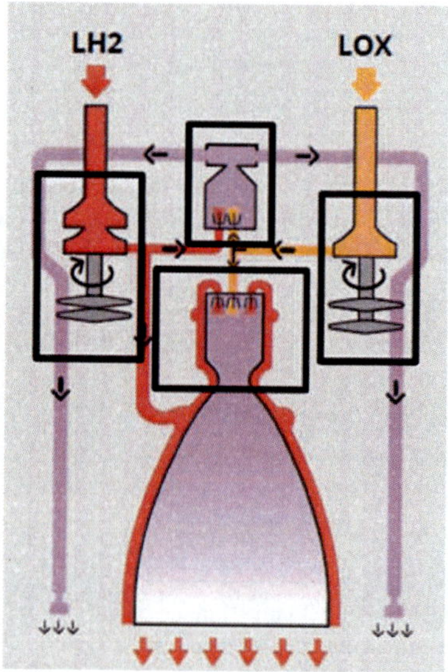

Fig. 1 Vulcain 2 rocket engine scheme [3]

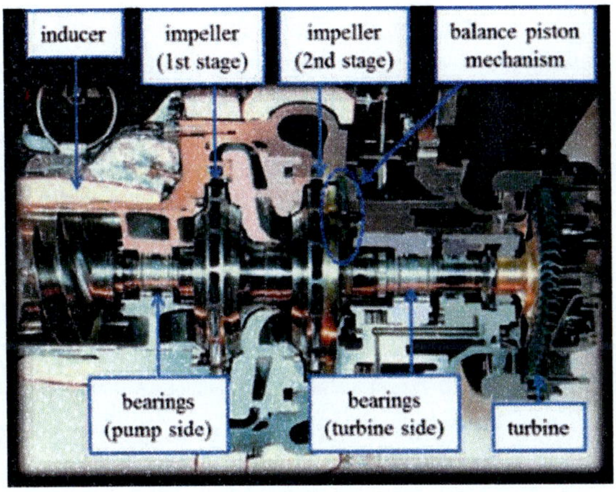

Fig. 2 Example of Rocket engine turbopump (hydrogen turbopump) [4]

As explained by El-Thaji et al. [8] and Luo et al. [9], bearings are among the most critical components of any rotating machinery. Failure of such component could result in irreversible damage to the turbo-pump. In the worst case, it could cause the loss of the system. In the domain of rocket engines, such generic problem results in explosions and loss of the mission leading to huge economical damages. Bearing RUL prediction has already been the object of many studies. As summarised by Lei et al. in [10], PHM prediction approaches can be categorised in three major families: physics model-based approaches, data driven approaches and hybrid approaches. In the first case, a theoretical model is used to describe the degradation evolution and approximate its trend with respect to time. Such methods are helpful when small quantities of data are available. On the other hand they may have a limited precision for very complex non-linear systems. In the second case, data driven approaches rely on statistical models and Artificial Intelligence (AI). In both cases, relevant information about the degradation is inferred from the available data and used to train the algorithm. These methods are easier to implement when intricate systems are studied but may prove to be ineffective if not enough data is available for training. Finally, hybrid methods are a combination of the two previous categories. By bringing together different methodologies, a more complete hybrid technique can be proposed. The domain of rocket engines presents two main issues: (i) lack of data due to small quantities of specimen produced and used, (ii) high-non-linearities coming from its strong complexity. The objective of the present work is to develop a data-driven PHM method for RUL prediction of rocket turbo-pump bearings. The proposed approach treats bearings vibration signals combining wavelets Multi Resolution Analysis (MRA) and polynomial approximation. The HI used to track the degradation evolution in time is the cumulative Root Mean Square (RMS) which provides an indication of the signal energy amount accumulated in time and thus an indication of the degradation stage. The proposed approach was applied to a database available on the NASA repository [11] in view of when some reusable rocket engine data will be available. It shows good capabilities of degradation detection and trend estimation. The contribution of this paper consists in proposing a cumulative health indicator with a physical meaning characterised by a strictly monotonic trend. Indeed such HI is constructed in such a way to maintain a direct link with the component physical degradation while enhancing and ensuring an increasing trend with respect to time.

The remainder of the paper is divided into the following sections: Sect. 2 describes the methodology. Section 3 shows the experimental data used in this study and the obtained results. Finally, Sect. 4 draws some concluding remarks and future perspectives.

2 Methodology of the Proposed Approach

RUL estimation of Rolling Elements Bearing (REB) is a difficult task. Indeed, during its operation, a bearing suffers an inevitable degradation process concerning different tribological phenomena such as wear and flaking [12]. Such complex damage phenomena present many non-linearites and a big variety of outcomes depending on the degradation environment. The aim of the present work is to define a model describing the turbopump bearings overall degradation and tracking its evolution in order to predict the RUL. Figure 3 shows the main steps of the proposed method in a synthetic way. The retrieved vibration data is pre-processed using the Maximum Overlap Discrete Wavelet Package Transform (MODWPT) which splits the signal frequency content into different frequency bands. After that, the RMS is computed and used as statistical indicator for degradation detection. Once the initial degradation instant is defined, a second indicator is computed and fitted with a polynomial model using the least squares method. Finally the obtained model is used to predict the bearing degradation and to compute the RUL. A detailed description of the different steps is given hereafter:

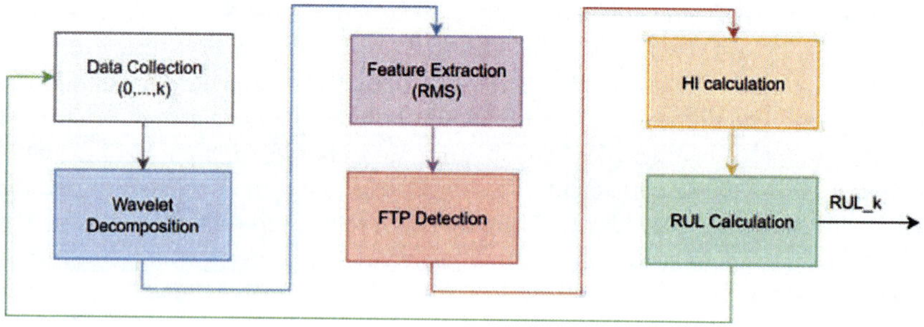

Fig. 3 Proposed algorithm steps

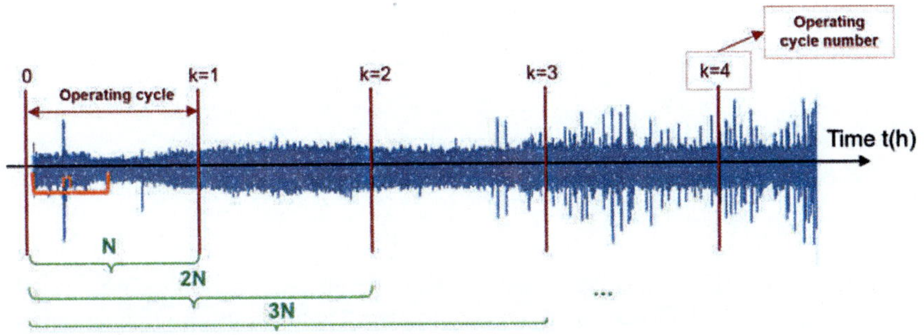

Fig. 4 Data collection process

1. **Data collection.** A vibration signal recorded with accelerometers is retrieved each time the component completed its operating cycle. During each cycle, N samples are acquired. The samples are denoted n and the sampling time p_s (Fig. 4). Thus $t = n \cdot p_s$. In this study the duration of one cycle, $N \cdot p_s$, is assumed to be constant and the transient phases at the beginning and at the end of each cycle are neglected.
2. **Wavelets Decomposition.** Each vibration signal is decomposed using the MODWPT [13]. The Discrete Wavelet Package Transform (DWPT) is a signal processing technique which allows to capture the signal behaviour at different frequency bands by projecting it on a basis of zero-mean oscillating function called wavelets. The MODWPT is a more complete technique which allows to process the signal without down-sampling and time delay making it more suitable for time series [14] . The mother wavelet *db5* was chosen for its orthogonality which results in the conservation of the signal energy through the decomposition process. The decomposition frequency bands can be computed using Eq. 1, where i is the index of the frequency band, fb, and L is the decomposition level.

$$fb_i = \left[(i-1) \cdot \frac{F_s}{2^{L+1}}; (i) \cdot \frac{F_s}{2^{L+1}} \right] \quad \text{with} \quad i = 1, \ldots, 2^L \tag{1}$$

   For example, in our case, these parameters are defined such that : $L = 3$, $F_s = 25.6$ kHz, $i = 1, \ldots, 8$.
3. **Feature extraction.** For each frequency band of the wavelet decomposition, a statistical indicator is computed. Here, the RMS is computed following Eq. 2. This indicator can alternatively be the Kurtosis, the energy or the relative energy, see Eqs. 3 to 5 respectively where $x_s(j, i)$ is a sub-signal of frequency band i, $1 < j < n$ is the vibration sample index, σ is the standard variation and $\bar{x}_s$ is the mean value of $x_s(j, i)$.

$$RMS(n,i) = \left(\frac{1}{n}\sum_{j=1}^{n} x_s(j,i)^2\right)^{1/2} \quad (2)$$

$$En_{tot}(n,i) = \sum_{j=1}^{n} |x_s(j,i)|^2 \quad (4)$$

$$KTS(n,i) = \frac{\sum_{j=1}^{n} x_s(j,i) - \bar{x}_s(j,i)}{n\sigma^4} \quad (3)$$

$$EN_{rel}(n,i) = \frac{En_{fb_i}}{En_{tot}} \quad (5)$$

The RMS and Kurtosis are general statistical indicators and have been widely used for bearing vibration analysis [10, 15, 16]. The first one provides an indication of the mean energy of the signal while the second one quantifies the signal impulsivity. The energy and relative energy, [17, 18], are also of particular interest when MODWPT is applied. Moreover, as pointed out by Luo et al. [9], the bearing degradation has a strong influence on the signal frequency energy distribution. Thus, its evolution with respect to time can be used to quantify the degradation in an indirect way.

4. **FPT detection.** The first step towards RUL computation is to detect the presence of a degradation. The indicators computed in step 3 for all frequency bands are monitored simultaneously using paired contiguous sliding windows that check for mean variation in each sub-signal with respect to a detection threshold.
5. **HI computation.** Once the First Predicting Time (FPT), denoted t_{fpt}, is obtained the analysis is carried on only in the frequency band of interest. A second health indicator is calculated from the first one. The proposed HI consists in the discrete cumulative integral of the RMS:

$$HI(n,i) = \sum_{l=1}^{n} RMS(l,i) \quad (6)$$

The motivation of this HI is as follows: when the bearing balls pass over a defect the signal amplitude increases due to the impact between the ball surface and the ring damaged surface. This magnitude variation results in energy variation. The classical RMS gives an indication of the instantaneous degradation rate. As a consequence, the integral of such quantity between the FPT and the current time quantifies the total bearing surface damage. The proposed approach was first introduced by Boness et al. in [19] for wear volume estimation of a ball-on-cylinder test apparatus.
6. **RUL estimation**. The HI profile is fitted with the polynomial Eq. 7 of order 2:

$$\hat{HI}(t) = a(t - t_{fpt})^2 + b(t - t_{fpt}) + c \quad \text{with} \quad t > t_{fpt} \quad (7)$$

A polynomial of order 2 was chosen as compromise between approximation accuracy and monotonicity. Indeed, a polynomial of order 1 is too simple and a polynomial of order 3, or higher, presents at least a maximum and a minimum. Writing down $t_k = kNp_s$ the time when the kth operating cycle is completed and where the samples records $[(k-1)N+1, kN]$ are available, the coefficients a, b and c are fitted from $HI(n,i)$ using all the samples n from times t_{fpt} to t_k and the least squares method. $\hat{HI}(t)$ is then used to predict the HI for $t > t_k$.

This polynomial model is used for HI forecast. Finally, the RUL is calculated as the time interval between the current time and the future instant when the HI exceeds a predefined threshold, Th:

$$RUL = \Delta t \quad \text{such that} \quad \hat{HI}(t_k + \Delta t) = Th \tag{8}$$

Defining the HI failure threshold is a sensitive step. Indeed finding a general limit is quite tough given the wide range of possible degradation profiles. In some cases, as in Jia et al. [20], the threshold is defined according to external factors like operational risk. Other authors set the failure limit using some training data to define a threshold for each degradation trend. Later the failure limit is chosen according to similarity between the tested signal and the training signals. This methodology, used in [9], is considered in the present paper.

3 Results and Discussion

The proposed algorithm is tested with run to failure vibration signals retrieved from the NASA prognostic repository, [11]. The data were obtained running experiments on the PRONOSTIA platform at the FEMTO-ST Institute, [21]. The platform, provided with two high frequencies accelerometers one the x and y axis, was used to test individually seventeen bearings of type NSK 6804RS. The experiments were conducted in three different operating conditions whose details can be found in Table 1. Each vibration signal was recorded with a sampling frequency of 25.6 kHz. The samples are recorded every 10 seconds and have a duration of 0.1 s. As an example, Fig. 5 shows the raw vibration signal for Bearing 2-1. The presented signal is strongly non-stationary and is characterised by amplitude variations at low and high frequencies.

Table 1 Experimental setting conditions

	Condition 1	Condition 2	Condition 3
Speed (RPM)	1800	1650	1500
Load (N)	4000	4200	5000
Bearings	Bearing 1_1 to 1_7	Bearing 2_1 to 2_7	Bearing 3_1 to 3_3

Fig. 5 Raw vibration signal

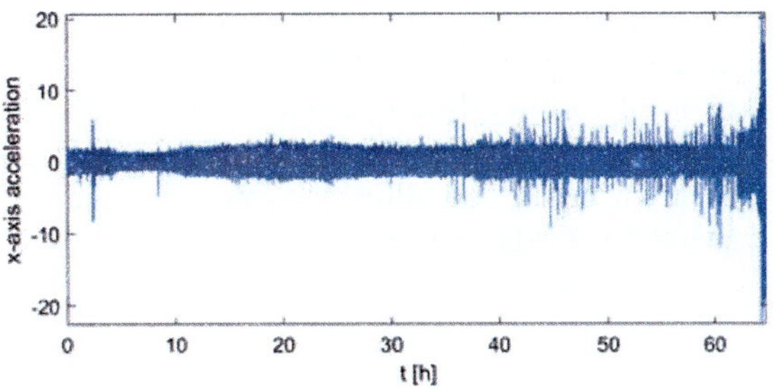

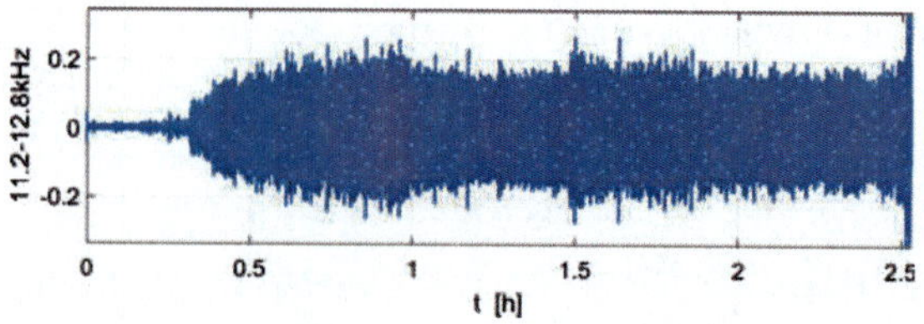

Fig. 6 FPT Detection on the 8 sub-signals

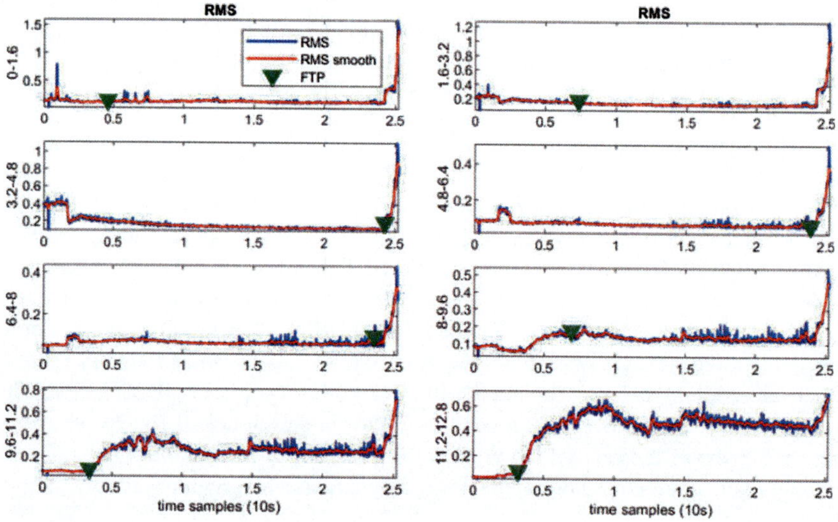

Fig. 7 RMS evolution of the decomposed signal

When the data collection phase is complete the vibration signal is treated with the MODWPT, step 2 in Fig. 3. In this case, the signal was split into 8 frequency sub-bands. The sub-signal of the frequency band [11.2–12.5 kHz] is reported in Fig. 6 as an example. As it can be seen, the initial amplitude variation are not present and the signal increases in amplitude starting from about t=0.3 hours. Such sub-signal reveals a clearer degradation trend. All the obtained sub-signals are used for feature extraction. The time evolution of the RMS in each sub-band is shown in Fig. 7. The blue curves represent the RMS values with respect to time. To ease the detection task, the RMS curves were smoothed.

The green markers show the position of the true FPT. Indeed some sub-signals present some initial instabilities before entering a steady state. Such phenomenon is explained by the fact that at the beginning of their life the bearing may present some fabrication defects. Such defects are reduced thanks to friction during the first period of functioning and must not be considered as actual bearing degradation. Thanks to the wavelet MRA, it is possible to separate the different vibration phenomena and to focus only on the ones provoked by damages on the bearing. The FPT detection, step

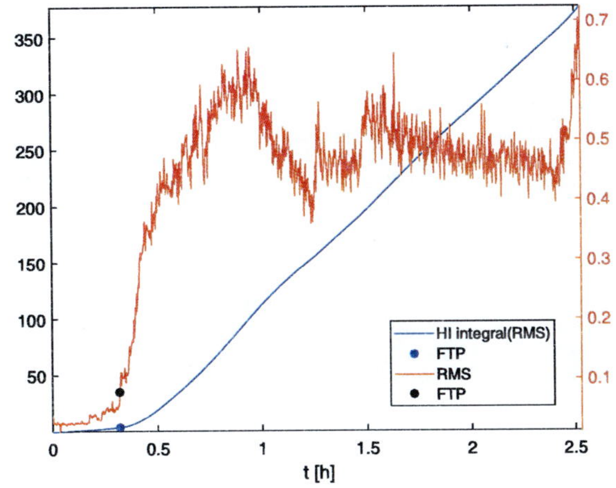

Fig. 8 RMS and cumul. RMS Bearing 2-1

4 in Fig. 3, is a crucial step in the RUL estimation. Not only it needs to be precise but also it has to detect the presence of a degradation as soon as possible. Given that the proposed algorithm is capable of estimating the true FPT in each frequency band, the first FPT to be computed defines the frequency band that is to be used for RUL estimation. In Fig. 6, it is possible to see that the frequency band number 8, [11.2–12.5 kHz], is the first one to show a sign of degradation. Figure 8 shows the RMS and cumulative RMS for frequency band [11.2–12.5kHz]. With a close look on the graph, it is possible to see that they show quite different profiles.

The RMS used for FPT detection shows a general monotonic increasing trend affected by fluctuation especially towards the end. This kind of curve is not ideal from the point of view of HI prediction since it could be interpreted as a reduction of the degradation level. Oppositely, the degradation process is believed to be irreversible. The cumulative RMS shows a strictly monotonic increasing trend and for this reason, it is more suitable for RUL estimation. At last, Fig. 9 shows the results of step 6: the estimated RUL in terms of life cycles. A life cycle of 5 minutes was considered. The obtained results look promising: the initial estimation falls within a 30% uncertainty interval and crosses it only twice during the entire calculation. Figures 10, 11, 12 and 13 show the RUL estimation for four other bearings: 2-2, 2-4, 3-1, 3-3 respectively.

The obtained results show good estimation of the RUL and a fast convergence towards the real RUL. Bearing 3-1 and 3-3 show a good result as well. In this case, a different failure threshold was needed to be defined since the degradation profile was too different from the previous cases. The threshold value was estimated using signal 3-1 as learning set and was later applied to signal 3-3. Lastly, the dashed line in Figs. 9, 10, 11, 12 and 13 represents the cycle duration and shows whether the component is reusable or not. Indeed, if the estimated RUL is less than the cycle duration, the bearing must be changed or at least repaired. For example, Bearing 2-2 in Fig. 10, could be reused for 9 cycles, Bearing 2-4 in Fig. 11, for 14 cycles and Bearing 3-3, in Fig. 13, for 2 cycles. This result is an example of how RUL estimation

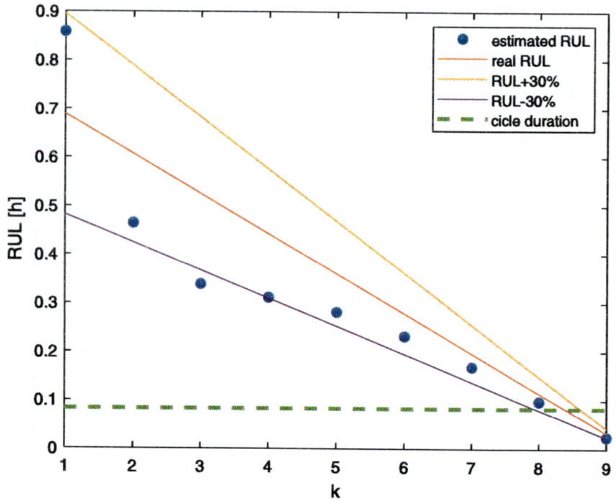

Fig. 9 RUL Estimation Bearing 2-1

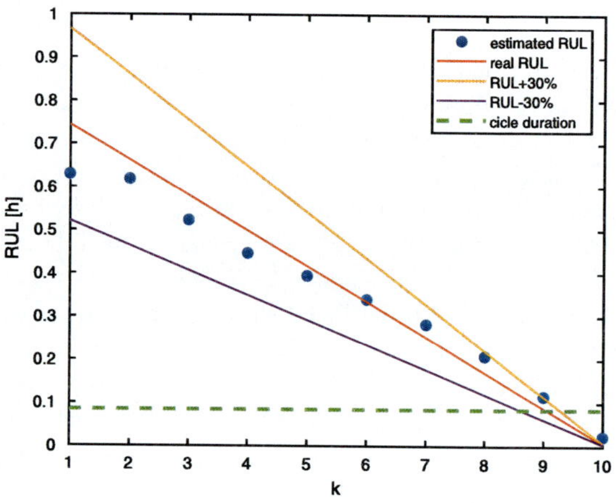

Fig. 10 RUL Estimation Bearing 2-2

can be used as aid to the decision process concerning maintenance planning. The performance of the algorithm was estimated as in [15] by calculating the percentage of RUL estimated values that fall in the confidential interval at +/-20% during the last 500 s of component life. In this case, [15], the bearing RUL was computed using the RMS of the vibration signal as HI and dynamic regression models (DRM) for HI prediction. The obtained results were rearranged in order to be comparable to the ones in literature. Table 2 shows the values of the chose metrics for Bearing 2-4 and 3-3. As it can be seen, the proposed approach has a better prognostic performance.

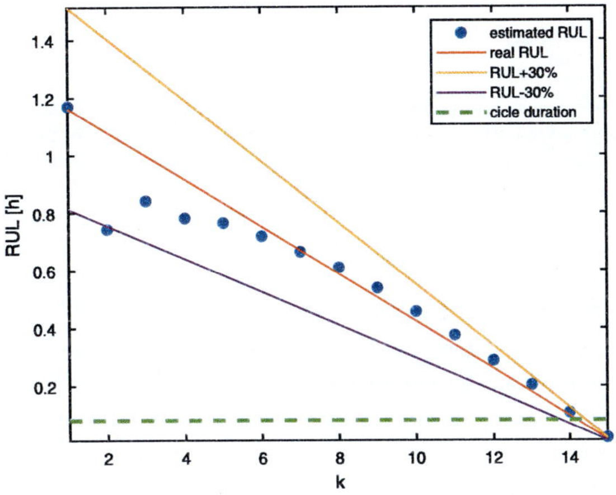

Fig. 11 RUL Estimation Bearing 2-4

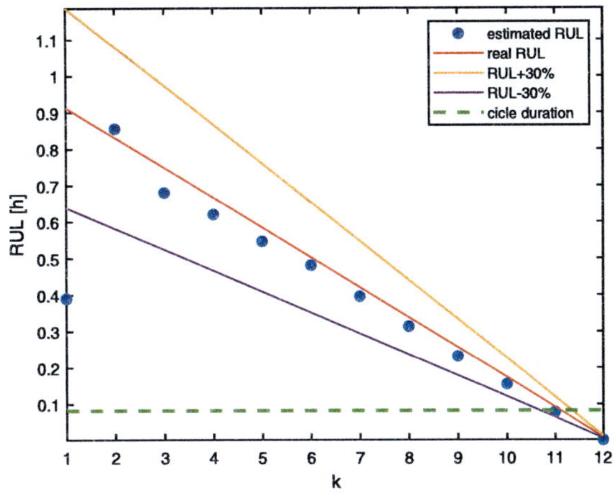

Fig. 12 RUL Estimation Bearing 3-1

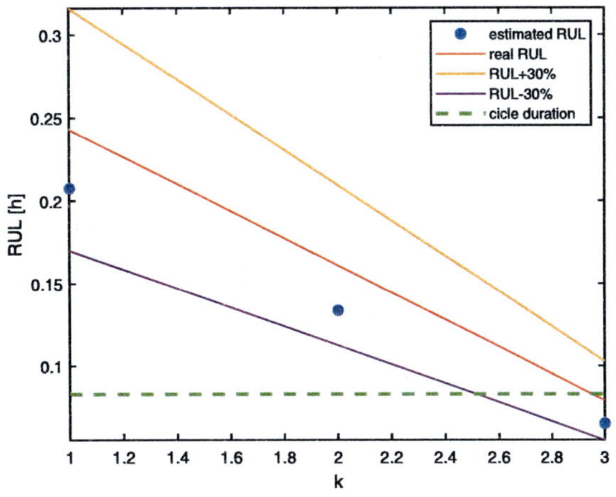

Fig. 13 RUL Estimation Bearing 3-3

Table 2 Performance metrics of the proposed approach and the approach in [15]

	Proposed method	DRM method		Proposed method	DRM method
Bearing 2_4	87.5%	38%	Bearing 3_3	87.5%	80%

4 Conclusion

In this paper, an approach for RUL estimation of rolling element bearings was proposed and tested with a set of vibration degradation data. The contribution of this methodology mainly relies on two aspects. The first one concerns the introduction of wavelets MRA analysis for degradation detection and feature extraction. The second one is the use of the cumulative RMS as health indicator. The proposed HI has proven to be able to effectively quantify the level of degradation and predict it for a certain group of degradation evolution profiles. The obtained results showed that this technique can be successfully used to separate and isolate vibration trend connected to progressive degradation. Further investigation is foreseen to improve the proposed algorithm. In particular, future work will focus on defining a more adaptive threshold tuning technique, real time approximation modelling and uncertainty analysis.

Acknowledgements This work is co-funded by CNES "Centre Nationale d'Etudes Spatiales" and the Normandy Region.

References

1. Kawatsu, K.: PHM by using multi-physics system-level modeling and simulation for EMAs of liquid rocket engine. In: IEEE Aerospace Conference, pp. 1–10. Big Sky, USA (2019)
2. Banerjee, B., Kraemer, L., Solano, W.: Particle filtering for diagnosis and prognosis of anomalies in rocket engine tests. In: AIAA Infotech at Aerospace Conf. and Exhibit, St. Louis, MO, USA (2011)
3. Vulcain, A.H.: ®2. 1. Technical report. https://www.ariane.group/wp-content/uploads/2020/06/VULCAIN2.1_2020_04_PS_EN_Web.pdf
4. Hiraki, H., Inoue, T., Yabui, S.: Influence of impeller's elastic deformation on the stability of balance piston mechanism of rocket engine turbopump. In: IOP Conference Series: Earth and Environmental Science, vol. 240, p. 052028. IOP Publishing (2019)
5. Kanso, S., Jha, M.S., Galeotta, M., Theilliol, D.: Remaining useful life prediction with uncertainty quantification of liquid propulsion rocket engine combustion chamber. In: SAFEPROCESS, vol. 55, pp. 96–101. Greece, Cyprus (2022)
6. Chelouati, M., Jha, M.S., Galeotta, M., Theilliol, D.: Remaining useful life prediction for liquid propulsion rocket engine combustion chamber. In: Conference on Control and Fault-Tolerant Systems, SysTol, pp. 225–230. Saint-Raphael, France (2021)
7. Hötte, F., Sethe, C.V., Fiedler, T., Haupt, M.C., Haidn, O.J., Rohdenburg, M.: Experimental lifetime study of regeneratively cooled rocket chamber walls. Int J Fatigue (2020)
8. El-Thalji, I., Jantunen, E.: A summary of fault modelling and predictive health monitoring of rolling element bearings. Mech. Syst. Signal Proc. (2015)

9. Luo, H., Bo, L., Liu, X., Zhang, H.: A novel method for remaining useful life prediction of roller bearings involving the discrepancy and similarity of degradation trajectories. Comput. Intell. Neurosci. (2021)
10. Lei, Y., Li, N., Guo, L., Li, N., Yan, T., Lin, J.: Machinery health prognostics: a systematic review from data acquisition to RUL prediction. Mech. Syst. Signal Proc. (2018)
11. Orzech, G.: Prognostics center of excellence data set repository (2022). http://www.nasa.gov/content/prognostics-center-of-excellence-data-set-repository
12. El-Thalji, I., Jantunen, E.: A descriptive model of wear evolution in rolling bearings. Eng. Fail. Anal. (2014)
13. Percival, D.B., Walden, A.T.: Wavelets Methods for Time Series Analysis. Cambridge University Press, Cambridge (2000)
14. Yang, Y., He, Y., Cheng, J., Yu, D.: A gear fault diagnosis using hilbert spectrum based on modwpt and a comparison with emd approach. Meas.: J. Int. Meas. Confed. (2009)
15. Ahmad, W., Khan, S.A., Islam, M.M.M., Kim, J.: A reliable technique for remaining useful life estimation of rolling element bearings using dynamic regression models. Reliab. Eng. Syst. Safety (2019)
16. Wang, D., Tsui, K.L., Miao, Q.: Prognostics and health management: a review of vibration based bearing and gear health indicators. IEEE Access (2017)
17. Tobon-Mejia, D., Medjaher, K., Zerhouni, N., Tripot, G., Tobon-Mejia, D.A., Medjaher, K., Zerhouni, N., Tripot, G.: Estimation of the remaining useful life by using wavelet packet decomposition and HMMs. In: Aerospace Conference. Big Sky, USA (2011)
18. Ocak, H., Loparo, K.A., Discenzo, F.M.: Online tracking of bearing wear using wavelet packet decomposition and probabilistic modeling: a method for bearing prognostics. J. Sound Vibr. (2007)
19. Boness, R.J., McBride, S.L.: Adhesive and abrasive wear studies using acoustic emission techniques, Wear (1991)
20. Jia, X., Li, W., Wang, W., Li, X., Lee, J.: Development of multivariate failure threshold with quantifiable operation risks in machine prognostics. In: Annual Conference of the PHM Society, vol. 12, pp. 9–9. PHM Society (2020)
21. Nectoux, P., Gouriveau, R., Medjaher, K., Ramasso, E., Morello, B., Zerhouni, N., Varnier, C.: PRONOSTIA: an experimental platform for bearings accelerated degradation tests. In: IEEE International Conference on PHM, PHM'12. Denver, CO, United States (2012)

Advanced Control

Adaptive Finite Horizon Degradation-Aware Regulator

Amirhossein Hosseinzadeh Dadash and Niclas Björsell

Abstract Predicting the failure and estimating the machine's state of health is information that supports the production planning and maintenance management systems to increase productivity and reduce maintenance and downtime costs. However, controlling the degradation in the machines will improve the system's reliability and resilience and make high-level decisions more accurate and reliable. To control the degradation in the machines, time should be included in the cost function as a variable, which alters the markovian properties of the system dynamic. In this article, we include the degradation cost in the quadratic cost function of the infinite horizon controller and calculate the optimal feedback according to the dynamics of the degradation using dynamic programming. It will be shown that the infinite horizon control will convert to the finite horizon, and the controller will be able to control the degradation according to the desired degradation at the desired time. In the end, with the help of simulation, we show that the degradation controller can control the degradation in the MIMO systems.

1 Introduction

In the era of Industry 4.0, the horizons of the definition of "optimal control" can include more variables than its classical definition. Keeping the system output close to the desired output, which was the optimal control's primary goal, is now among many other goals that must be achieved simultaneously. For instance, the Linear Quadratic Regulator (LQR) is considered an optimal controller when the only parameters to

A. Hosseinzadeh Dadash (✉) · N. Björsell
Department of Electronics, Mathematics and Natural Sciences, University of Gävle, 801 76 Gävle, Sweden
e-mail: amirhossein.hosseinzadeh.dadash@hig.se

N. Björsell
e-mail: niclas.bjorsell@hig.ses

D. Theilliol et al. (eds.), *Recent Developments in Model-Based and Data-Driven Methods for Advanced Control and Diagnosis*, Studies in Systems, Decision and Control 467,
https://doi.org/10.1007/978-3-031-27540-1_11

control are the input and output, but being optimal in controlling the degradation simultaneously needs some new information from the system and modifications to the method [1, 2].

Fault-tolerant controllers have been the subject of research [3]. However, degradation tolerant controllers, which are more helpful in long-term production planning and maintenance management systems [4, 5], have not been discussed much. For instance, one of the methods for reducing the maintenance cost is to synchronize the machines' degradation so they can reach the maintenance time simultaneously. This simultaneity is achieved in [6] using advanced high-level controllers and has the limitation that it is only applicable to systems with similar machines working on the same process.

To reduce the calculation and data storage cost and make the degradation control applicable to machines that are not identical and do not work on the same process, the degradation controller should be integrated with the controller or work on its side without a need to exchange much information with high-level controller (production management) [7, 8]. For doing this integration, the degradation model should be identified. This identification should be in a way that the model can be related to the system dynamics. This identification can be made using process-aware neural networks [9, 10] or sparse regressions [11, 12]. In both cases, the identification of the degradation might not be as accurate as model identification without the limitation of considering the process parameters, but it can be used for control purposes which have their own benefits.

Assuming that the machine and degradation models are known, it is possible to control the degradation in two ways: including the degradation as the system's state or including it in the feedback loop. The traditional way of including degradation inside the model has been tested successfully [13, 14], but having a closed form of including the degradation in the feedback loop will be helpful in the future and can change the controller design process. Also, an adaptive feedback loop can be configured faster with the degradation model change than the adaptive controller.

This paper will introduce the adaptive feedback loop for controlling the degradation at the same time as the output. In the first section, the method for calculating different parts of this feedback loop will be shown, and in the second section, the method will be validated using a simulation model.

2 Method

The method for calculating the adaptive degradation-aware feedback comprises two parts, the formulation of the feedback, which is explained first, and mapping from the recorded degradation into the system's states.

2.1 Formulation

The state-space model of the system can be written as follows:

$$\begin{cases} x(t+1) = Ax(t) + Bu(t) + Mv_1(t) \\ y(t) = Cx(t) + v_2(t) \end{cases} \quad \Omega = \begin{bmatrix} v_1 \\ v_2 \end{bmatrix}, \tag{1}$$

where $x(t)$ includes system state(s) at time t, $u(t)$ is system input(s) at time t. A and B represent the physical system parameters, and C defines the relationship between the output(s) and state(s) of the system. v_1 and v_2 are the input and sensor noise, respectively.

The quadratic cost that the LQ controller minimizes for the infinite horizon is [15]

$$J = \| e \|_{Q_1}^2 + \| u \|_{Q_2}^2, \tag{2}$$

where $e = x - r$ and r is the reference signal, and e is the error; and Q_1 and Q_2 are penalty matrices for the error and input signal, respectively. However, adding the third term for controlling the degradation converts (2) to

$$J(t = 1:T) = E\left\{ \sum_{k=0}^{N} \| e_k \|_{Q_1}^2 + \sum_{k=0}^{N-1} \| u_k \|_{Q_2}^2 + \sum_{k=0}^{N} \| G_{x^f,u^f}(t) \|_{Q_3}^2 \right\}, \tag{3}$$

$$G_{x^f,u^f}(t) = D_{x^f,u^f} - D_d(t), \tag{4}$$

where x^f, u^f are feature of x and feature of u respectively, D_{x^f,u^f} is the degradation at state x^f with the input u^f, and $D_d(t)$ is the desired degradation at time t. To better understand x^f and u^f, some explanation is needed. First, the working cycle should be defined as the interval where the machine starts from its standby state and settles on the desired output (usually the step response of the machine till it fully settles). For example, a working cycle of a wood-cutting saw starts when it starts rotating from the stop position until it reaches the desired speed (full load working). Second, in reality, the degradation rate is lower than the rate of change in the system state (the degradation happens over tens or hundreds, or even thousands of cycles of operation while the controller works on the scale of a fraction of a second). So, in order to be able to detect the degradation from the recorded signal, a fixed point (in time) or a feature from one machine's working cycle (for each state and input) should be recorded for different cycles. This way, the degradation trend can be detected and mapped to the system state. These recorded features are defined to be x^f and u^f. Figure 1 shows this process more clearly. Finally, as the features are the variables that are going to be used in the last part of (3), t will be defined as a cycle; this means that t is not an instance and instead is a time duration. However, as only one feature from each recorded signal during each cycle is recorded in (3), it can be considered a time instance.

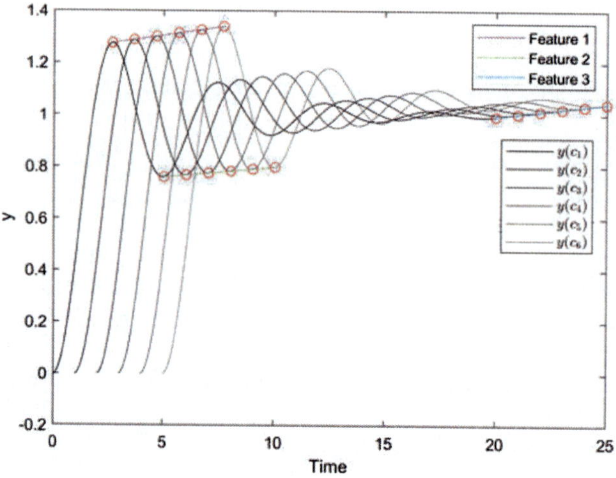

Fig. 1 Relation of the features per cycle and recorded signal (red circles are the features e.g., x^f or u^f that are recorded at each cycle, and their trend is used to map the degradation to the system states)

After these adaptations of the cost function with the limitation of the function estimation methods, the whole optimization becomes a finite horizon optimization that tries to control the degradation of the machine. Hence, it reaches the maintenance condition (x_N) at the desired time (T). Note that the degradation of the machine is not time-dependent (Markovian). However, the desired degradation is only the function of the time (because the optimization goal is to make the machine reach the desired maintenance time while keeping the output within acceptable limits).

Applying the dynamic programming for this optimization, we have

$$J_N(T) = e_N(T)'\, Q_1\, e_N(T) \tag{5}$$

$$J_{k=1:N-1}(t = 1:T) = \min_{u_k} E\left\{e' Q_1 e + u' Q_2 u + G' Q_3 G + J_{k+1}(Ax_k + Bu_k + v_{1_k})\right\}, \tag{6}$$

where J_k is the cost-to-go from state x_k at time t to x_N at time T.

Now for calculating the best trajectory, the cost function should be calculated backward from x_N at time T to x_1 at time 0 [16]. For now, we assume that the degradation of the machine can be identified as a function of the machine's states and inputs. In this case, we have

$$D_{x,u} = \begin{bmatrix} W_x & W_u \end{bmatrix} \begin{bmatrix} x_k \\ u_k \end{bmatrix}. \tag{7}$$

Now expanding (6) gives

$$J_k(t) = e'Q_1e + \min_{u_k} E\left\{u'Q_2u + G'Q_3G + (Ax_k + Bu_k + w_k)'Q_1(Ax_k + Bu_k + w_k)\right\}. \tag{8}$$

Expanding this statement and differentiating it in order to find the u_k^* (optimal input) and assuming that $E\{v_1\} = 0$ will result in

$$u_x(t)^* = -(Q_2 + B'Q_1B)^{-1}(B'Q_1Ax_k + (W_x'x_k(t-1) + W_u'u_k(t-1) - D_d(t-1))W_uQ_3). \tag{9}$$

This equation is composed of two parts, the first part

$$u_x^* = -(Q_2 + B'Q_1B)^{-1}(B'Q_1Ax_k), \tag{10}$$

which is the infinite horizon optimal feedback, and only depends on x_k, and second part

$$u(t)^* = -(Q_2 + B'Q_1B)^{-1}(W_x'x_k(t-1) + W_u'u_k(t-1) - D_d(t))W_uQ_3, \tag{11}$$

which depends on t, x and u and is the adaptive degradation compensation feedback.

2.2 Relevance Vector Machine

In (7), it was assumed that the degradation of the system could be calculated as a function of the system's state and input. There are many methods to do this mapping. One of these methods used for this research is the Relevance Vector Machine (RVM). The relevance vector machine (RVM) was first introduced in [17]. The RVM structure is very similar to the support vector machine, which is given as follows:

$$\hat{y}(x) = \sum_{\ell=1}^{L} w_\ell k(\mathbf{x}, \mathbf{x}_\ell) + c, \tag{12}$$

where w_ℓ is a coefficient in $\mathbf{w}$, which is the vector of coefficients, $k(\cdot, \cdot)$ is the kernel function, and c is the bias parameter. RVM defines the conditional distribution for the target value $\mathbf{y}$ given the vector of covariates $\mathbf{x}$, prediction outcome $\hat{\mathbf{y}}$, regression coefficients $\mathbf{w}$, and a precision parameter called ψ as

$$p(y_\ell \,|\mathbf{x}_\ell, \mathbf{w}, \psi) = \mathcal{N}(y_\ell \,|\hat{y}(\mathbf{x}_\ell),\ \psi^{-1}), \tag{13}$$

$$\hat{\mathbf{y}} = \Phi(\mathbf{X})\mathbf{w}, \tag{14}$$

$$\Phi(\mathbf{X}) = [\phi(\mathbf{x}_1), \phi(\mathbf{x}_2), \ldots, \phi(\mathbf{x}_L)]^T, \tag{15}$$

$$\phi(\mathbf{x}_\ell) = [1, k(\mathbf{x}_\ell, \mathbf{x}_1), k(\mathbf{x}_\ell, \mathbf{x}_2), \ldots, k(\mathbf{x}_\ell, \mathbf{x}_L)]. \tag{16}$$

The likelihood function for $\mathbf{y}$ can be written as

$$p(\mathbf{y}|\mathbf{X}, \mathbf{w}, \psi) = \prod_{\ell=1}^{L} p(t_\ell|\mathbf{x}_\ell, \mathbf{w}, \psi^{-1}). \tag{17}$$

RVM introduces a prior distribution for each w in $\mathbf{w}$ as a hyperparameter α

$$p(\mathbf{w}|\alpha) = \prod_{i=1}^{M} \mathcal{N}(w_i|\,0, \alpha_i^{-1}), \tag{18}$$

where M is the number of covariates (bias included). The hyperparameter α measures the precision of each w_i. Following Bayesian inference, the distribution of the weights becomes Gaussian and takes the following form:

$$p(\mathbf{w}|\mathbf{y}, \mathbf{X}, \boldsymbol{\alpha}, \psi) = \mathcal{L}(\mathbf{w}|\, m, \Sigma), \tag{19}$$

in which

$$\mathbf{m} = \psi\Sigma\Phi^T\mathbf{y}, \tag{20}$$

$$\Sigma = (\Delta_\alpha + \psi\Phi^T\Phi)^{-1}, \tag{21}$$

and $\Delta_\alpha = diag(\alpha_i)$.

2.2.1 Results

For this proof of idea, the model used is the simple mass and spring model with two inputs. The reason for considering two inputs is that, for SISO systems, with a correct configuration of Qs, the optimal input of the degradation controller would be the same as the optimal input calculated by solving the Ricatti equation for LQR because the only parameter available to control both output and degradation would be a single input which prioritizing the output quality will make the system a regular controller, not a degradation controller. However, having more than one input will make the controller capable of controlling two outputs (the system's output and system degradation).

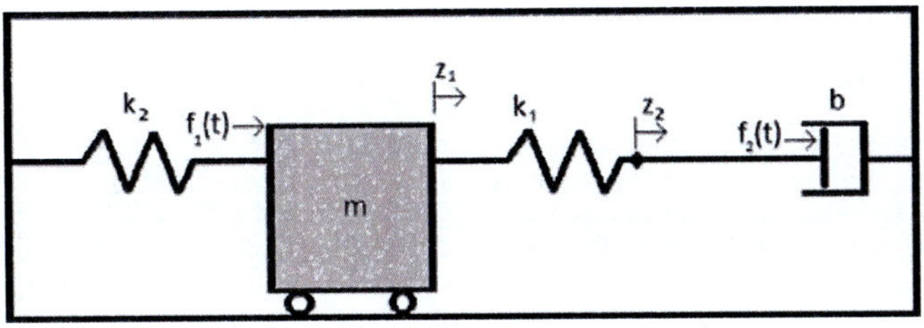

Fig. 2 Mass and Spring Model

The model used for this simulation is shown in Fig. 2, and its respective state-space is shown in (22) and (23) [18].

$$\begin{bmatrix}\dot{z}_1\\ \ddot{z}_1\\ \dot{z}_2\end{bmatrix} = \begin{bmatrix}0 & 1 & 0\\ \frac{k2-k1}{m} & 0 & -\frac{k_1}{m}\\ \frac{k_1}{b} & 0 & -\frac{k_1}{b}\end{bmatrix}\cdot\begin{bmatrix}z_1\\ \dot{z}_1\\ z_2\end{bmatrix} + \begin{bmatrix}0 & 0\\ \frac{1}{m} & 0\\ 0 & \frac{1}{b}\end{bmatrix}\cdot\begin{bmatrix}f_1\\ f_2\end{bmatrix}, \tag{22}$$

$$y = Ix, \tag{23}$$

where z_1 and z_2 are displacements shown in Fig. 2, m is the mass, k_1 and k_2 are respective spring constants, b is the damping ratio and f_1 and $f_2$are respective input forces. The parameter to control or the desired output for this simulation is z_1, which is the position of the mass. Although, according to different conditions, the degradation might differ in real situations, to reduce the complexity of the result and be able to focus on the research idea, the degradation parameter chosen to be k_1. The degradation model for k_1 considered to be exponential [19]:

$$k_1(t+1) = k_1(t) + 2 \times 10^{-5} \times \exp(t * 5 * 10^{-5}). \tag{24}$$

Two failure thresholds were defined for this simulation:

$$|\hat{z_1}(t) - z_1(t)| > 0.01, \tag{25}$$

$$k_1 > 0.02, \tag{26}$$

where $\hat{z_1}(t)$ is defined as the desired output at time t, and $z_1(t)$ is the system output at time t, and the system is considered as failed (from the maintenance point of view) when the difference between actual output and desired output passes this threshold. The closed-loop step response of the system is shown in Fig. 3. It can be seen that the controller is controlling the system according to the desired output. Figure 4 shows the same controller under degradation explained in (24) and failure threshold defined in (25) and (26). The left part of the figure shows the normal controller and on the right is the response of the degradation controller. Note that the x-axis of the Fig. 4

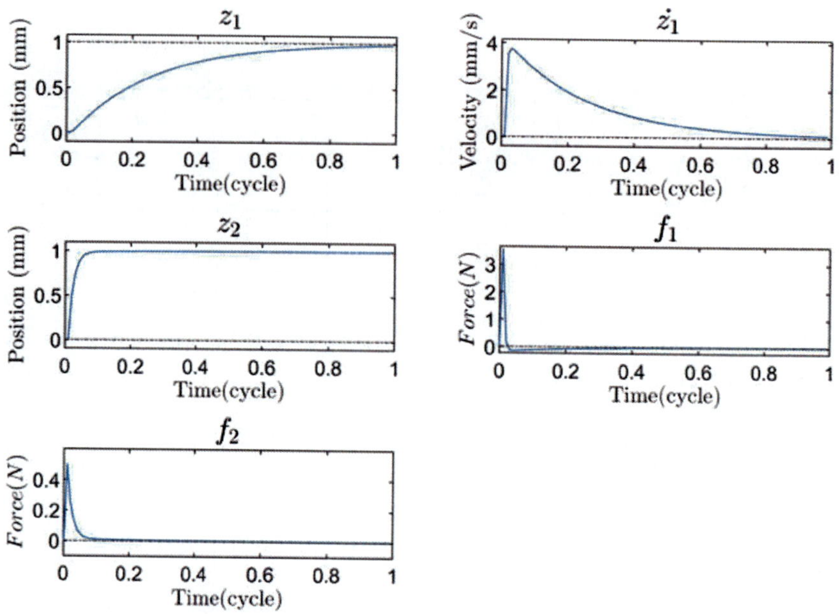

Fig. 3 Closed-loop step response of the infinite horizon controller

is in cycles, which means that only features of the output (in this case, maximum) are recorded from each cycle and plotted in the figure, so each point in the figure is a feature recorded from a whole cycle as shown in Fig. 1.

The degradation coefficients W_x and W_u are calculated according to recorded data shown in the left part of Fig. 4.

It can be seen that the regular controller loses control, the system output deviates from the desired output by an amount more than the failure threshold, and the system is considered as failed after around 90 cycles. However, the degradation controller keeps the output inside the acceptable threshold for around 500 cycles. In the end, the simulation finishes not because of the output deviation but because of the failure threshold mentioned in (26).

3 Discussion

Controlling the degradation at the same time as the output will impact the industry and reduce the maintenance cost. Unlike most existing control methods, the degradation control will depend upon machine learning methods because identifying the physical laws governing the degradation is costly and environment-dependent. Although machine learning methods are considered the best solutions for solving modern con-

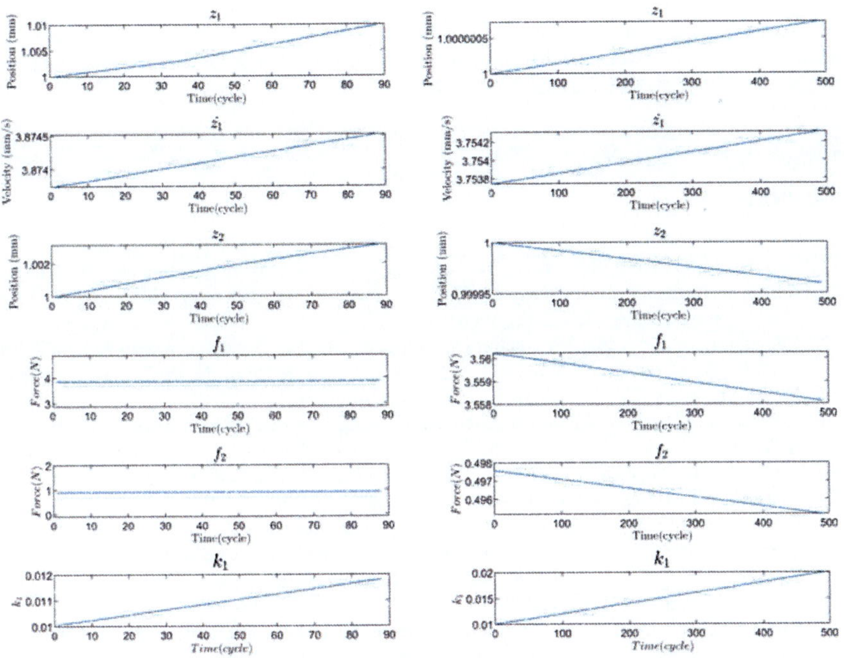

Fig. 4 Controller without degradation awareness (left) versus degradation controller (right)

trol problems, adapting them to existing methods will be challenging. This article proved that with the correct choice of machine learning method, it is possible to adapt the traditional control method with machine learning. The stability of the closed loop is among the most critical questions that answering it is outside the scope of this article. However, the short answer is that another cost function can be used to find the optimal point if the feedback makes the system unstable.

4 Conclusion

In this article, the degradation-aware adaptive feedback loop was introduced. First, the formulation of the quadratic cost function of the infinite horizon controller was updated, and the third term for penalizing the controller was introduced. Then using dynamic programming, the optimal feedback was calculated, which is a combination of the infinite horizon optimal feedback and time, state, and output dependent function. Then the limitation of identifying the degradation was considered in the formulation and adapted to the limitations, and RVM was used as the machine learning method compatible with the formulation and limitations. Finally, with the help of simulation, the functionality of the degradation-aware feedback loop was proved.

Acknowledgements The research project is financed by the European Commission within the European Regional Development Fund, Swedish Agency for Economic and Regional Growth, Region Gävleborg, and the University of Gävle.

References

1. Hua, C., Li, L., Ding, S.X.: Reinforcement learning-aided performance-driven fault-tolerant control of feedback control systems. IEEE Trans. Autom. Control **67**(6), 3013–3020 (2022). https://doi.org/10.1109/TAC.2021.3088397
2. Chemali, E., et al.: Minimizing battery wear in a hybrid energy storage system using a linear quadratic regulator. In: IECON 2015 - 41st Annual Conference of the IEEE Industrial Electronics Society, pp. 003265–003270 (2015). https://doi.org/10.1109/IECON.2015.7392603
3. Abbaspour, A., et al.: A survey on active fault-tolerant control systems. Electronics **9**(9), 1513 (2020)
4. Hu, J., Sun, Q., Ye, Z.-S.: Condition-based maintenance planning for systems subject to dependent soft and hard failures. IEEE Trans. Reliab. **70**, 1468–1480 (2020)
5. Omshi, E.M., Grall, A., Shemehsavar, S.: A dynamic auto-adaptive predictive maintenance policy for degradation with unknown parameters. Eur. J. Oper. Res. **282**, 81–92 (2020)
6. Björsell, N., Dadash, A.H.: Finite horizon degradation control of complex interconnected systems. IFAC-PapersOnLine **54**, 319–324 (2021)
7. Salazar, J.C., et al.: Health-aware and fault-tolerant control of an octorotor UAV system based on actuator reliability. Int. J. Appl. Math. Comput. Sci. **30**(1), 47–59 (2020)
8. Ahmed, Ibrahim, Quiñones-Grueiro, Marcos, Biswas, Gautam: Fault-tolerant control of degrading systems with on-policy reinforcement learning. IFAC-PapersOnLine **53**(2), 13733–13738 (2020)
9. Read, J.S., Jia, X., Willard, J., Appling, A.P., Zwart, J.A., Oliver, S.K., Karpatne, A., Hansen, G.J., Hanson, P.C., Watkins, W., et al.: Processguided deep learning predictions of lake water temperature. Water Resour. Res. **55**, 9173–9190 (2019)
10. Daw, A., Karpatne, A., Watkins, W., Read, J., Kumar, V.: Physics-guided neural networks (pgnn): an application in lake temperature modeling. arXiv:1710.11431
11. Rudy, S.H., Brunton, S.L., Proctor, J.L., Kutz, J.N.: Data-driven discovery of partial differential equations. Sci. Adv. **3** (2017)
12. Brunton, S.L., Proctor, J.L., Kutz, J.N.: Discovering governing equations from data by sparse identification of nonlinear dynamical systems. Proc. Natl. Acad. Sci. **113**, 3932–3937 (2016)
13. Sun, J., Zuo, H., Wang, W., Pecht, M.G.: Application of a state space modeling technique to system prognostics based on a health index for condition-based maintenance. Mech. Syst. Signal Proc. **28**, 585–596 (2012)
14. Storvik, G.: Particle filters for state-space models with the presence of unknown static parameters. IEEE Trans. Signal Proc. **50**, 281–289 (2002)
15. Glad, T., Ljung, L.: Control Theory. CRC Press, Boca Raton (2018)
16. Bertsekas, D.: Dynamic Programming and Optimal Control: Volume I, vol. 1. Athena Scientific, Nashua (2012)
17. Tipping, M.E.: Sparse Bayesian learning and the relevance vector machine. J Mach Learn Res **1**, 211–244 (2001)
18. Close, C.M., Frederick, D.K., Newell, J.C.: Modeling and Analysis of Dynamic Systems. Wiley, New York (2001)
19. Gebraeel, N.: Sensory-updated residual life distributions for components with exponential degradation patterns. IEEE Trans. Autom. Sci. Eng. **3**(4), 382–393 (2006)

Set-Membership Fault Detection Approach for a Class of Nonlinear Networked Control Systems with Communication Delays

Afef Najjar and Jean-Christophe Ponsart

Abstract In this paper, a Fault Detection (FD) problem for a class of Nonlinear Networked Control Systems (NNCS) in a set-membership framework is investigated. Under the assumption of bounded network-induced delays and process uncertainties (i.e. process disturbances and measurement noises), a residual generator is constructed based on a set-membership estimation-based predictor approach. Finally, a numerical example illustrating the performances of the proposed method is given.

Keywords Fault detection · Nonlinear Networked Control System (NNCS) · Unknown network delay · Interval observer · Predictor

1 Introduction

The NCS are systems wherein some or all signals are transmitted among the system's components as information flows through a shared network [7]. Compared with conventional point-to-point architectures, the advantages of NCS are lighter wiring, lower installation costs and greater abilities in diagnostic, reconfigurability and maintenance [7]. Thanks to these distinctive benefits, application of NCS ranges over various industry's fields nowadays [7, 13]. However, using a shared network for data exchange make system control [10], monitoring [9] or diagnostic [11] more difficult where some communication constraints should be considered such as packets losses, sampling problems and network-induced delays [7]. The last mentioned is one of the most common problem in literature [6], especially in NCS FD. Intensive research addresses this challenging subject, one can see for example [6, 11] and the references therein. Filtering method, Markovian jump approach and observer-based approach are ones of the most used approaches dealing with this problem. In this

A. Najjar (✉) · J.-C. Ponsart
CNRS, CRAN, Université de Lorraine, F-54000 Nancy, France
e-mail: afef.najjar@univ-lorraine.fr

J.-C. Ponsart
e-mail: Jean-Christophe.Ponsart@univ-lorraine.fr

D. Theilliol et al. (eds.), *Recent Developments in Model-Based and Data-Driven Methods for Advanced Control and Diagnosis*, Studies in Systems, Decision and Control 467,
https://doi.org/10.1007/978-3-031-27540-1_12

paper, a FD technique is proposed in a set-membership framework for the NNCS with unknown communication delays and unknown external disturbances [5] by only knowing their bounds. The main idea consists in detect faults by generating interval residual signals. Then, FD is ensured through a belonging test of the zero signal to the interval delimited by upper and lower residual signals.

This paper is structured as follows. Section 2 introduces some preliminaries. Section 3 represents the studied system architecture. In Sect. 4, we develop the proposed fault detection technique. Then, Sect. 5 is devoted to simulation results proving the proposed FD strategy. Conclusions are given in Sect. 6.

2 Preliminaries

$\mathbb{R}$ and $\mathbb{N}$ represent the sets of real and natural numbers, respectively. The eigenvalues set of a matrix $A \in \mathbb{R}^{n\times n}$ is named $\lambda(A)$ and $\mathrm{Re}(z)$ is the real part of the complex number z. The set of Hurwitz matrices from the set $\mathbb{R}^{n\times n}$ is denoted by $\mathbb{H}$, i.e. $R \in \mathbb{H} \Leftrightarrow \mathrm{Re}(\lambda) < 0, \forall\lambda \in \lambda(R)$. We denote $\mathbb{M}$ as the set of Metzler matrices from the set $\mathbb{R}^{n\times n}$, i.e. $R = \{r_{ij}\}_{i,j=1}^{n} \in \mathbb{M} \Leftrightarrow r_{i,j} \geq 0$ for $i \neq j$. For a variable $x(t) \in \mathbb{R}^n$, the upper and lower bounds are denoted by $\overline{x}(t) \in \mathbb{R}^n$ and $\underline{x}(t) \in \mathbb{R}^n$, respectively, such that $\underline{x}(t) \leq x(t) \leq \overline{x}(t)$ and the relation $\leq$ should be interpreted elementwise for vectors and matrices, i.e. $A = (a_{i,j}) \in \mathbb{R}^{n\times m}$ and $B = (b_{i,j}) \in \mathbb{R}^{n\times m}$ such that $A \geq B$ if and only if $a_{i,j} \geq b_{i,j}$ $\forall\, i \in \{1, \ldots, n\}$ and $j \in \{1, \ldots, m\}$, $i, j \in \mathbb{N}$. For a matrix $R \in \mathbb{R}^{n\times m}$, define $R^+ = \max\{0, R\}$ and $R^- = R^+ - R$. The matrix of absolute values of all elements of a matrix $M \in \mathbb{R}^{n\times m}$ is $|M| = M^+ + M^-$. The vector E_p is stated for $(p \times 1)$ vector with unit elements, and I_n denotes the identity matrix of $n \times n$ dimension. Superscript T denotes the transpose of a matrix or a vector. $\mathcal{K}$ is the set of continuous increasing functions $\gamma : \mathbb{R}_+ \to \mathbb{R}_+$ with $\gamma(0) = 0$. We refer by $\beta \in \mathcal{KL}$ if $\beta(\cdot, t) \in \mathcal{K}$ for all $t \geq 0$ and $\beta(r, \cdot)$ is continuous and strictly decreasing to zero for all $r > 0$. $\|.\|$ is the standard 2-norm.

Lemma 1 ([3]) *Let $\underline{x}, x, \overline{x} \in \mathbb{R}^n$ be vectors satisfying $\underline{x} \leq x \leq \overline{x}$ and $A \in \mathbb{R}^{n\times m}$ be a time-invariant matrix. Then, the inequalities below hold:*

$$A^+\underline{x} - A^-\overline{x} \leq Ax \leq A^+\overline{x} - A^-\underline{x}. \tag{1}$$

Lemma 2 ([2]) *Consider the following system:*

$$\begin{cases} \dot{x}(t) = Ax(t) + \psi(t), \\ y(t) = Cx(t), \end{cases} \tag{2}$$

where $\psi(t)$ is a continuous function and A, C are known matrices. Suppose that there exist two known continuous-time functions $\underline{\psi}(t)$ and $\overline{\psi}(t) : \mathbb{R} \to \mathbb{R}^n$ satisfied $\underline{\psi}(t) \leq \psi(t) \leq \overline{\psi}(t)$, $\forall\, t \geq 0$.

If there exists a gain L such that $(A - LC) \in \mathbb{H} \cap \mathbb{M}$ *and* $\underline{x_0}, x_0, \overline{x_0} \in \mathbb{R}^n$, $\underline{x}_0 \leq x_0 \leq \overline{x}_0$, *then the system:*

$$\begin{cases} \dot{\overline{x}}(t) = A\overline{x}(t) + \overline{\psi}(t) + L(y(t) - C\overline{x}(t)) \\ \dot{\underline{x}}(t) = A\underline{x}(t) + \underline{\psi}(t) + L(y(t) - C\underline{x}(t)) \end{cases} \tag{3}$$

is an interval observer for (2) and $\underline{x}(t) \leq x(t) \leq \overline{x}(t)$, $\forall t \geq 0$.

Definition 1 Consider the nonlinear system

$$\dot{x} = f(x, u), \tag{4}$$

with $f(x, u) \in \mathbb{R}^n$, the system (4) is input-to-state Stable (ISS) if for any input $u \in \mathbb{R}^m$ and $x_0 \in \mathbb{R}^n$ there exist functions $\beta \in \mathcal{KL}$ and $\gamma \in \mathcal{K}$ such that

$$|x(t, x_0, u)| \leq \beta(x_0, t) + \gamma(\|u\|), \quad \forall t \geq 0. \tag{5}$$

3 NCS Architecture Description and Problem Formulation

Consider the following NNCS:

$$\begin{cases} \dot{x}(t) = Ax(t) + F(u(t), y(t)) + w(t) + Hf(t), \\ y(t) = Cx(t) + v(t), \end{cases} \tag{6}$$

where $x \in \mathbb{R}^n$ denotes the state vector, $y \in \mathbb{R}^p$ the measurable output vector $u \in \mathbb{R}^m$ the known input vector, where $w \in \mathbb{R}^n$, $v \in \mathbb{R}^p$ and $f \in \mathbb{R}^n$ are the external disturbances and the additive faults to be detected. The functions u, v, w are continuous. The function $F(u(t), y(t)) \in \mathbb{R}^n$ is a globally Lipschitz nonlinear function. The matrices A, C and H are known matrices of compatible dimensions.

Before proceeding further, we make some assumptions on the process matrices.

Assumption 1 The pair (A, C) is detectable. □

Assumption 2 $w(t)$ and $v(t)$ are unknown but bounded functions with a priori known bounds, for $\overline{w}$, $\underline{w} \in \mathbb{R}^n$, $\overline{V} \in \mathbb{R}_+$:

$$\underline{w} \leq w(t) \leq \overline{w}, \quad |v(t)| \leq \overline{V} E_p, \quad \forall t \geq 0 \tag{7}$$

The network induces two uncertain delays: $d_a(t)$ and $d_m(t)$ which refer to the actuation and the measurement channels delay, respectively. These delay functions are unknown but bounded:

$$0 \leq \underline{d_a} \leq d_a(t) \leq \overline{d_a}, \quad 0 \leq \underline{d_m} \leq d_m(t) \leq \overline{d_m}, \quad \forall t, \tag{8}$$

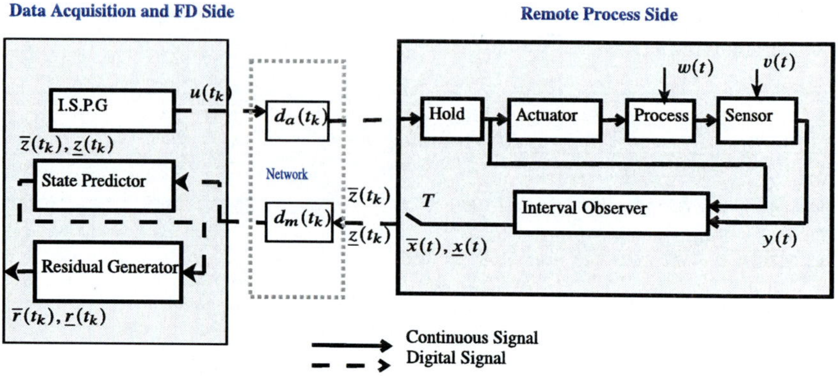

Fig. 1 NCS architecture

where $\overline{d_a}$, $\underline{d_a}$ are respectively the upper and lower bounds of $d_a(t)$ and $\overline{d_m}$, $\underline{d_m}$ are respectively the upper and lower bounds of $d_m(t)$. Suppose that $\overline{d}$ and $\underline{d}$ are respectively the upper and lower bounds of the communication delays such that:

$$\overline{d} = \max\{\overline{d_a}, \overline{d_m}\}, \underline{d} = \min\{\underline{d_a}, \underline{d_m}\}. \tag{9}$$

Considering the communication delay in the actuation channel, the process (6) can be modeled as an input delayed system:

$$\begin{cases} \dot{x}(t) = Ax(t) + F(u(t - d_a(t)), y(t)) + w(t) + Hf(t) \\ y(t) = Cx(t) + v(t). \end{cases} \tag{10}$$

Assumption 3 We assume that $u(t - d_a(t))$ is bounded. □

Let $\{t_k, k \in \mathbb{N}\}$ be the sequence of sampling instants such that $t_{k+1} - t_k = T$ and $\lim_{k\to\infty} t_k = \infty$, T is the sampling period and t_k is an increasing sequence such that $t_k = kT$. In a network environment, data sampling is needed. Therefore, the next assumptions are required.

Assumption 4 ([1]) The sampling communication delays $d_a(t_k)$ and $d_m(t_k)$ are unknown but bounded with a priori known bounds and the upper bound $\overline{d}$ is assumed to be a multiple of the sampling period T. □

Assumption 5 ([1]) The information on the control signal $u(t_k)$, the information on the output $y(t_k)$ and the information on the sampling upper and lower bounds $\underline{z}(t_k)$, $\overline{z}(t_k)$ could be stored and used $\forall\, t_k \in \left[t_k - \overline{d}, t_k\right)$. □

Each block of the NCS architecture shown in Fig. 1 performs the same function as described in [5]. However, the added Residual Generator block is implemented in the calculator for residual generation; after receiving predictor outputs; $\overline{z}(t_k)$ and $\underline{z}(t_k)$, Residual Generator calculates interval residual signals $\overline{r}(t_k)$ and $\underline{r}(t_k)$ which will be used for FD test detailed in Sect. 4.

3.1 Interval Observer Structure

In this section, the interval observer developed in [5] is tackled to estimate the unavailable states of the process under unknown input delay $d_a(t)$ and process uncertainties. In free faulty case, the system (10) is:

$$\begin{cases} \dot{x}(t) = Ax(t) + F(u(t - d_a(t)), y(t)) + w(t) \\ y(t) = Cx(t) + v(t). \end{cases} \tag{11}$$

Since it is not always possible to compute a gain L for the system (11) such that $A - LC \in \mathbb{H} \cap \mathbb{M}$, a change of coordinates $\xi = Sx$ with a nonsingular matrix S such that the matrix $S(A - LC)S^{-1} \in \mathbb{H} \cap \mathbb{M}$ is used to relax this restriction [8].

Theorem 1 ([5]) *Let Assumptions 1–3 be satisfied and* $\underline{x_0} \leq x_0 \leq \overline{x_0}$. *If there exists a change of coordinates* $\xi = Sx$ *satisfying* $\mathcal{E} = S(A - LC)P \in \mathbb{H} \cap \mathbb{M}$, $P = S^{-1}$ *so that the following system*

$$\begin{cases} \dot{\hat{x}}^+(t) = \mathcal{E}\hat{x}^+(t) + \overline{SF}(u(t - \underline{d}), y(t)) + SLy(t) + S^+\overline{w} - S^-\underline{w} + |SL|\, E_p\overline{V}, \\ \dot{\hat{x}}^-(t) = \mathcal{E}\hat{x}^-(t) + \underline{SF}(u(t - \underline{d}), y(t)) + SLy(t) + S^+\underline{w} - S^-\overline{w} - |SL|\, E_p\overline{V}, \end{cases} \tag{12}$$

where

$$\begin{cases} \overline{SF}(u(t - \underline{d}), y(t)) = \max\limits_{\alpha \in [0, \overline{d} - \underline{d}]} \{SF(u(t - \underline{d} - \alpha), y(t))\}, \\ \underline{SF}(u(t - \underline{d}), y(t)) = \min\limits_{\alpha \in [0, \overline{d} - \underline{d}]} \{SF(u(t - \underline{d} - \alpha), y(t))\}, \end{cases} \tag{13}$$

and the initial conditions are calculated as follows:

$$\hat{x}^+(0) = S^+\overline{x}_0 - S^-\underline{x}_0, \quad \hat{x}^-(0) = S^+\underline{x}_0 - S^-\overline{x}_0, \tag{14}$$

is input-to-state stable (ISS) interval observer for the system (11) satisfying [4]

$$\hat{x}^-(t) \leq \xi(t) \leq \hat{x}^+(t), \quad \forall t \geq 0 \tag{15}$$

where the bounds of the solution $x(t)$ *are:*

$$\overline{x}(t) = P^+\hat{x}^+(t) - P^-\hat{x}^-(t), \underline{x}(t) = P^+\hat{x}^-(t) - P^-\hat{x}^+(t). \tag{16}$$

such that

$$\underline{x}(t) \leq x(t) \leq \overline{x}(t), \quad \forall t \geq 0. \tag{17}$$

Proof The proof of the above theorem is detailed in [5].

3.2 State Predictor Design

To compensate the large unknown communication delays $d_m(t_k)$, an interval predictor introduced in [5] is used. Based on the delayed data $\overline{z}(\frac{t_k}{T} - \frac{\overline{d}}{T})$, $\underline{z}(\frac{t_k}{T} - \frac{\overline{d}}{T})$, we will reconstruct $\overline{z}(\frac{t_k}{T})$, $\underline{z}(\frac{t_k}{T})$ after a finite time $t_k = \overline{d}$.

Assumption 6 T is selected such that the matrix $\Phi = I_n + T\mathcal{E}$ is positive. □

Theorem 2 ([5]) *If Assumptions 4–6 hold* $\forall\, t_k \geq \overline{d}$, *we get:*

$$\hat{z}^+(k) = \Phi^{k_2}\hat{z}^+(k-k_2) + \sum_{j=1}^{k_2} \Phi^{k_2-j} T\overline{SF}(u(k-k_1-k_2+j-1), y(k-k_2+j-1)) + \sum_{j=1}^{k_2} \Phi^{k_2-j} L_1 y(k-k_2+j-1) + \sum_{j=1}^{k_2} \Phi^{k_2-j}\overline{\beta}, \tag{18}$$

$$\hat{z}^-(k) = \Phi^{k_2}\hat{z}^-(k-k_2) + \sum_{j=1}^{k_2} \Phi^{k_2-j} T\underline{SF}(u(k-k_1-k_2+j-1), y(k-k_2+j-1)) + \sum_{j=1}^{k_2} \Phi^{k_2-j} L_1 y(k-k_2+j-1) + \sum_{j=1}^{k_2} \Phi^{k_2-j}\underline{\beta}, \tag{19}$$

where $j = 1, 2, 3, \ldots, k_2$, $k = \frac{t_k}{T}$, $k_1 = \left\lfloor \frac{d}{T} \right\rfloor$ *and* $k_2 = \frac{\overline{d}}{T}$, $\Phi = (I_n + T\mathcal{E})$, $L_1 = TSL$, $\overline{\beta} = T(S^+\overline{w} - S^-\underline{w} + |SL|\, E_p \overline{V})$, $\underline{\beta} = T(S^+\underline{w} - S^-\overline{w} - |SL|\, E_p\overline{V})$ *and*

$$\overline{z}(k) = S^+\hat{z}^+(k) - S^-\hat{z}^-(k), \underline{z}(k) = S^+\hat{z}^-(k) - S^-\hat{z}^+(k) \tag{20}$$

are a predictor from the sampling instant of time $k - k_2$ *to* k *for the process* (11) *i.e.* $\overline{z} \to \overline{x}$ *and* $\underline{z} \to \underline{x}$ $\forall\, t_k \geq \overline{d}$ *and the following inclusion holds*

$$\underline{z}(\frac{t_k}{T}) \leq x(\frac{t_k}{T}) \leq \overline{z}(\frac{t_k}{T}), \quad t_k \geq \overline{d}. \tag{21}$$

Proof Please see the proof detailed in [5].

4 Fault Detection

In this section, a procedure of fault detection is developed thanks to Residual Generator block as shown in Fig. 1. The residual evaluation in a set-membership context is ensured via the following belonging test [12]: If the zero signal is enclosed by upper and lower bounds of the residual signal, it is a fault-free case. Otherwise, a fault is occurred. Two steps are required to indicate the presence of faults:

- **Step1: Residuals generation**: in this first step, the Residual Generator calculates upper and lower residuals defined as the gap between measured outputs and estimated outputs using stored information.
  From Lemma 1, the upper and lower bounds of the estimated output are then computed as follows:

$$\begin{cases} \overline{y}(t_k) = C\overline{z}(t_k) + \overline{V}E_p = C^+\overline{z}(t_k) - C^-\underline{z}(t_k) + \overline{V}E_p \\ \underline{y}(t_k) = C\underline{z}(t_k) - \overline{V}E_p = C^+\underline{z}(t_k) - C^-\overline{z}(t_k) - \overline{V}E_p \end{cases} \tag{22}$$

  Then, upper and lower residuals are :

$$\overline{r}(t_k) = \overline{y}(t_k) - y(t_k), \underline{r}(t_k) = \underline{y}(t_k) - y(t_k)\,. \tag{23}$$

- **Step 2: Residuals evaluation:** this second step is detailed as follows. When a fault is occurred, an inconsistency is detected shown that the estimated outputs are no more compatible with the measurements where:

$$y(t_k) \notin [\underline{y}(t_k), \overline{y}(t_k)] \tag{24}$$

  The above belonging test is rewritten as follows:

$$\begin{cases} 0 \notin [\underline{y}(t_k), \overline{y}(t_k)] - y(t_k) \\ 0 \notin [\underline{y}(t_k) - y(t_k), \overline{y}(t_k) - y(t_k)] \\ 0 \notin [\underline{r}(t_k), \overline{r}(t_k)] \end{cases} \tag{25}$$

  Then, the zero signal is enclosed by $\overline{r}$ and $\underline{r}$ in the fault free case. Otherwise, a fault is detected.

From (23) and using the fact that $C = C^+ - C^-$, $\overline{r}$ and $\underline{r}$ are computed as follows:

$$\overline{r}(t_k) = \overline{y}(t_k) - y(t_k) = C^+(\overline{z}(t_k) - z(t_k)) + C^-(z(t_k) - \underline{z}(t_k)) - v(t_k) + \overline{V}E_p \tag{26}$$

$$\underline{r}(t_k) = \underline{y}(t_k) - y(t_k) = -C^+(z(t_k) - \underline{z}(t_k)) - C^-(\overline{z}(t_k) - z(t_k)) - v(t_k) - \overline{V}E_p \tag{27}$$

An augmented system is then defined as:

$$\begin{bmatrix}\overline{r}\\ \underline{r}\end{bmatrix}=\begin{bmatrix}C^{+} & C^{-}\\ -C^{-} & -C^{+}\end{bmatrix}\begin{bmatrix}\overline{e_z}\\ \underline{e_z}\end{bmatrix}+\begin{bmatrix}-v+\overline{V}E_p\\ -v-\overline{V}E_p\end{bmatrix},$$

$$\overline{e_z}(t_k)=\overline{z}(t_k)-x(t_k),\ \underline{e_z}(t_k)=x(t_k)-\underline{z}(t_k) \tag{28}$$

Or we have $\overline{z}\rightarrow\overline{x}$ and $\underline{z}\rightarrow\underline{x}\ \forall\ t_k\geq\overline{d}$ (see Sect. 3.2), then

$$\begin{cases}\overline{e_z}(t_k)=\overline{z}(t_k)-x(t_k)\simeq\overline{x}(t_k)-x(t_k)=\overline{e}(t_k)\\ \underline{e_z}(t_k)=x(t_k)-\underline{z}(t_k)\simeq x(t_k)-\underline{x}(t_k)=\underline{e}(t_k)\end{cases} \tag{29}$$

Known that the measurement noise v is bounded, stability analysis of upper and lower residual signals is equivalent to assure the stability of the estimation errors.

From Theorem 1, we have upper and lower bound estimation errors $\overline{e}$ and $\underline{e}$ are ISS [5]. Therefore, from (29), one can prove that $\overline{e_z}$ and $\underline{e_z}$ are ISS and then the residual signals $\overline{r}(t_k)$ and $\underline{r}(t_k)$ are ISS $\forall\ t_k\geq\overline{d}$.

5 Numerical Example

To prove the efficiency of the proposed FD strategy, the next system (6) is considered:

$$A=\begin{bmatrix}-3.5000 & 0 & 0.5000\\ 0 & -2.7540 & 0\\ 0 & 0 & -1.2000\end{bmatrix},\ F(u(t),y(t))=\begin{bmatrix}\sin(u(t))\\ 1.5\sin(u(t)y(t))\\ 2\sin(u(t))\end{bmatrix},$$
$$C=\begin{bmatrix}0\,0\,1\end{bmatrix},\ H=[-3\quad 2\quad 1]^T$$

$$w(t)=[0.1\cos(2t)\quad 0.1sin(3t)\quad 0.1\cos(4t)]^T,\ \overline{w}=[0.1\quad 0.1\quad 0.1]^T,\ \underline{w}=-\overline{w}\text{ and}$$
$$v(t)=0.2\cos(t)\cos(5t)\sin(10t)\sin(20t)\text{ with }\overline{V}=0.2.$$

On can see that the function $F(u(t),y(t))$ is globally Lipschitz.

Considering communication delays, system (6) will be modeled as (10); its output signal and the input signal delivered by the I.S.P.G are depicted in Fig. 2.

Network proprieties: The network induces unknown but bounded delays as described by (8). These bounds are $\underline{d}=0.7s$, $\overline{d}=0.1s$ and the distribution of delays is shown in Fig. 3 with the disturbances and the measurement noise. The sampling period is given by $T=0.01s$.

Interval observer design: A gain $L=\begin{bmatrix}-10\,0\,0\end{bmatrix}^T$ is computed satisfying $A-LC\in\mathbb{H}$. However, the matrix $A-LC$ is Hurwitz but is not Metzler. Then, a transformation of coordinates as described in Sect. 3.1 is needed. We propose

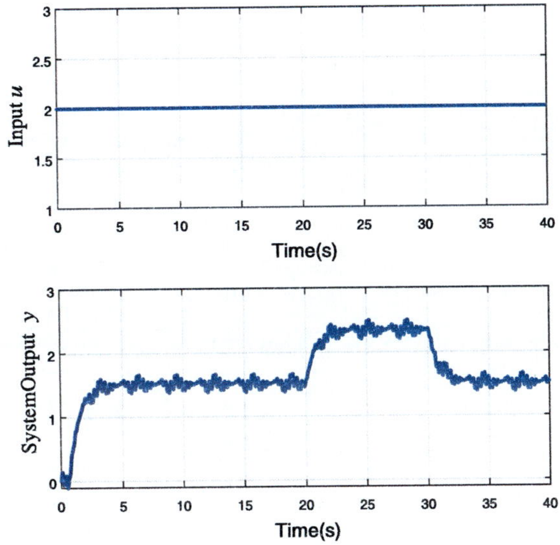

Fig. 2 System input and output

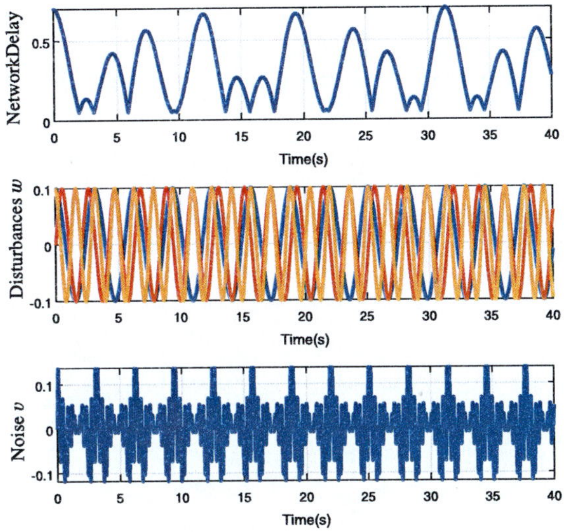

Fig. 3 Network delay, disturbances and noise

$S = \begin{bmatrix} 0.5005 & 0.0000 & -2.2800 \\ 0.000 & 1.0064 & 0.0000 \\ -0.0010 & -0.0012 & 5.5701 \end{bmatrix}$, and we can easily verify that $\mathcal{E} = S(A - LC)S^{-1} = \begin{bmatrix} -3.5000 & 0.0000 & 0.0020 \\ 0.0000 & -2.7540 & 0.0000 \\ 0.0049 & 0.0019 & -1.2000 \end{bmatrix} \in \mathbb{H} \cap \mathbb{M}$. Therefore, an interval observer as (12) can be designed. The initial conditions of system (10) are chosen such that $\underline{x}_0(t) \leq x_0(t) \leq \overline{x}_0(t)$; $\overline{x}(0) = \left[0.5\ 0.5\ 0.5\right]^T$ and $\underline{x}(0) = \left[-0.5\ -0.5\ -0.5\right]^T$. Also the estimator (12) is initialized by $\hat{x}^+(0)$ and $\hat{x}^-(0)$ which are defined in (14).

State predictor design: For the sampling period T = 0.01s, we obtain $\Phi = \begin{bmatrix} 0.9650 & 0.0000 & 0.0000 \\ 0.0000 & 0.9724 & 0.0000 \\ 0.00005 & 0.00002 & 0.9880 \end{bmatrix}$. The predictor described by (18) and (19) can be computed with Φ, $\tilde{L_1} = [-0.0500 \quad 0.0000 \quad 0.0001]^T$ and $\overline{\beta} = [0.0127 \quad 0.0010 \quad 0.0010]^T$, $\underline{\beta} = -\overline{\beta}$.

Fault evolution: The system is affected by an additive fault $f(t)$:

$$\begin{cases} f(t) = 1 & 20s \leq t \leq 30s \\ f(t) = 0 & else \end{cases} \tag{30}$$

The fault evolution is presented in Fig. 4. The evolution of the residual signals in faulty case is plotted in Fig. 4 where black lines are the zero signal and red, blue lines represent the upper and lower residuals, respectively. As shown in Fig. 5 upper and lower residual signals are sensitive to the fault f i.e. $0 \notin [\underline{r}, \overline{r}]$ when $20s \leq t \leq 30s$. Then we can conclude that the fault detection strategy is ensured and validated despite unknown communication delays and external process uncertainties.

Fig. 4 Fault evolution

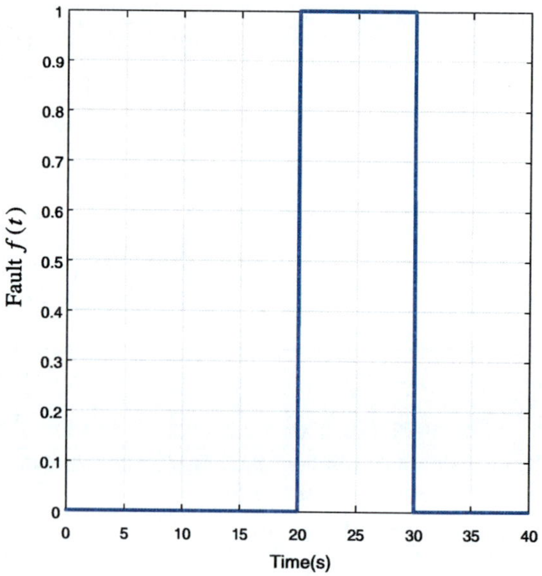

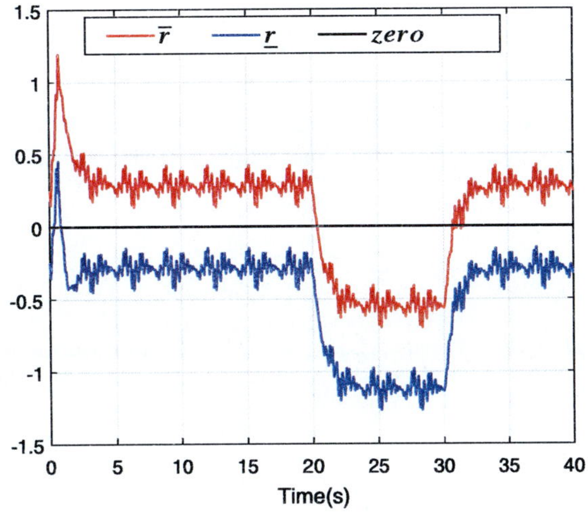

Fig. 5 Evolution of upper and lower residual signals

6 Conclusion and Future Work

In this work, a set-membership FD technique is developed for NNCS subject to network delays and uncertainties. These uncertainties are assumed to be unknown but bounded with a priori known bounds. The main contribution consists in using upper and lower residuals for FD decision. A belonging test of the zero signal to the interval delimited by the upper and lower residuals is used to ensure the detection of faults. Theoretical results have been validated through the numerical example.

Nevertheless, the fault isolation problem for such NNCS is not investigated in this contribution. It will be the subject of a future work.

References

1. Yang, R., Liu, G.P., Shi, P., Thomas, C., Basin, M.V.: Predictive output feedback control for networked control systems. IEEE Trans. Ind. Electron. **61**(1), 512–520 (2013). https://doi.org/10.1109/TIE.2013.2248339
2. Gouzé, J.L., Rapaport, A., Hadj-Sadok, M.Z.: Interval observers for uncertain biological systems. Ecol. Model. **133**(1–2), 45–56 (2000). https://doi.org/10.1016/S0304-3800(00)00279-9
3. Efimov, D., Raïssi, T.: Design of interval observers for uncertain dynamical systems. Autom. Remote. Control. **77**(2), 191–225 (2016). https://doi.org/10.1134/S0005117916020016
4. Dashkovskiy, S., Efimov, D.V., Sontag, E.D.: Input to state stability and allied system properties. Autom. Remote. Control. **72**(8), 1579–1614 (2011). https://doi.org/10.1134/S0005117911080017
5. Najjar, A., Dinh, T.N., Amairi, M., Raïssi, T.: Interval observer-based supervision of nonlinear networked control systems. Turk. J. Electr. Eng. Comput. Sci. **30**(4), 1219–1234 (2022). https://doi.org/10.55730/1300-0632.3845

6. Ren, C., Park, J.H., He, S.: Positiveness and finite-time control of dual-switching poisson jump networked control systems with time-varying delays and packet drops. IEEE Trans. Control Netw. Syst. **9**(2), 575–587 (2022)
7. Zhang, X.M., Han, Q.L., Ge, X., et al.: Networked control systems: a survey of trends and techniques. IEEE/CAA J. Autom. Sin. **7**(1), 1–17 (2019)
8. Najjar, A., Dinh, T.N., Amairi, M. et al.: Supervision of nonlinear networked control systems under network constraints. In: 2019 4th Conference on Control and Fault Tolerant Systems (SysTol), pp. 270–275. IEEE
9. Keller, J.Y., Sauter, D.: Monitoring of stealthy attack in networked control systems. In: 2013 Conference on Control and Fault-Tolerant Systems (SysTol), pp. 462–467. IEEE
10. Wu, C., Liu, J., Jing, X., et al.: Adaptive fuzzy control for nonlinear networked control systems. IEEE Trans. Syst. Man Cybern.: Syst. **47**(8), 2420–2430 (2017)
11. Gao, Z., Cecati, C., Ding, S.X.: A survey of fault diagnosis and fault-tolerant techniques-part i: fault diagnosis with model-based and signal-based approaches. IEEE Trans. Industr. Electron. **62**(6), 3757–3767 (2015)
12. Chevet, T., Dinh, T.N., Marzat, J. et al.: Robust sensor fault detection for linear parameter-varying systems using interval observer. In: 31st European Safety and Reliability Conference. Proceedings of the 31st European Safety and Reliability Conference, Angers, France, pp. 1486–1493. https://doi.org/10.3850/978-981-18-2016-8_380-cd. https://hal-cnam.archives-ouvertes.fr/hal-03239385
13. Heijmans, S., Postoyan, R., Nešić, D., et al.: An average allowable transmission interval condition for the stability of networked control systems. IEEE Trans. Autom. Control **66**(6), 2526–2541 (2021). https://doi.org/10.1109/TAC.2020.3012526. https://hal.archives-ouvertes.fr/hal-02909161

Demanded Power Point Tracking for Urban Wind Turbines

Felix Dietrich, Lukas Jobb, and Horst Schulte

Abstract In addition to increasing the rated power of wind turbines in the multiple megawatt range, research is advancing into suitable small compact wind turbines for urban areas. In particular, the requirements for their flexible automated commissioning and plug-and-play solutions are growing. Up to now, small wind turbines have been operated in such a way that these provide the maximum achievable power from the wind. In the future, however, it will be essential to integrate them into a power system as a flexibly controllable current or voltage source. A demanded power point tracking algorithm based on a model-free extremum-seeking control scheme is proposed to get closer to this aim. As a logical extension of the well-known maximum power point algorithm, the method is presented in detail, applied to a real wind turbine of a manufacturer, and validated in simulation studies for direct application in the field.

Keywords Active power control · Wind energy · Extremum-seeking control · MPPT · Variable-speed control

1 Introduction

Wind energy has seen rapid market share growth over the past few years. Furthermore, several countries, such as the U.S., have pledged to develop this field in the coming years further [1]. Given this development, it has become more important for wind turbines to be able to react to (sudden) changes in load-side energy demand. For grid-connected turbines, this means adjusting to the grid's frequency variations. One

F. Dietrich · H. Schulte (✉)
School of Engineering-Energy and Information, Control Engineering Group,
University of Applied Sciences (HTW) Berlin, Wilhelminenhofstraße 75A, 12459 Berlin, Germany
e-mail: schulte@htw-berlin.de

L. Jobb
MOWEA - Modulare Windenergieanlagen GmbH, Storkower Str. 115A, 10407 Berlin, Germany

D. Theilliol et al. (eds.), *Recent Developments in Model-Based and Data-Driven Methods for Advanced Control and Diagnosis*, Studies in Systems, Decision and Control 467,
https://doi.org/10.1007/978-3-031-27540-1_13

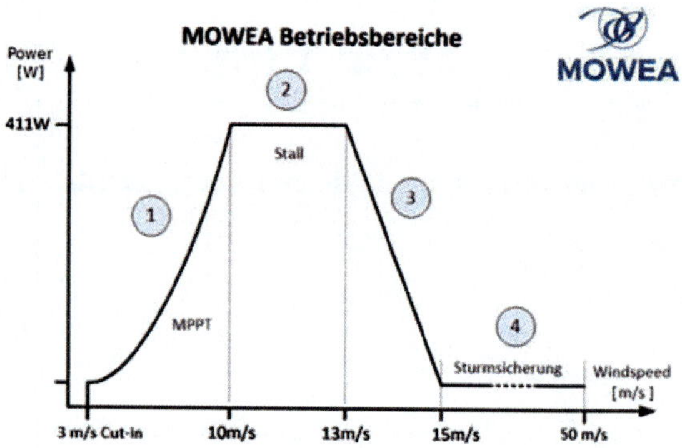

Fig. 1 Power curve of a variable speed wind turbine divided into Region 1 to Region 4

should note that over the years, variable-speed wind turbines have replaced fixed wind turbines as the most common newly installed wind turbine type [8].

With a full-frequency converter, variable-speed wind turbines decouple the inertial turbine dynamics from the power grid through their power electronics. Therefore, a variable speed wind turbine without an additional control scheme has no inertial response like conventional power generators when the power demand and grid frequency vary [3]. However, an advantage of variable-speed wind turbines is that they can increase their overall power efficiency by applying a power curve, as depicted in Fig. 1. When they operate below their rated power (Region 1), a Maximum Power Point Tracking (MPPT) control methodology can be applied to obtain the highest possible power output at different wind speeds. It is achieved with a wind turbine controller using torque control to optimize the turbine's power output [2]. However, once the rated wind speed is surpassed, the objective is to maintain the wind turbine's power output at the rated power level while minimizing the structural loads on the turbine (Region 2). Wind turbines in the standard megawatt class commonly achieve this through pitch control while the torque is constant. The pitch control changes the rotor blades' angles to reduce the turbine's aerodynamic efficiency to operate below its maximum power point.

This paper focuses on small wind turbines with a rated power of up to 2 kW without a blade pitch mechanism. Therefore, all control is carried out exclusively by adjusting the generator torque without pitch control. When the wind speed reaches the cut-out speed, the turbine is shut down to prevent structural damage. In some cases, the cut-out is not carried out abruptly, but the power output is continuously reduced according to the wind speed (Region 3). That way, some energy can still be harvested while the load on the mechanical parts of the turbine is still manageable. To follow external power commands in Region 1 and Region 2 and internal reference for smooth power shutdown in Region 3, this paper proposes a demanded power point tracking (DPPT).

To achieve DPPT functionality, various control strategies have been developed and proposed in the literature. First, we give an overview of the wind turbines of the megawatt-class, which are controlled by the generator torque and rotor blade pitch angle: To cover the whole wind speed spectrum in [4] the authors proposed a linear parameter variable (LPV) control design method for variable-pitch wind turbines. DPPT methods, also called active power tracking control (APC), to support grid integration with pitch-based and torque-based power reserve control strategies are studied and compared in [9]. One proposed control strategy was to obtain rotor speed set-points according to demanded power points. It was validated through simulations and field tests with a 550 kW wind turbine. In [12], a transition region between the partial load region (Region 1) and full load region (Region 2) is established, see Fig. 1. Here, mode switches enable or disable the DPPT control ability. The proposed algorithm is validated through simulation and field tests with a down-scaled wind turbine in a wind tunnel. In [5], a model-based multi-variable controller design without switching mechanisms for power following behavior in the range without and with pitch adjustment is presented and experimentally validated.

In this work, the focus is on small wind turbines. Due to the large model uncertainties in the aerodynamic map, a model-free approach such as a modified extremum-seeking (ES) algorithm controller is appropriate. On the other hand, the ES method is unsuitable for megawatt-class turbines because the manipulated variable constraints are significant, and the ES algorithm often reaches saturation, making this method not well suited. The extremum-seeking control algorithms belong to the model-free real-time optimization control strategy class. The literature proposes ES algorithms to solve the maximum power point tracking (MPPT) problem for small wind turbines and photovoltaic systems. In [6], the authors used an ES controller and combined it with field-oriented generator control to optimize the power output of a grid-connected wind turbine. Logarithmic power feedback as input is proposed in [14] for the ES algorithm to cover a wider spectrum of wind speeds.

The generalization of the ES algorithm to track a given external slope reference input has been presented in [11] and extends the results to the multivariable case of gradient seeking. A similar approach is taken in [13]. Here, instead of using the extremum seeking to optimize a measurable system output, it will be used to optimize a suitable auxiliary cost function. This cost function is selected, so the system output reaches a certain percentage of its optimal value.

To the best of the authors' knowledge, this paper presents a DPPT algorithm for a real industrial small wind power plant[1] based on a modified ES method using a similar cost function as presented in [13] for the first time. It is shown that the ES control can be used for DPPT purposes as an extension of the existing MPPT method. Therefore, the proposed method provides a simple, industrial robust control strategy for the entire power range (Region 1-4) shown in Fig. 1 without the need for mode switching or transition control between two control regions.

The paper is structured as follows: Sect. 2 explains the technical scheme of the small wind power plant under consideration, derives the mathematical first-principle model, and presents the internal control structure. In Sect. 3, the operation of an extremum-seeking algorithm for maximum power point tracking is first described, and second, as a natural extension, the modified ES is presented to achieve the

[1] MOWEA String turbine, Modulare Windenergieanlagen GmbH, Berlin, Germany.

required power point tracking. Afterward, the simulation results, which demonstrate the performance of the demanded power point tracking algorithm using the ES method, are presented in Sect. 4. Section 5 concludes the paper with a discussion and an outlook.

2 Control Objectives and Process Description

The power P_r captured from the wind by the turbine rotor is defined by

$$P_r = \frac{1}{2}\,\rho\, v^3\, \pi\, R^2\, c_P\,(\lambda)\ , \tag{1}$$

where ρ denotes the air density [kg/m^3], v the wind speed [m/s] far in front of the rotor, and R the rotor radius in [m]. Therefore $\pi\, R^2$ is the area perpendicular to the airflow swept by the rotor blades. A wind turbine can only harness a fraction of the wind power. The amount of power the turbine captures depends on the aerodynamic efficiency of the rotor. The latter is described by the power coefficient c_P. This paper investigates a wind turbine whose power coefficient depends on the tip speed ratio λ. The power coefficient would also be a function of the pitch angle β [deg] for wind turbines with pitch angle control. The tip-speed ratio is defined by:

$$\lambda = \frac{\omega_r\, R}{v}\ , \tag{2}$$

where ω_r denotes the rotor speed [rad/s]. The Fig. 2 shows the $c_P(\lambda)$ curve of the studied wind turbine of the manufacturer MOWEA. In this case, the power coefficient reaches its maximum of $c_{P,max} = 0.4$ at a tip-speed ratio of $\lambda_{opt} = 7.5$ (see the upper part of Fig. 2), which implies that the power output can be set dynamically by varying the rotor speed. The control objectives of wind turbines are highly dependent on the current operating region shown in Fig. 1: In Region 1, the objective would be to reach λ_{opt} to operate at the maximum power point for a given wind speed. In Region 2 and 3, the objective would be to keep the power output constant or reduce it, with the turbine operating below its maximum power point. Since the rotor power is $P_r = T_r\, \omega_r$ the rotor torque T_r is described by

$$T_r(v, \omega_r) = \frac{1}{2}\,\rho\, v^2\, \pi\, R^3\, \frac{c_P(\lambda(v, \omega_r))}{\lambda(v, \omega_r)}\ . \tag{3}$$

The dynamics of the drive train of a small wind turbine is given by the 1-DOF equation of motion

$$\dot{\omega}_r = \frac{1}{J}\left(T_r(v, \omega_r) - T_g\right)\ , \tag{4}$$

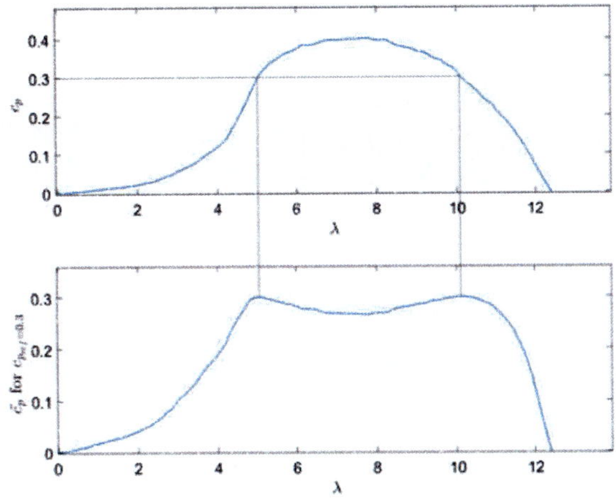

Fig. 2 $c_P(\lambda)$ curve of the MOWEA Champ 400 wind turbine (upper figure) and the transformed $\tilde{c}_P(\lambda)$ curve for a $c_{P_{ref}}$ of 0.3 (lower figure)

where J is the total inertia of the drive with the main components, rotor, and generator related to the rotation around the longitudinal axis of the drive train. Since a gearbox is usually not required for small wind turbines because the rotor speed is high enough to drive the generator directly.

This paper studies the MOWEA wind turbine using a brushless DC (BLDC) motor as a generator. Its power flow is reversed, so the BLDC motor is a generator. At its output, the AC power generated by the generator is rectified by a controllable inverter that feeds a DC load. The rectifier is also used as a boost DC-DC converter. This simplifies the design, and thus the generator model is used as follows:

$$T_g = \frac{1}{2\pi} k_M \, \psi \, i_{DC} \tag{5}$$

where k_M denotes the motor parameter, and ψ is the constant flux parameter caused by the rotor's permanent magnets. The variable i_{DC} represents the output current after being rectified and passed through the DC-DC converter. The control system structure is cascaded with an inner rotor speed controller K_{sc}, and an outer control loop using the ES control method is denoted as K_{ES}. The input variables of the ES controller are the setpoint resp. tracking power value and the instantaneous electrical power with $P_{DC} = u_{DC}\, i_{DC}$, measured at the load. The extreme value controller passes on a setpoint speed to the rotor speed controller's inner control loop. The latter controls the setpoint by adjusting the inverter's duty cycle signal u_d at its output depending on the control error. The duty cycle determines the DC-DC voltage ratio. It is thus inversely proportional to the current ratio, which is used to set the speed via the generator torque according to (5) and the motion Eq. 4. As quantified by Eqs. (1) and (2), a change in rotor speed affects the power output of the turbine, closing the causal chain of the outer loop (Fig. 3).

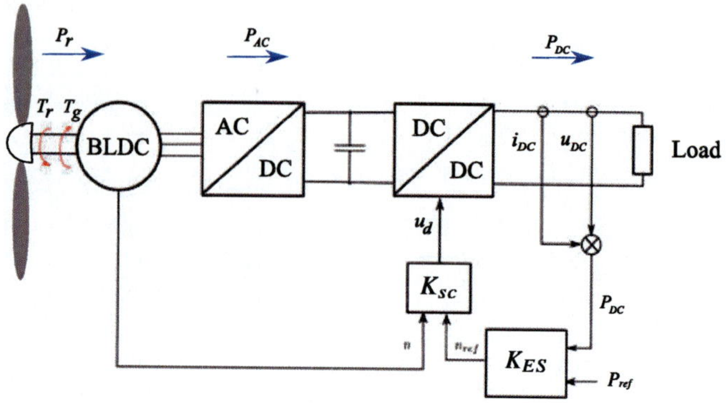

Fig. 3 Drive train, power electronics, and control scheme with Extremum seeking for DPPT used in small wind turbines

3 Methods

It is now investigated how the ES control algorithm can be used for MPPT control in Region 1 and DPPT control in Region 2 and 3 without structure switching. Specifically, in Region 2, the turbine is used in stall mode, and Region 3, it is used for power control to zero to reach Region 4. The overall objectives are

1. Implementing the ES algorithm as an MPPT control strategy by adapting its control output n_{ref} where $2\pi\ n_{ref} = \omega_{r,ref}$ so that the system output P_{DC} converges to the maximum power point of the $c_P(\lambda)$ curve;
2. Transforming the $c_P(\lambda)$ curve so that its transformed maximum lies at the reference power point P_{ref} with a respective $c_{P_{ref}}(\lambda)$.

In the following three subsections, we first discuss the original operation of the ES optimizer, then present the modified ES optimization method to be used as a power tracking control scheme.

3.1 *Original Extremum Seeking Optimizer*

Due to the unknown constantly changing effective wind speed in front of the rotor, the actual maximum power point is unknown in advance. The maximum power point is shifting continuously. For these reasons, the extremum-seeking optimization method is advantageous for finding its maximum power point. The method is a model-free, real-time optimization with rigorously provable stability and convergence [10].

In Fig. 4, an already modified ES scheme is depicted. The control block D is omitted from the original ES optimization method scheme, and the plant's output signal is directly fed to the ES algorithm. ES uses a continuous oscillatory perturbation signal $i(t) = \hat{i} \sin \omega_p t$ superimposed onto the ES output, the rotor speed set point

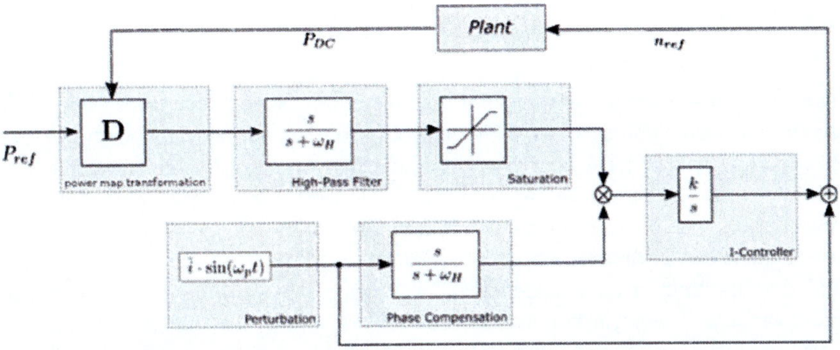

Fig. 4 Schematic of the extended extremum-seeking controller for DPPT of small wind turbines (denoted as a plant) without pitch-control

n_{ref}. The system response and electrical power output are continuously measured, and its gradient is extracted using a highpass filter. The gradient is then assessed by multiplying it by the phase-compensated perturbation signal. Finally, the resulting gradient is amplified and controlled to zero by an integrator. In this way, the ES continuously updates its output, the rotor speed reference value in proportion to the gradient. A unique maximum for a given wind speed is observed, comparable to the $c_p(\lambda)$ curve depicted in the upper part of Fig. 2. The plants' output signal P_{DC} eventually converges to this maximum power point. The convergence time depends on the Hessian matrix of the underlying plant model of the unknown $c_p(\lambda)$ curve therefore, it is impossible to determine it analytically. However, there is the possibility of building an estimation by measuring the convergence time at different set points. Alternatively, the ES algorithm can be changed to a so-called Newton version, rendering the convergence time to be user-assignable [7]. For the sake of simplicity, the Newton version has not been implemented.

The parameters of the extremum-seeking algorithm must be well chosen. The dynamics of the system and the characteristics of the disturbances (i.e., the wind changes) have to be considered to ensure fast convergence to the maximum power point and overall robustness. The largest time constant of our system can be approximated to $\tau_n \approx 4s$. Therefore, the perturbation frequency is $\omega_p = 1s^{-1}$. It is sufficiently slow so that no dynamic mode of the turbine is excited. The perturbation amplitude is chosen as small as possible ($\hat{n} = 15$ rpm), considering a noisy systems output signal and requiring a clear system response in the operational space. The gradient of the system output can be obtained with the help of a high-pass filter. Its cut-off frequency, $\omega_H = 0.2s^{-1}$ is chosen to let the perturbation frequency through but to cut out lower frequencies.

Additionally, a saturation block is used to limit the effects of disruptive changes in wind velocity on the algorithm. It is chosen to have a maximum level of ± 5 W. Finally, the gain of the integral controller k is selected to ensure fast convergence for all available urban wind scenarios.

3.2 Modified Extremum Seeking Control for DPPT

The objective of demand power point tracking control is not the convergence to the maximum power point as in MPPT control but rather the convergence to the referenced power. The $c_P(\lambda)$ curve in the upper part of Fig. 2 is a concave function. Therefore, two λ_1 and λ_2 values exist for any power value, except for its maximum power point. The control objective is to transform this $c_p(\lambda)$ curve so that its maximum point lies at the reference value. To achieve this goal, a control block D is added to the input of the ES optimizer, as shown in Fig. 4. Block D transforms the input variable P_{DC} according to

$$P^* = \frac{P_{ref}^2 - (P_{ref} - P_{DC})^2}{P_{ref}} \tag{6}$$

The transformed $c_p(\lambda)$ curve is shown in the lower part of Fig. 2. Assuming P_{ref} is constant for a period of time and P_{DC} is a concave function of λ, the expression $-(P_{ref} - P_{DC})^2$ is a quadratic-shaped function with two local maxima at λ_1 and λ_2 and a local minimum at its original maximum power point at λ_{opt}. The expression P_{ref}^2 is added to this expression to be positive definite over most of its operating range. This is important because the ES configured in this paper can only operate with positive definite input functions. Finally, it is divided by the reference value so that when $P_{DC} = P_{Ref}$, which is to say when the system output is converging to its reference value, $P^* = P_{ref}$. The lower part of Fig. 2 shows a transformation of the $c_P(\lambda)$ curve for a reference power equalling $c_{P_{ref}} = 0.3$. The two maxima can be observed for λ-values of 5.02 and 10.07, respectively. λ_{opt} of $c_P(\lambda)$ curve of the upper figure is now a local minimum in the transformed $c_P(\lambda)$ curve since the transformed function is concave on the intervals $[-\inf, \lambda_{opt}]$ and $[\lambda_{opt}, +\inf]$ the ES algorithm will converge to one of the two new maxima.

4 Results

To validate our extremum-seeking approach for DPPT, a wind turbine subjected to different wind speeds and power reference values is simulated. The simulations were performed using Simulink®Mathworks R2022a. The simulation is visualized for a total time frame of 120 s. The results are shown in Fig. 5. The upper simulated time series shows the wind power plant's power reference value and electrical power output. The second depicts changes in the turbine speed, and the third shows generator and rotor torque variations. The lower time series represents the wind speed, which is the disturbance in our controlled system.

The power plant undergoes a cold start, and its power output convergences after $t = 7s$ to the reference value. The ripple in the turbine speed values, the power output signal, and torque values are characteristic of the extremum-seeking optimizer. Although the turbine is subjected to disruptive wind changes, a step at $t = 25$ s from

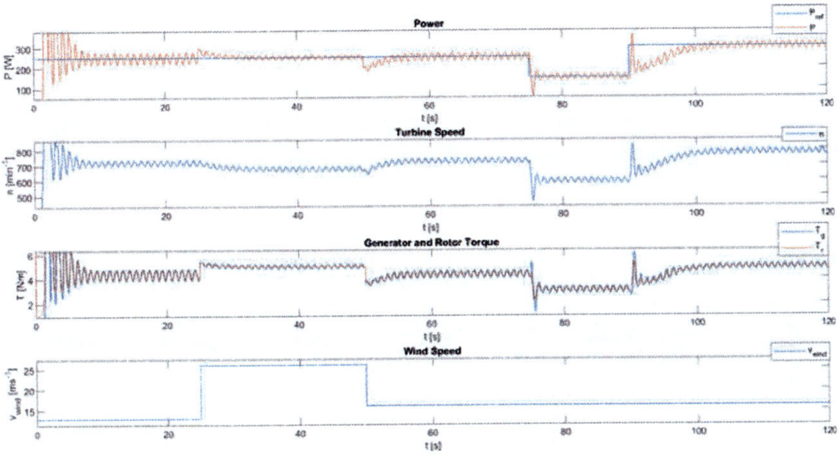

Fig. 5 Simulation with artificial wind velocity

$v_{wind} = 13\,\text{ms}^{-1}$ to $v_{wind} = 26\,\text{ms}^{-1}$ and back to $v_{wind} = 16\ \text{ms}^{-1}$ at $t = 50$ s, the DPPT-controller can converge to the reference value after a maximum of 8 s. From $t = 75$ s, changes in the power reference value are applied, but the controller can track them in less than 10 s. Note that convergence time varies because we deal with a strongly nonlinear system.

5 Discussion and Outlook

Wind turbines capable of active power variation by an external reference signal (i.e., superior control) require different control systems than those that feed in as much power as possible at partial load. The latter requires an MPPT control scheme, while the former requires limiting or reducing power output. This paper proposed a demanded power point tracking control scheme to satisfy the requirements of rated and derated wind regions as well as the power demands of the load side. To avoid switching in the controller during operation, an MPPT algorithm was modified as an extremum-seeking optimizer that works as a power controller. It was shown that the proposed control scheme provides fast convergence to power reference signals and robustness when faced with strong disturbances (wind speed changes). As necessary in practice, the control system can perform a cold start. Furthermore, it does not need error-prone and costly wind speed sensors to operate. It is computationally efficient and can be easily implemented in a microcontroller.

Future work will address its implementation onto a wind turbine using the Hardware-In-The-Loop methodology. Analysis of the response time of the control system will be necessary to assess if it complies with the requirements of the existing wind turbine system. Furthermore, its application for the adaptive control of a multi-rotor system will be investigated.

References

1. U.S. Department of Energy: 20% Wind Energy by 2030: Increasing Wind Energy's Contribution to U.S. Electricity Supply. DOE/GO-102008-2567 (2008)
2. Bossanyi, E.A.: The design of closed loop controllers for wind turbines. Wind Energy **3**(3), 149–163 (2000)
3. Callegari, G., Capurso, P., Lanzi, F., Merlo, M., Zaottini, R.: Wind power generation impact on the frequency regulation: study on a national scale power system. In: 14th International Conference on Harmonics and Quality of Power (ICHQP), pp. 1–6 (2010)
4. Inthamoussou, F.A., De Battista, H., Mantz, R.J.: LPV-based active power control of wind turbines covering the complete wind speed range. Renew. Energy **99**, 996–1007 (2016)
5. Pöschke, F., Petrovic, V., Berger, F., Neuhaus, L., Hölling, M., Kühn, M., Schulte, H.: Model-based wind turbine control design with power tracking capability: a wind-tunnel validation. Control Eng. Pract. **120** (2022). https://doi.org/10.1016/j.conengprac.2021.105014
6. Ghaffari, A., Krstic, M., Seshagiri, S.: Power optimization and control in wind energy conversion systems using extremum seeking. IEEE Trans. Control Syst. Technol. **22**(5) (2014)
7. Ghaffari, A., Krstic, M., Nesic, D.: Multivariable Newton-based extremum seeking. Automatica **48**, 1759–1767 (2012)
8. Hansen, A.D., Hansen, L.H.: Market penetration of wind turbine concepts over the years (2006). https://doi.org/10.1002/we.210
9. Jeong, Y., Johnson, K., Fleming, P.: Comparison and testing of power reserve control strategies for grid-connected wind turbines. Wind Energy **17**(3), 343–358 (2014)
10. Krstic, M., Wang, H.-H.: Stability of extremum seeking feedback for general nonlinear dynamic systems. Automatica **36**, 595–601 (2000)
11. Krstic, M., Ariyur, K.B.: Slope seeking: a generalization of extremum seeking. Int. J. Adapt. Control Signal Process **18**(1) (2004). https://doi.org/10.1002/acs.777
12. Kim, K., Kim, H.G., Kim, C.J., Paek, I., Bottasso, C.L., Campagnolo, F.: Design and validation of demanded power point tracking control algorithm of wind turbine. Int. J. Precis. Eng. Manufact.-Green Technol. **5**(3), 387–400 (2018). July
13. Labar, Christophe, Garone, Emanuele, Kinnaert, Michel: Sub-optimal extremum seeking control. IFAC-Papers Online **50**(1), 7762–7768 (2017)
14. Rotea, M.A.: Logarithmic power feedback for extremum seeking control of wind turbines. IFAC-Papers Online **50**(1), 4504–4509 (2017)

Degradation Simulator for Infinite Horizon Controlled Linear Time-Invariant Systems

Amirhossein Hosseinzadeh Dadash and Niclas Björsell

Abstract Diagnosis, fault prediction, and Remaining Useful Life (RUL) estimation are among the predictive maintenance research subjects used for maintenance cost reduction. Using the available data with different machine learning methods, especially deep learning methods, the accuracy of estimation and prediction of faults and RUL have increased dramatically. However, due to the statistical nature of the machine learning methods and the limitations of available datasets, physically interpreting this information might be impossible. On the other hand, controlling the degradation and faults in the machines as the optimum predictive maintenance solution needs the physical interpretation of the method's outcome. In order to test the new process-based methods for degradation and fault control, datasets with more information are required (compared to available datasets). In this article, we introduce an open-source degradation simulator for linear systems. This simulator can simulate the degradation in closed-loop machines whose dynamics are known. It is also possible to simulate different degradation models for different system parts simultaneously by adding different processes and output noise to the system. This simulator can generate enough data to test new machine learning-based predictive maintenance methods.

1 Introduction

The classic definition of predictive maintenance premises that monitoring the system condition using different indicators will provide the necessary data for preventing unscheduled outages and maximizing the Mean Time To Failure (MTTF) [1]. Using

A. Hosseinzadeh Dadash (✉) · N. Björsell
Department of Electronics, Mathematics and Natural Sciences, University of Gävle, 801 76 Gävle, Sweden
e-mail: amirhossein.hosseinzadeh.dadash@hig.se

N. Björsell
e-mail: niclas.bjorsell@hig.ses

D. Theilliol et al. (eds.), *Recent Developments in Model-Based and Data-Driven Methods for Advanced Control and Diagnosis*, Studies in Systems, Decision and Control 467,
https://doi.org/10.1007/978-3-031-27540-1_14

different machine learning methods, the amount and quality of the information about the systems' state of health have increased dramatically. This has led to a more precise prediction of the system lifetime [2, 3].

However, in the era of Industry 4.0 and complex industrial processes that exploit distributed systems, the expectations have improved from mere prediction to the need for control and synchronization using prediction. Having enough information about the system and its respective degradation, it is possible to control the complex interconnected systems so sub-systems can reach their maintenance time simultaneously [4]. Controlling the degradation in complex systems makes it possible to save more downtime costs compared to the cases that only exploit prediction.

Considering that most of the research for failure prediction is focused on machine learning methods such as deep learning [2], Support Vector Machine [5, 6], and k-nearest neighbor [7, 8], regardless of the outcome precision, their outcome might not be interpretable in the real world [9, 10]. On the other hand, from the control point of view, the structure of the available datasets has many limitations, like the absence of input and any information about the noise. Although these shortages make these datasets more applicable for research on machine learning methods, they are not useful from the control point of view.

In order to exploit the diagnosis and fault prediction information derived using machine learning methods in the control scheme and improve the system performance, it is necessary to adapt the machine learning method and explore system dynamics [11–14]. For being able to test the new methods for degradation control, special datasets with different characteristics are needed. These datasets should include more information compared to available datasets. Firstly, the noise should be configurable (or at least known), so in the end, the effect of noise can be calculated precisely. Secondly, controlling the degradation means diverting it from one part to the other inside the system; these datasets should include enough information so that this diversion can be traced and its consequences can be studied. Thirdly, the input to the system (the desired system output and control inputs) should be known. The input is necessary for the outcome to be interpretable physically and be helpful for control. Finally, the effect of the feedback loop should be studied. This is because the closed-loop system always compensates for the degradation (system parameters deviation) by manipulating the system states using control inputs. This might degrade the system's more critical and expensive part(s), which might not be the intention of the maintenance management system.

This article introduces the open-access simulator for simulating the degradation in infinite horizon closed-loop linear time-invariant systems. This simulator is fully configurable and can be used for research on controlling the degradation in a closed-loop system according to system dynamics. It can also be used for the identification of the dynamics of the system.

2 Method

The prerequisite for designing the degradation simulator is to study how the degradation happens and how it appears in the system dynamics.

2.1 Degradation in the Closed-Loop System

The state-space model (SSM) of the system can be written as follows:

$$\begin{cases} \dot{x} = Ax + Bu + Nv_1 \\ z = Mx \\ y = Cx + v_2 \end{cases}, \quad v = \begin{bmatrix} v_1 \\ v_2 \end{bmatrix}, \quad \Phi_v(w) = \begin{bmatrix} R_1 & R_{12} \\ R_{12}^T & R_2 \end{bmatrix}, \tag{1}$$

where x is a vector containing system states, u is a vector of inputs. A and B are the system parameters, C contains the output parameter, and M defines the states to control. Also, v_1 is the input noise, v_2 is the sensor noise, and Φ is the disturbances covariance matrix.

The degradation can be detected as a change in single or multiple signals recorded from the system [15]. The effect of degradation can be detected in variations of input, output, or both [16], meaning the degradation is a function of monotonic change in the y, u, or both (or their feature) over time. However, the change in the u or y is a result of the degradation in the system's parameters (A, B or C from (1)), but, as controllers control the system's output in a way that it follows the reference signal, and the system is considered time-invariant the controller continuously manipulates the system variables (states) using the system's inputs in order to keep the output close to the desired output. Eventually, at some point, the deviation between the system's parameters' real values and the system's parameters' initial values increases so much that either the controller becomes unstable or the output will go outside the acceptable range. This is when the system fails, and maintenance is needed.

2.2 Degradation Simulation Algorithm

The whole process of simulating the degradation is shown in Fig. 1. Four inputs need to be known: system state-space model, degradation model, disturbance model, and failure or maintenance criterion.

2.2.1 Degradation and Disturbance Model

As mentioned in Sect. 2.1, although the degradation appears in the systems' output, it is the result of the controller actions for compensating for degradation system

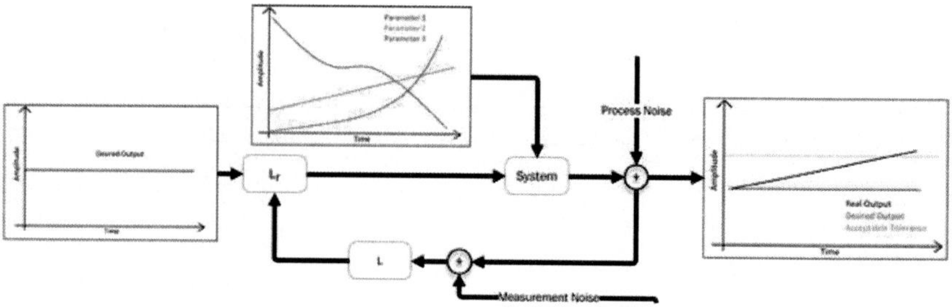

Fig. 1 Degradation simulation

parameters (A, B, or C). In the second step of the simulation, the degradation should be defined. The degradation can be only time-dependent or depend on different internal or external parameters, or it might come from the neural network or other function estimators. Like the degradation models, the disturbances can come from any source. They can also include process noise, measurement noise, or both.

2.2.2 Failure or Maintenance Criterion

When the degradation increases, the output deviates from the desired output. To some system dependant tolerance, this deviation is acceptable, and after that, this will consider a failure. The failure tolerance or the maximum deviation for the desired output should be defined so the maintenance time can be calculated.

Also, no system can work for an infinite time without needing maintenance. This maintenance might not be considered a response to a failure, but it is necessary to keep the system going. For example, the system's oil change is required when the oil degrades. This can also be configured as the maximum MTTF of the system in the simulator.

2.2.3 Controller Design

The final stage before the algorithm is the linear quadratic regulator (LQR) design, which should be calculated according to system state space [17]. The quadratic criterion that the LQ controller minimizes is given as follows:

$$J = \int e^T Q_1 e + u^T Q_2 u, \tag{2}$$

where $e = z - r$ and r is the reference signal, z is the controlled output, and e is the error; u is the control input, and Q_1 and Q_2 are penalty matrices for the error and input signal, respectively. The optimal control signal for this controller can be written as follows:

$$u = -Lx + L_r r, \tag{3}$$

$$\dot{x} = Ax + Bu, \tag{4}$$

where L is the optimal feedback gain and L_r keeps the static gain of the closed-loop equal to the identity matrix.

2.2.4 Algorithm

The following pseudo-code shows how the degradation simulator works. The simulator is designed based on the working cycles. This means that the parameters are considered constant during each working cycle and get updated at the end of the cycle. The working cycle here refers to the time from the beginning of one process until the time that the process is done. For instance, the wood-cutting machine might cut thousands of wood pieces until its saw needs to change. Each working cycle will be from the beginning to the end of cutting one piece of wood.

2.3 Sample Simulation

For demonstration, two models are used for degradation simulation. The first model is the simple mass and spring model with three states and one input, and the second model is a more complicated linearized model of a rolling mill with five states and two inputs. All simulation parameters can be found in the appendix.

Pseudo-code

Inputs: *system state-space model*
degradation model(s)
disturbance model(s)
failure criterion

Outputs: *Time series of system outputs*
Time series of system inputs
Record of a configurable feature for each of the working cycles

Calculate L and L_r
While the failure criterion or maintenance criterion is not met
  For each working cycle:
    Update the model parameters according to their degradation model
    Calculate inputs and outputs and add the disturbances while keeping the controller (L and L_r) the same
    Record the control inputs, outputs and the desired features

2.3.1 Mass and Spring

Figure 2 shows the model used for the first degradation simulation. The model state-space is

$$\begin{bmatrix} \dot{x} \\ \ddot{x} \\ \dot{z} \end{bmatrix} = \begin{bmatrix} 0 & 1 & 0 \\ \frac{k2-k1}{m} & 0 & -\frac{k_1}{m} \\ \frac{k_1}{b} & 0 & \frac{k_1}{b} \end{bmatrix} \cdot \begin{bmatrix} x \\ \dot{x} \\ z \end{bmatrix} + \begin{bmatrix} 0 \\ \frac{1}{m} \\ 0 \end{bmatrix} \cdot [f], \tag{5}$$

$$M = [1\,0\,0]x, \tag{6}$$

$$y = Ix + Iv_2. \tag{7}$$

Two degradation models with different disturbances were considered for this system. The first degradation for k_1 and its respective disturbances are modeled according to

$$k_1(t+1) = k_1(t) + C_1 exp(C_2 t),\ v_2 = \begin{bmatrix} \mathcal{N}(\mu_1, \sigma_1) \\ \mathcal{N}(\mu_2, \sigma_2) \\ \mathcal{N}(\mu_3, \sigma_3) \end{bmatrix}. \tag{8}$$

The second degradation for k_1 is modeled according to

$$k_1(t+1) = k_1(t) + \mathcal{N}(\mu_4, \sigma_4) \times exp(C_3 t), \tag{9}$$

with the same disturbances as (8).

The result of the degradation simulation can be seen in Fig. 3.

2.3.2 Rolling Stand

The second degradation model is the linearized version of a rolling stand (as shown in the left part of Fig. 2) that was introduced in [18]. The state-space model of the mill is

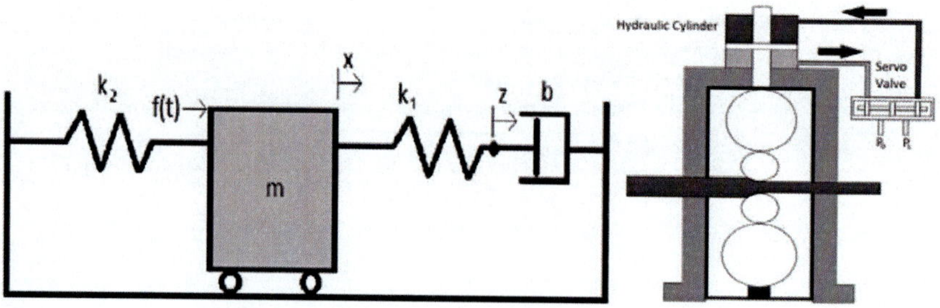

Fig. 2 (Left) mass and spring model, (right) rolling stand model

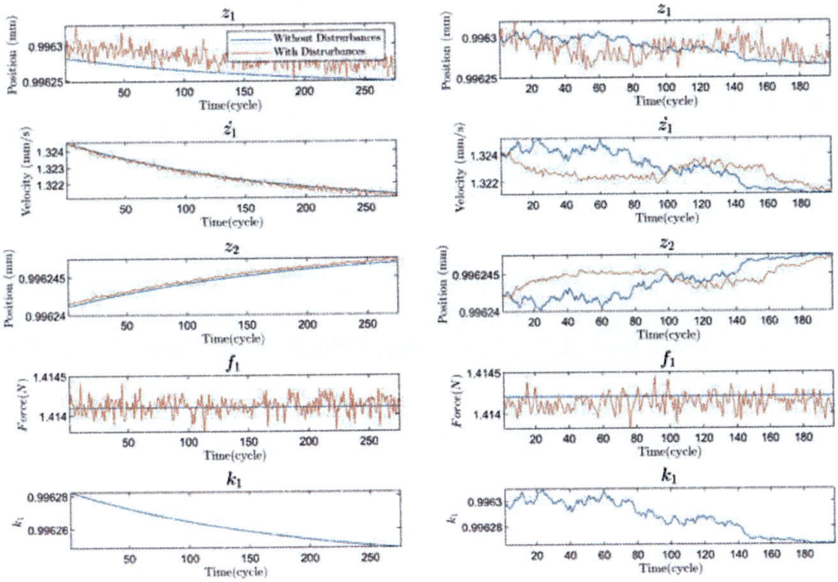

Fig. 3 Degradation simulation in mass and spring model (left) degradation model 1, (right) degradation model 2

$$\begin{bmatrix} \dot{h} \\ \ddot{h} \\ \dot{P}_1 \\ \dot{P}_2 \\ \dot{x}_v \end{bmatrix} = \begin{bmatrix} 0 & 1 & 0 & 0 & 0 \\ 0 & -\frac{F_c}{m} & \frac{A_1}{m} & -\frac{A_2}{m} & 0 \\ 0 & -\frac{\beta A_1}{V1} & -\frac{\beta(C_i+C_e)}{V1} & \frac{\beta C_i}{V1} & \frac{\beta P_s}{C_l V1} \\ 0 & \frac{\beta A_2}{V2} & \frac{\beta C_i}{V2} & -\frac{\beta(C_i+C_e)}{V2} & \frac{2\beta P_s}{C_l V2} \\ 0 & 0 & 0 & 0 & -\frac{1}{\tau_v} \end{bmatrix} \cdot \begin{bmatrix} h \\ \dot{h} \\ P_1 \\ P_2 \\ x_v \end{bmatrix} + \begin{bmatrix} 0 & 0 \\ \frac{-1}{m} & 0 \\ 0 & 0 \\ 0 & 0 \\ 0 & \frac{k_v}{\tau_v} \end{bmatrix} \cdot \begin{bmatrix} F_L \\ u_v \end{bmatrix} + \begin{bmatrix} 0 \\ 1 \\ 0 \\ 0 \\ 0 \end{bmatrix} .v_1, \tag{10}$$

$$y = Ix + Iv_2. \tag{11}$$

where h is the displacement of the rollers, $\dot{h}$ is the vertical velocity of the rollers displacement, P_s is the valve's supply pressure, P_1 and P_2 are the pressures inside the hydraulic cylinder; x_v is the hydraulic valve spool displacement, F_L is the input force, F_c is the viscous force, m is the mass, A_1 and A_2 are the areas inside hydraulic cylinder; V_1 and V_2 are the volumes of the primary or secondary side of the hydraulic cylinder, respectively; β is the oil effective bulk modulus, C_i is the internal oil leakage, C_e is the external oil leakage, C_l is the constant of linearization, τ_v is the hydraulic valve time constant, k_v is the gain of the hydraulic valve, and u_v is the input voltage to the valve.

The first degradation for C_e and its respective disturbances are modeled according to

$$C_e(t+1) = C_e(t) + C_4 exp(C_5 t), \;\; v_2 = \left[\mathcal{N}(\mu_i, \sigma_i),\right], i = 1 \ldots 5, \tag{12}$$

For the second degradation model, while the disturbances are kept the same, the degradation is defined as:

$$\rho(t+1) = \rho(t) - (C_6 exp(\rho(t)C_7 t) - 0.1 - C_e, \quad (13)$$

$$C_e(t+1) = C_e(t) + C_8 exp(C_e(t)C_9 t) + \mathcal{N}(C_{10} exp(C_e(t)C_{11} t), C_{12} exp(C_e(t)C_{13}), \quad (14)$$

$$F_c(t+1) = F_c e(t) + C_{14} exp(F_c(t)C_{15} t), \quad (15)$$

$$\beta(t+1) = \beta(t) - C_{16}\rho(t) + C_e(t)C_{17}\rho(t) - F_c(t)C_{18}, \quad (16)$$

where ρ is the hydraulic oil density. The result of the degradation simulation for the two mentioned degradation models on the rolling stand can be seen in Fig. 4.

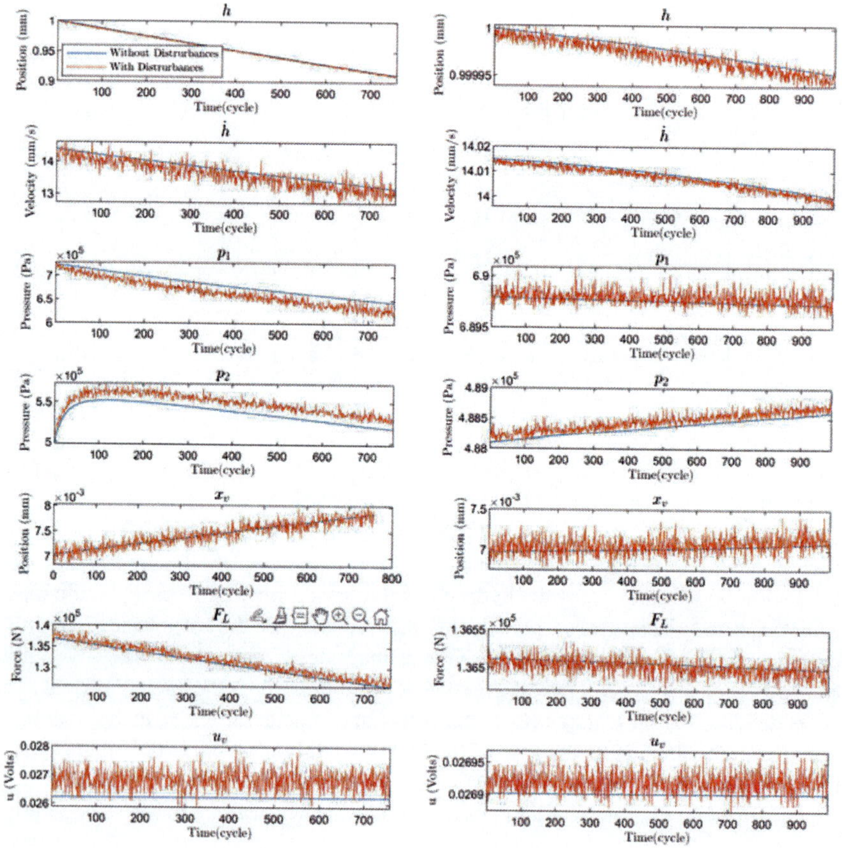

Fig. 4 Degradation simulation in rolling mill model (left) degradation model 1, (right) degradation model 2

3 Conclusion

In this article, the degradation simulator, which is accessible through [19], was introduced. Detailed information about the degradation simulation and its relation to the system's state space was discussed, and the algorithm used in this simulator was explained. Finally, with the help of two models number of degradation was simulated, and the results were shown. This simulator can be used to further investigate the degradation control problems and degradation detection and RUL estimation in closed-loop systems according to the system's model.

Acknowledgements The research project is financed by the European Commission within the European Regional Development Fund, Swedish Agency for Economic and Regional Growth, Region Gävleborg, and the University of Gävle.

Appendix

Degradation models parameters:

$A_1 = 100e^{-3}$	$A_2 = 100e^{-3}$	$P_s = 1e^9$	$F_c = 10$	$\beta = 2e^5$	$V_1 = 0.5$	$V_2 = 0.65$
$C_i = 1e^{-6}$	$C_e = 1e^{-6}$	$m = 1e^2$	$C_l = 1.1$	$\tau_v = 1e^{-3}$	$k_v = 1$	
$C_1 = 0.005$	$C_2 = 0.005$	$C_3 = 0.005$	$\mu_1 = 1e^{-5}$	$\mu_2 = 1e^{-4}$	$\mu_3 = 1e^{-5}$	$\mu_4 = 5e^{-3}$
$\sigma_1 = 1e^{-5}$	$\sigma_2 = 1e^{-4}$	$\sigma_3 = 1e^{-5}$	$\sigma_4 = 1e^{-2}$			
$C_4 = 1e^{-8}$	$C_5 = 5e^{-7}$	$C_6 = 1e^{-20}$	$C_7 = 1e^{-4}$	$C_8 = 5e^{-11}$	$C_9 = 1e^{-7}$	$C_{10} = 5e^{-11}$
$C_{11} = 1e^{-7}$	$C_{12} = 5e^{-11}$	$C_{13} = 1e^{-8}$	$C_{14} = 5e^{-4}$	$C_{15} = 1e^{-4}$	$C_{16} = 1e^{-4}$	$C_{17} = 2$
$C_{18} = 50$	$\mu_5 = 1e^{-3}$	$\mu_6 = 1e^{-1}$	$\mu_7 = 1e^4$	$\mu_8 = 1e^4$	$\mu_9 = 1e^4$	$\sigma_5 = 1e^{-3}$
$\sigma_6 = 1e^{-1}$	$\sigma_7 = 1e^4$	$\sigma_8 = 1e^{-4}$	$\sigma_9 = 1e^{-5}$			

References

1. Mobley, R.K.: An Introduction to Predictive Maintenance. Elsevier (2002)
2. Vollert, S., Theissler, A.: Challenges of machine learning-based RUL prognosis: a review on NASA's C-MAPSS data set. In: 2021 26th IEEE International Conference on Emerging Technologies and Factory Automation (ETFA). IEEE (2021)
3. Ge, M.-F., et al.: A review on state of health estimations and remaining useful life prognostics of lithium-ion batteries. Measurement **174**, 109057 (2021)
4. Björsell, N., Dadash. A.H.: Finite horizon degradation control of complex interconnected systems. IFAC-PapersOnLine **54**(1), 319–324 (2021)
5. Han, H., et al.: Least squares support vector machine (LS-SVM)-based chiller fault diagnosis using fault indicative features. Appl. Therm. Eng. **154**, 540–547 (2019)
6. Zhu, X., Xiong, J., Liang, Q.: Fault diagnosis of rotation machinery based on support vector machine optimized by quantum genetic algorithm. IEEE Access **6**, 33583–33588 (2018)
7. Madeti, S.R., Singh, S.N.: Modeling of PV system based on experimental data for fault detection using KNN method. Solar Energy **173**, 139–151 (2018)

8. Xiong, J., et al.: An information fusion fault diagnosis method based on dimensionless indicators with static discounting factor and KNN. IEEE Sens. J. **16**(7), 2060–2069 (2015)
9. Kahana, A., et al.: Obstacle segmentation based on the wave equation and deep learning. J. Comput. Phys. **413**, 109458 (2020)
10. Jin, K.H., et al.: Deep convolutional neural network for inverse problems in imaging. IEEE Trans. Image Process. **26**(9), 4509–4522 (2017)
11. Read, J.S., Jia, X., Willard, J., Appling, A.P., Zwart, J.A., Oliver, S.K., Karpatne, A., Hansen, G.J., Hanson, P.C., Watkins, W., et al.: Process-guided deep learning predictions of lake water temperature. Water Resour. Res. **55**, 9173–9190 (2019)
12. Daw, A., Karpatne, A., Watkins, W., Read, J., Kumar, V.: Physics-guided neural networks (PGNN): an application in lake temperature modeling. arXiv:1710.11431
13. Rudy, S.H., Brunton, S.L., Proctor, J.L., Kutz, J.N.: Data-driven discovery of partial differential equations. Sci. Adv. **3** (2017)
14. Brunton, S.L., Proctor, J.L., Kutz, J.N.: Discovering governing equations from data by sparse identification of nonlinear dynamical systems. Proc. Natl. Acad. Sci. **113**, 3932–3937 (2016)
15. Isermann, R.: Process fault detection based on modeling and estimation methods-a survey. Automatica **20**, 387–404 (1984)
16. Zagorowska, M., Wu, O., Ottewill, J.R., Reble, M., Thornhill, N.F.: A survey of models of degradation for control applications. Annu. Rev. Control **50**, 150–173 (2020)
17. Glad, T., Ljung, L.: Control Theory. CRC Press (2018)
18. Yan, J., Li, B., Guo, G., Zeng, Y., Zhang, M.: Nonlinear modeling and identification of the electro-hydraulic control system of an excavator arm using BONL model. Chin. J. Mech. Eng. **26**, 1212–1221 (2013)
19. Dadash, A.H.: Degradation Simulator (2022). Available via Github. http://github.com/ahhzd86/degradation_simulator

Distributed Observer-Based Leader-Following Consensus Control Robust to External Disturbance and Measurement Sensor Noise for LTI Multi-agent Systems

Jesus A. Vazquez Trejo, Jean-Christophe Ponsart, Manuel Adam-Medina, Guillermo Valencia-Palomo, and Juan A. Vazquez Trejo

Abstract A robust observer-based leader-following consensus control for linear multi-agent systems subject to external disturbance and measurement noise is developed. The H_∞ criterion is implemented to guarantee stability and robustness of the synchronization error and estimation error through the Lyapunov approach. A set of linear matrix inequalities are obtained to compute the control and observer gain matrices. To show its effectiveness, the proposed strategy is carried out to solve the leader-following formation control consensus-based problem in a fleet of unmanned aerial vehicles under the effect of wind turbulence and measurement noise.

1 Introduction

In recent years, multi-agent systems have attracted the attention of researchers [1]. Multi-agent systems in this work are handled as found in the control community, where despite the name being shared with the computer science community, the meanings, the objectives, and the tools of common use are adapted to the control area [2]. An agent is defined as an autonomous dynamical system, e.g., Unmanned

J. A. Vazquez Trejo (✉) · J.-C. Ponsart · J. A. Vazquez Trejo
Université de Lorraine, CNRS, CRAN, 54000 Nancy, France
e-mail: jesus-avelino.vazquez-trejo@univ-lorraine.fr

J.-C. Ponsart
e-mail: Jean-Christophe.Ponsart@univ-lorraine.fr

J. A. Vazquez Trejo
e-mail: juan-antonio.vazquez-trejo@univ-lorraine.fr

M. Adam-Medina
TecNM/CENIDET, Cuernavaca, Morelos 62490, Mexico
e-mail: manuel.am@cenidet.tecnm.mx

G. Valencia-Palomo
TecNM/IT Hermosillo, Av. Tec y Per Poniente S/N, 83170 Hermosillo, Mexico
e-mail: gvalencia@hermosillo.tecnm.mx

D. Theilliol et al. (eds.), *Recent Developments in Model-Based and Data-Driven Methods for Advanced Control and Diagnosis*, Studies in Systems, Decision and Control 467,
https://doi.org/10.1007/978-3-031-27540-1_15

Aerial Vehicles (UAVs), or satellites, among others. Nevertheless, multi-agent systems can be subject to external disturbance and measurement noise which compromise the synchronization and coordination of the agents. It is therefore necessary to consider these exogenous inputs in the dynamic modeling of the multi-agent system in order to be able to develop a robust control of these external disturbances and measurement noise. This is why it is essential to add robustness to the distributed control and observer because external disturbance (and sensor noise) affect their performance [3]. Different works in the literature have addressed this problem. A robust finite-time problem for a second-order multi-agent system is addressed in [4] for rejecting external disturbances. Also, a sliding mode leader-following control for multiple UAVs is proposed in [5], where the system remains insensitive to external disturbances. Alternatively, in [6], a leader-following sliding mode formation control approach for underactuated surface vehicles considering model uncertainties and external disturbance is developed. In [7], a robust controller using H_2 performance is implemented in a system with communication and input time delays presented in the frequency domain minimizing the error and the disturbance effect. In [8], the robust optimal formation control problem for heterogeneous multi-agent systems considering external disturbances is addressed based on reinforcement learning. An adaptive semi-global bipartite consensus assuming a connected switching topology graph under input saturation and external disturbance is proposed in [9].

In this paper, the main contribution is the design of the distributed leader-following control law based on state estimates of the neighboring agents to coordinated linear multi-agent systems, where the robust control and observer gains are computed simultaneously based on linear matrix inequalities (LMIs). The main difference concerning previous work is the distributed robust observer-based leader-following control for linear multi-agent systems under external disturbance and measurement noise, where the main objective is to follow the trajectories described by a virtual leader and maintain consensus between followers despite the external disturbance and the sensor noise.

2 Problem Statement and System Description

Consider a linear homogeneous multi-agent system, which means that A, B, and C matrices are identical for each agent, as follows:

$$\begin{aligned} \dot{x}_i(t) &= Ax_i(t) + Bu_i(t) + D_u d_{u_i}(t), \\ y_i(t) &= Cx_i(t) + D_y d_{y_i}(t), \end{aligned} \tag{1}$$

where $i = 1, 2, \ldots, N$; N is the total number of agents; $x_i(t) \in \mathbb{R}^n$, $u_i(t) \in \mathbb{R}^m$, $y_i(t) \in \mathbb{R}^p$, $d_{u_i}(t) \in \mathbb{R}^m$, $d_{y_i}(t) \in \mathbb{R}^p$, are the state, input, output, external disturbance and noise vectors respectively; $A \in \mathbb{R}^{n\times n}$, $B \in \mathbb{R}^{n\times m}$, $C \in \mathbb{R}^{p\times n}$, $D_u \in \mathbb{R}^{n\times m}$, and $D_y \in \mathbb{R}^{p\times n}$ are constant matrices of the system. Graph theory [2] can be used for the communication description of multi-agent systems. In this way, the Laplacian matrix $\mathcal{L}$ describes the dynamics of the multi-agent system through the adjacency matrix $\mathcal{A}$. Let us consider a directed graph $\mathcal{G}(\mathcal{V}, \mathcal{E}, \mathcal{A})$ where $\mathcal{V} = \{v_1, v_2, \ldots, v_N\}$ is a set of nodes (agents), $\mathcal{E} = \{(i, j) : i, j \in \mathcal{V}\} \subseteq \mathcal{V} \times \mathcal{V}$ is a set of edges. The adjacency matrix $\mathcal{A} = [a_{ij}] \in \mathbb{R}^{N \times N}$ is defined as $a_{ii} = 0$, $a_{ij} = 1$ if and only if the pair $(i, j) \in \mathcal{E}$ otherwise $a_{ij} = 0$. When the graph is undirected, also, $a_{ij} = a_{ji}$, $\forall i \neq j$ and $\mathcal{A} = \mathcal{A}^T$. The Laplacian matrix $\mathcal{L} \in \mathbb{R}^{N \times N}$ is defined as $\mathcal{L}_{ii} = \sum_{j \neq i} a_{ij}$ and $\mathcal{L}_{ij} = -a_{ij}$. The neighbors of i-th agent are denoted as $j \in \mathcal{N}_i$. The following assumptions are held in this paper:

Assumption 1 The pair (A, B) is stabilizable.

Assumption 2 The pair (A, C) is observable.

Assumption 3 The graph $\mathcal{G}$ is undirected and connected.

Lemma 1 ([10]) *The matrix $\bar{\mathcal{L}}$ has non negative eigenvalues. The matrix $\bar{\mathcal{L}}$ is positive definite if and only if the graph is connected and undirected.*

The main objective of this work is the design of a robust leader-following control for multi-agent systems consensus-based, where the virtual leader dynamics is presented as follows:

$$\dot{x}_l(t) = Ax_l(t), \tag{2}$$

where $x_l(t) \in \mathbb{R}^n$ is the state vector of the virtual leader. Let $\delta_i = x_i - x_l$, then, the dynamics of the synchronization error between each agent i and the leader is:

$$\dot{\delta}_i(t) = A\delta_i(t) + Bu_i(t) + D_u d_{u_i}(t). \tag{3}$$

In order to estimate the states of each agent, the following distributed observer is proposed:

$$\begin{aligned} \dot{\hat{x}}_i(t) &= A\hat{x}_i(t) + Bu_i(t) + L(y_i(t) - C\hat{x}_i(t)), \\ \hat{y}_i(t) &= C\hat{x}_i(t), \end{aligned} \tag{4}$$

where $\hat{x}_i(t) \in \mathbb{R}^n$ is the estimated state vector, $L \in \mathbb{R}^{n \times p}$ is the observer gain to be designed, and $\hat{y}_i(t) \in \mathbb{R}^p$ is the estimated output vector. The dynamics of the estimation error $e_i = x_i - \hat{x}_i$ is computed as follows:

$$\dot{e}_i(t) = (A - LC)e_i(t) + D_u d_{u_i}(t) - LD_y d_{y_i}(t). \tag{5}$$

The considered observer-based leader-following consensus control law is [11]:

$$u_i(t) = K\left[\sum_{j \in \mathcal{N}_i} a_{ij}(\hat{x}_i(t) - \hat{x}_j(t)) + \alpha_i(\hat{x}_i(t) - x_l(t))\right], \tag{6}$$

where $K \in \mathbb{R}^{n \times p}$ is the control gain to be designed, a_{ij} are the elements of the adjacency matrix, $\hat{x}_i(t)$ is the estimated state vector of the i-th agent, and $\hat{x}_j(t)$ is the estimated state vector of the neighboring agents, α_i represents the communication between the leader and the followers, where $\alpha_i > 0$ if there is a directed edge from the leader to the i-th agent, otherwise $\alpha_i = 0$. The considered problem is the design of a robust control gain K and a robust observer gain L to steer the multi-agent system (1) subject to external disturbance and sensor noise, to follow the virtual leader's trajectories.

3 Observer-Based Leader-Following Consensus Robust Controller

In this section, the LMI conditions to guarantee the existence of the robust control and observer gains based on the Lyapunov stability analysis are presented. Considering the following new variables $e(t) = [e_1(t)^T, e_2(t)^T, \ldots, e_N(t)^T]^T$, the disturbance vector $d_u(t) = [d_{u_1}(t)^T, d_{u_2}(t)^T, \ldots, d_{u_N}(t)^T]^T$, and noise vector $d_y(t) = [d_{y_1}(t)^T, d_{y_2}(t)^T, \ldots, d_{y_N}(t)^T]^T$, then, using the Kronecker product $\otimes$ the dynamic estimation error can be expressed as:

$$\dot{e}(t) = (I_N \otimes (A - LC))e(t) + (I_N \otimes D_u)d_u(t) - (I_N \otimes LD_y)d_y(t). \quad (7)$$

Replace (6) in (3) and let $\delta(t) = \left[\delta_1(t)^T, \delta_2(t)^T, \ldots, \delta_N(t)^T\right]^T$, and $\overline{\mathcal{L}} = \mathcal{L} + \Lambda$ where $\mathcal{L}$ is the Laplacian matrix and $\Lambda = diag(\alpha_1, \alpha_2, \ldots, \alpha_N)$ is the communication exchange between the virtual leader and the followers. Then, the synchronization error of the multi-agent system is rewritten as:

$$\dot{\delta}(t) = (I_N \otimes A + \overline{\mathcal{L}} \otimes BK)\delta(t) - (\overline{\mathcal{L}} \otimes BK)e(t) + (I_N \otimes D_u)d_u(t). \quad (8)$$

Let $z(t) = [\delta(t)^T, e(t)^T]^T$ then, system (1) can be expressed by:

$$\dot{z}(t) = \widetilde{A}z(t) + \widetilde{D}_u d_u(t) - \widetilde{D}_y d_y(t), \quad (9)$$

where

$$\widetilde{A} = \begin{bmatrix} I_N \otimes A + \overline{\mathcal{L}} \otimes BK & -\overline{\mathcal{L}} \otimes BK \\ 0 & I_N \otimes (A - LC) \end{bmatrix}, \widetilde{D}_u = \begin{bmatrix} I_N \otimes D_u \\ I_N \otimes D_u \end{bmatrix}, \widetilde{D}_y = \begin{bmatrix} 0 \\ I_N \otimes LD_y \end{bmatrix}. \quad (10)$$

The distributed H_∞ criterion is considered as in [11], in order to guarantee the existence of robust control and observer gains able to reject the effect of the external disturbances and sensor noise, then, the following theorem is proposed.

Theorem 1 *Consider the closed-loop system* (9). *Given the eigenvalues* $\lambda_j(\bar{\mathcal{L}})$, $j = 1, 2, \ldots, N$; *if there exist symmetric matrices* $\bar{P}_1 > 0 \in \mathbb{R}^{n \times n}$ *and* $P_2 > 0 \in \mathbb{R}^{n \times n}$, *the tuning scalar variable $\mu > 0$, and minimizing γ by the H_∞ criterion satisfying the following condition:*

$$\begin{bmatrix} \eta_{1_j} & 0 & D_u & 0 & -\lambda_j B N_c & 0 \\ * & \eta_2 & P_2 D_u & -M_o D_y & 0 & I \\ * & * & -\gamma^2 I_N & 0 & 0 & 0 \\ * & * & * & -\gamma^2 I_N & 0 & 0 \\ * & * & * & * & -\mu^{-1}\overline{P}_1 & 0 \\ * & * & * & * & * & -\mu\overline{P}_1 \end{bmatrix} < 0, \tag{11}$$

the synchronization error (8) and estimation error (7) are stable; where $\eta_{1_j} = He\{A\overline{P}_1 + \lambda_j B N_c\}$, and $\eta_2 = He\{P_2 A - M_o C\} + I_N$, then, the robust control law can be computed with $K = N_c \overline{P}_1^{-1}$, and the observer gain with $L = P_2^{-1} M_o$.

Proof Consider the candidate Lyapunov function as follows:

$$V(t) = z(t)^T \begin{bmatrix} I_N \otimes P_1 & 0 \\ 0 & I_N \otimes P_2 \end{bmatrix} z(t), \tag{12}$$

where, $P_1 \in \mathbb{R}^{n \times n} = P_1^T > 0$, and $P_2 \in \mathbb{R}^{n \times n} = P_2^T > 0$. The derivative of the candidate Lyapunov function along the trajectories of (9) is given by:

$$\begin{aligned} \dot{V}(t) =& 2\delta(t)^T (I_N \otimes P_1 A + \overline{\mathcal{L}} \otimes P_1 B K)\delta(t) - 2\delta(t)^T (\overline{\mathcal{L}} \otimes P_1 B K) e(t) \\ &+ 2e(t)^T (I_N \otimes P_2(A - LC)) e(t) + 2\delta(t)^T (I_N \otimes P_1 D_u) d_u(t) \\ &+ 2e(t)^T (I_N \otimes P_2 D_u) d_u(t) - 2e(t)^T (I_N \otimes P_2 L D_y) d_y(t). \end{aligned} \tag{13}$$

Let us perform a spectral decomposition of the matrix $\overline{\mathcal{L}}$, such that $\overline{\mathcal{L}} = TJT^{-1}$ with an invertible matrix $T \in \mathbb{R}^{N \times N}$ and a diagonal matrix $J = diag(\lambda_1, \lambda_2, \ldots, \lambda_N) \in \mathbb{R}^{N \times N}$. Defining the change of coordinates as $\xi = (T^{-1} \otimes I_N)\delta$, $\epsilon = (T^{-1} \otimes I_N) e$, $\omega_u = (T^{-1} \otimes I_N) d_u$, and $\omega_y = (T^{-1} \otimes I_N) d_y$. For reasons of space, the notation of the time dependence of certain variables is omitted. Replacing the new coordinates in (13) leads to:

$$\begin{aligned} \dot{V} =& 2\xi^T (I_N \otimes P_1 A + J \otimes P_1 B K)\xi - 2\xi^T (J \otimes P_1 B K)\epsilon + 2\epsilon^T (I_N \otimes P_2(A - LC))\epsilon \\ &+ 2\xi^T (I_N \otimes P_1 D_u)\omega_u + 2\epsilon^T (I_N \otimes P_2 D_u)\omega_u - 2\epsilon^T (I_N \otimes P_2 L D_y)\omega_y. \end{aligned} \tag{14}$$

Using Lemma 1, (14) can be rewritten as follows:

$$\begin{aligned} \dot{V} =& \textstyle\sum_{j=1}^N \xi_j^T He\{P_1 A + \lambda_j P_1 B K\}\xi_j - 2\sum_{j=1}^N \xi_j^T \lambda_j P_1 B K \epsilon_j + \sum_{j=1}^N \epsilon_j^T He\{P_2 A - P_2 L C\}\epsilon_j \\ &+ 2\textstyle\sum_{j=1}^N \xi_j^T He\{P_1 D_u\}\omega_{u_j} + 2\sum_{j=1}^N \epsilon_j^T He\{P_2 D_u\}\omega_{u_j} + 2\sum_{j=1}^N \epsilon_j^T He\{P_2 L D_y\}\omega_{y_j}, \end{aligned} \tag{15}$$

then, considering (15) with H_∞ criterion, $\varphi_j(t) = [\xi_j(t)^T, \epsilon_j(t)^T, \omega_{u_j}(t)^T, \omega_{y_j}^T]^T$, and $J_T \leq \sum_{j=1}^N \varphi(t)_j^T \Omega_j \varphi(t)_j$, where Ω_j is defined as follows:

$$\Omega_j = \begin{bmatrix} Q_{1_j} + I & -\lambda_j P_1 BK & P_1 D_u & 0 \\ * & Q_2 + I & P_2 D_u & -P_2 L D_y \\ * & * & -\gamma^2 I & 0 \\ * & * & * & -\gamma^2 I \end{bmatrix} < 0, \quad (16)$$

where $Q_{1_j} = He\{P_1 A + \lambda_j P_1 BK\}$, and $Q_2 = He\{P_2 A - P_2 LC\}$. Then, (16) is pre and post multiplied by $\overline{P}_1$, where $\overline{P}_1 = \overline{P}_1^T = P_1^{-1} > 0$

$$\begin{bmatrix} \Upsilon_{1_j} & 0 & D_u & 0 \\ * & \Upsilon_2 & P_2 D_u & -P_2 L D_y \\ * & * & -\gamma^2 I & 0 \\ * & * & * & -\gamma^2 I \end{bmatrix} + He\left\{ \begin{bmatrix} -\lambda_j BK \\ 0 \\ 0 \\ 0 \end{bmatrix} \begin{bmatrix} 0 & I & 0 & 0 \end{bmatrix} \right\} < 0, \quad (17)$$

where $\Upsilon_{1_j} = He\{A\overline{P}_1 + \lambda_j BK\overline{P}_1\} + \overline{P}_1\overline{P}_1$, and $\Upsilon_2 = He\{P_2 A - \; P_2 LC\} + I$. Applying the Young relation [12], the following inequality is obtained:

$$\begin{aligned} &\begin{bmatrix} \Upsilon_{1_j} & 0 & D_u & 0 \\ * & \Upsilon_2 & P_2 D_u & -P_2 L D_y \\ * & * & -\gamma^2 I & 0 \\ * & * & * & -\gamma^2 I \end{bmatrix} + \mu \begin{bmatrix} -\lambda_j BK \\ 0 \\ 0 \\ 0 \end{bmatrix} \overline{P}_1^{-1} \begin{bmatrix} (-\lambda_j BK)^T & 0 & 0 & 0 \end{bmatrix} \\ &+ \mu^{-1} \begin{bmatrix} I & 0 & 0 & 0 \end{bmatrix}^T \overline{P}_1^{-1} \begin{bmatrix} 0 & I & 0 & 0 \end{bmatrix} < 0, \end{aligned} \quad (18)$$

where $\mu > 0$. Using the Schur complement in (18), choosing $N_c = K\overline{P}_1$, and $M_o = P_2 L$, the proof is complete.

4 Simulation Results: Application to a Team of UAVs

In this section, the simulation results for an application to leader-following robust consensus control of a team of homogeneous UAVs are presented. The dynamic model of each UAV is:

$$\begin{aligned} \ddot{x}_i &= (c_{\psi_i} s_{\theta_i} c_{\phi_i} + s_{\psi_i} s_{\phi_i}) \frac{1}{m_i} T_i, & \ddot{\phi}_i &= \dot{\theta}_i \dot{\psi}_i \left(\frac{J_y - J_z}{J_x} \right) + \frac{1}{J_x} R_i, \\ \ddot{y}_i &= (s_{\psi_i} s_{\theta_i} c_{\phi_i} - c_{\psi_i} s_{\phi_i}) \frac{1}{m_i} T_i, & \ddot{\theta}_i &= \dot{\phi}_i \dot{\psi}_i \left(\frac{J_z - J_x}{J_y} \right) + \frac{1}{J_y} P_i, \\ \ddot{z}_i &= -g + (c_{\theta_i} c_{\phi_i}) \frac{1}{m_i} T_i, & \ddot{\psi}_i &= \dot{\theta}_i \dot{\phi}_i \left(\frac{J_x - J_y}{J_z} \right) + \frac{1}{J_z} Y_i, \end{aligned} \quad (19)$$

where the shorthand notation $c_{\{\cdot\}}$ and $s_{\{\cdot\}}$ is $\cos(\cdot)$ and $\sin(\cdot)$ respectively; x_i, y_i, and z_i are the position of the *ith* UAV in the euclidean space, ϕ_i, θ_i, and ψ_i, are the Euler angles *roll*, *pitch*, and *yaw* respectively, R_i, P_i, and Y_i are the rotor torques, m_i is the total mass of the UAVs, g is the gravitational acceleration, J_x, J_y, and J_z represents the moments of inertia in their respectively axis, T_i is the thrust of the UAVs rotors. The parameters used in the simulations were taken from [13].

The main objective of the multi-agent system, where each UAV is associated to one agent, is to follow the trajectories of a virtual leader and maintain a desired shape, where $H = [h_1^T, h_2^T, \ldots, h_N^T]^T$ contains the desired distance column vectors h_i of every agent. To solve the formation control problem, a second order representation of each agent [14] is used to find a solution for the LMIs presented in Theorem 1:

$$\dot{p}_i(t) = v_i(t), \quad \dot{v}_i(t) = u_i(t). \tag{20}$$

However, it is noted that for the simulations the complete nonlinear model (19) is used. Considering $p_i(t)$, $v_i(t)$, $u_i(t)$ the position, velocity and acceleration respectively of agent i, $\hat{\bar{s}}_i(t) = [\hat{p}_i(t)^T - h_i^T, \hat{v}_i(t)^T]^T$ the estimated states of agent i, $\bar{s}_l(t) = [p_l(t)^T, v_l(t)^T]^T$ the virtual leader states, h_i is a vector of matrix H of the respective agent, then, the control law (6) is rewritten as:

$$u_i(t) = K\sum_{j\in\mathcal{N}_i} a_{ij}(\hat{\bar{s}}_i(t) - \hat{\bar{s}}_j(t)) + \alpha_i K(\hat{\bar{s}}_i(t) - \bar{s}_l(t)), \tag{21}$$

where K is the control gain to be designed. Considering $\hat{\bar{\delta}}_i(t) = \hat{\bar{s}}_i(t) - \bar{s}_l(t)$ Eq. (21) is rewritten as follows:

$$u_i(t) = K\sum_{j\in\mathcal{N}_i} a_{ij}(\hat{\bar{\delta}}_i(t) - \hat{\bar{\delta}}_j(t)) + \alpha_i K\hat{\bar{\delta}}_i(t). \tag{22}$$

Since the dynamic of each UAV is considered as a particle, the desired Euler angles θ_{di}, ϕ_{di}, and ψ_{di} are computed as in [15] to control the position of the UAVs, where the consensus protocol (22) and the desired Euler angles are related as follows:

$$\phi_{di} = \arctan\left(\frac{-u_{2i}}{\sqrt{u_{1i}^2 + (u_{3i}+g)^2}}\right), \quad \theta_{di} = \arctan\left(\frac{u_{1i}}{u_{3i}+g}\right), \quad \psi_{di} = 0, \tag{23}$$

and the thrust of the UAVs rotors is computed by $T_i = m_i(u_{1i}^2 + u_{2i}^2 + (u_{3i}+g)^2)^{1/2}$. The communication topology of the multi-agent system is depicted in Fig. 1 (where a UAV represents an agent and the arrows the flow of information), and the Laplacian matrix $\mathcal{L}$:

$$\mathcal{L} = \begin{bmatrix} 5 & -1 & -1 & -1 & -1 & -1 \\ -1 & 2 & -1 & 0 & 0 & 0 \\ -1 & -1 & 2 & 0 & 0 & 0 \\ -1 & 0 & 0 & 2 & -1 & 0 \\ -1 & 0 & 0 & -1 & 3 & -1 \\ -1 & 0 & 0 & 0 & -1 & 2 \end{bmatrix} \tag{24}$$

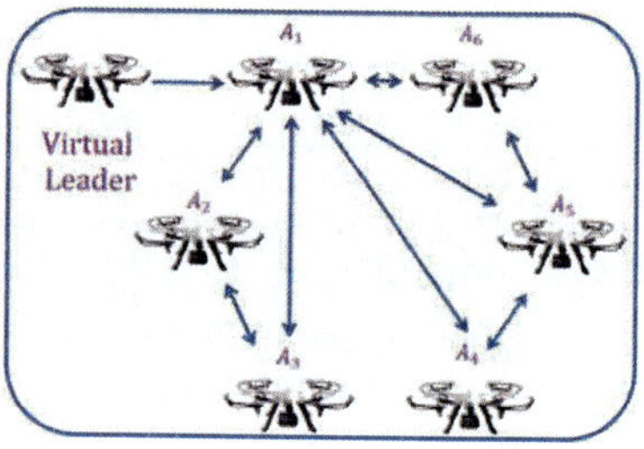

Fig. 1 Directions of the communication links between agents

H and Λ matrices are the following:

$$H = \begin{bmatrix} 0 & 4 & 6 & 4 & 0 & -2 \\ 0 & 0 & 2\sqrt{3} & 4\sqrt{3} & 4\sqrt{3} & 2\sqrt{3} \\ 0 & 0 & 0 & 0 & 0 & 0 \end{bmatrix}, \quad \Lambda = \begin{bmatrix} 1 & 0 & 0 & 0 & 0 & 0 \\ 0 & 0 & 0 & 0 & 0 & 0 \\ 0 & 0 & 0 & 0 & 0 & 0 \\ 0 & 0 & 0 & 0 & 0 & 0 \\ 0 & 0 & 0 & 0 & 0 & 0 \\ 0 & 0 & 0 & 0 & 0 & 0 \end{bmatrix}. \tag{25}$$

The first simulation result in Fig. 2, shows that the formation control problem is solved and the multi-agent system follows the trajectories of the virtual leader. All the UAVs reach the trajectories of the virtual leader while maintaining the rigid formation of a hexagon. This first result does not include external disturbance or measurement sensor noise.

The considered external disturbance is modeled as the Dryden wind turbulence presented in [16], and the sensor noise was added as a random function with normal distribution, mean value of zero, and standard deviation of $0.8\frac{\mathrm{m}}{\mathrm{s}}$, in order to verify the robustness of the control law (22). If the measurement noise or the external disturbance magnitudes increase, the control law (22) is not able to maintain the desired formation or to follow the virtual leader. To overcome the disturbance and noise problem, the LMI presented in (11) is computed with the Sedumi Toolbox [17], where the LMI variable γ value obtained with $\mu = 0.1$ is $\gamma = 1$. The computed control and observer gains are:

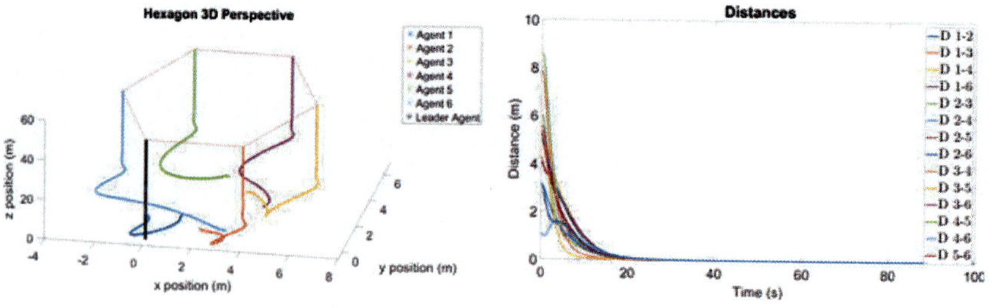

Fig. 2 Formation without disturbance or sensor noise and distances between followers

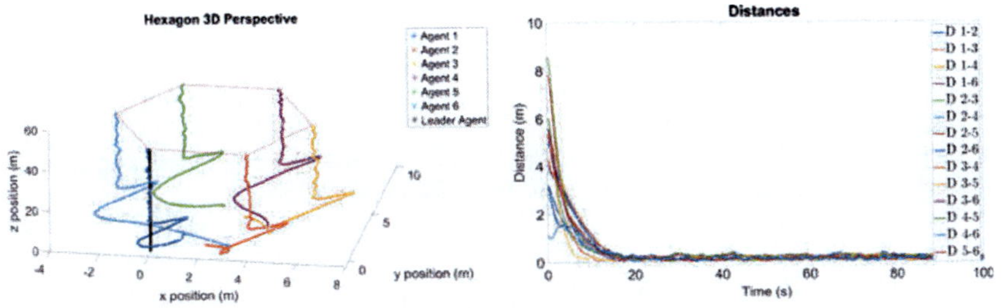

Fig. 3 Formation with disturbance and sensor noise using the control law (22) and observer (4) with gains K, and L (26) computed using Theorem 1

$$K = \begin{bmatrix} -0.756 & 0 & 0 & -2.651 & 0 & 0 \\ 0 & -0.756 & 0 & 0 & -2.651 & 0 \\ 0 & 0 & -0.756 & 0 & 0 & -2.651 \end{bmatrix}, L = \begin{bmatrix} 3.208 & 0 & 0 \\ 0 & 3.208 & 0 \\ 0 & 0 & 3.208 \\ 0.836 & 0 & 0 \\ 0 & 0.836 & 0 \\ 0 & 0 & 0.836 \end{bmatrix}. \quad (26)$$

The robustness of the distributed control and estimation implemented in the team of UAVs has been verified by solving the leader-follower formation control problem. The effectiveness of the proposed work is shown in Fig. 3, where each UAV achieves its desired position in the rigid formation of a hexagon while following the trajectory of the virtual leader because the external disturbance and the measurement sensor noise are rejected.

5 Conclusions

A distributed robust leader-following consensus control observer-based for LTI multi-agent systems was presented. The objective of the multi-agent system is to follow the trajectories described by a virtual leader's dynamics and to maintain a desired rigid formation between followers. This was exemplified with the simulation of a team of UAVs where the agents were able to reject external disturbance and measurement sensor noise. As further work, this approach will be extended to LPV multi-agent systems.

References

1. Skorobogatov, G., Barrado, C., Salamí, E.: Multiple UAV systems: a survey. Unmanned Syst. 8(02), 149–169 (2020)
2. Li, Z., Duan, Z.: Cooperative control of multi-agent systems: a consensus region approach. CRC Press (2017)

3. Trejo, J.A.V., Adam-Medina, M., Garcia-Beltran, C.D., Ramírez, G.V.G., Yolanda lópez Zapata, B., Sanchez-Coronado, E.-M., Theilliol, D.: Robust formation control based on leader-following consensus in multi-agent systems with faults in the information exchange: application in a fleet of unmanned aerial vehicles. IEEE Access **9**, 104940–104949 (2021)
4. Tian, X., Liu, H., Liu, H.: Robust finite-time consensus control for multi-agent systems with disturbances and unknown velocities. ISA Trans. **80**, 73–80 (2018)
5. Redrovan, D.V., Kim, D.: Multiple quadrotors flight formation control based on sliding mode control and trajectory tracking. In: 2018 International Conference on Electronics, Information, and Communication (ICEIC), pp. 1–6. IEEE (2018)
6. Sun, Z., Guoqing Zhang, Y.L., Zhang, W.: Leader-follower formation control of underactuated surface vehicles based on sliding mode control and parameter estimation. ISA Trans. **72**, 15–24 (2018)
7. Ahmed, Z., Khan, M.M., Saeed, M.A., Zhang, W.: Consensus control of multi-agent systems with input and communication delay: a frequency domain perspective. ISA Trans. **101**, 69–77 (2020)
8. Lin, W., Zhao, W., Liu, H.: Robust optimal formation control of heterogeneous multi-agent system via reinforcement learning. IEEE Access **8**, 218424–218432 (2020)
9. Haichuan, X., Housheng, S., Wang, Q., Chengjie, X.: Semi-global adaptive bipartite output consensus of multi-agent systems subject to input saturation and external disturbance under switching network. Int. J. Control Autom. Syst. **19**(9), 3037–3048 (2021)
10. Ni, W., Cheng, D.: Leader-following consensus of multi-agent systems under fixed and switching topologies. Syst. Control Lett. **59**(3–4), 209–217 (2010)
11. Chen, J., Zhang, W., Cao, Y.-Y., Chu, H.: Observer-based consensus control against actuator faults for linear parameter-varying multiagent systems. IEEE Trans. Syst., Man, Cybern.: Syst. **47**(7), 1336–1347 (2016)
12. Boyd, S., El Ghaoui, L., Feron, E., Balakrishnan, V.: Linear matrix inequalities in system and control theory. SIAM (1994)
13. Luis, C., Le Ny, J.: Design of a trajectory tracking controller for a nano quadcopter. Polytechnique Montreal, Technical report, Mobile Robotics and Autonomous Systems Laboratory (2016)
14. Yang, H., Zhang, Z., Zhang, S.: Consensus of second-order multi-agent systems with exogenous disturbances. Int. J. Robust Nonlinear Control **21**(9), 945–956 (2011)
15. Guerrero-Castellanos, J.-F., Vega-Alonzo, A., Marchand, N., Durand, S., Linares-Flores, J., Mino-Aguilar, G.: Real-time event-based formation control of a group of VTOL-UAVs. In: 2017 3rd International Conference on Event-Based Control, Communication and Signal Processing (EBCCSP), pp. 1–8. IEEE (2017)
16. Rodríguez-Mata, A.E., Flores, G., Martínez-Vásquez, A.H., Mora-Felix, Z.D., Castro-Linares, R., Amabilis-Sosa, L.E.: Discontinuous high-gain observer in a robust control UAV quadrotor: real-time application for watershed monitoring. Math. Probl. Eng. (2018)
17. Sturm, J.F.: Using SeDuMi 1.02, a MATLAB toolbox for optimization over symmetric cones. Optim. Methods Softw. **11**(1–4), 625–653 (1999)

Nonlinear Systems, Localization and FDI

Contact-Less Sensing and Fault Detection/Localization in Long Flexible Cantilever Beams via Magnetoelastic Film Integration and AR Model-Based Methodology

Robert-Gabriel Sultana and Dimitrios Dimogianopoulos

Abstract The integration of magnetoelastic film into a long, flexible polymer beam for obtaining sensing and fault diagnosis capabilities is investigated. Metglas® 2826 MB film is attached onto the clamped end of the beam (itself fixed as cantilever), with its free end connected to a mini-exciter. Vibrating the structure causes emission of variable magnetic flux by the integrated magnetoelastic film, intercepted in a contact-less manner by a low-cost reception coil suspended above the film. The resulting voltage is linked to the beam's vibratory motion and is recordable via conventional oscilloscopes without sophisticated/dedicated circuitry (e.g. power-amplifiers), making for a cost-efficient and low-complexity setup. Hence, the long, flexible beam may act as a sensing device with contact-less data transmission. The innovation is that *different* structural faults (represented by loads of various magnitudes and locations on the beam) are shown to affect accordingly the recorded signal's spectral characteristics. Furthermore, fault-induced frequency shifts and changed damping can be reliably estimated and traced back to specific faults via the currently proposed stochastic model-based algorithmic methodology. Tests involving two different loads applied on three distinct positions on the beam successfully conclude on the proposed setup's potential for vibration sensing and detection/localization of different faults (loads) on the beam.

1 Introduction

The need for wireless monitoring of hazardous substances in often hostile environments is one of the main reasons that magnetoelastic sensors marked a net progress over the past years. The concentration of dangerous chemical, environmental and bio-

R.-G. Sultana · D. Dimogianopoulos (✉)
Department of Industrial Design and Production Engineering, University of West Attica, Campus of Ancient Olive Grove, 250 Thivon Av., 12241 Athens, Greece
e-mail: dimogian@uniwa.gr

R.-G. Sultana
e-mail: auto47101@uniwa.gr

D. Theilliol et al. (eds.), *Recent Developments in Model-Based and Data-Driven Methods for Advanced Control and Diagnosis*, Studies in Systems, Decision and Control 467,
https://doi.org/10.1007/978-3-031-27540-1_16

logical factors could be remotely monitored without human intervention by means of magnetoelastic (magnetostrictive) sensors. These are made of (ferromagnetic) materials which alter their shape when subjected to external magnetic fields, or emit magnetic flux in accordance with externally imposed loading [1–3]. Variable changes of shape of a suitably fixed (for instance, clamped) magnetoelastic element (or *resonator*) will set it to vibrate. Vibratory dynamics change when biological [4] or environmentally polluting [5] substances accumulate onto the resonator's surface, due to the resulting change in its mass distribution. Again, a suitably coated resonator will exhibit changes in its mass distribution due to the coating's reaction with volatile organic compounds [6], or its adsorption of H_2O [7] or H_2O_2 [8]. Observing changes in the resonator's vibratory dynamics (shifted resonant frequencies) directly points to increased concentration of substances on it and, hence, the environment.

The majority of these resonators are set into vibration via an interrogation coil which, under suitable electrical excitation, produces variable magnetic flux towards the magnetoelastic material (usually in form of a ribbon or thin film), thus driving it to resonance. A reception coil is placed above the vibrating resonator and transforms the flux emitted into electrical signal in a contact-less manner. This signal is, then, monitored for frequency shifts enabling detection of the substance under consideration. The operational concept of these applications, based on the use of two coils and referred to as the active setup, is comprehensively reviewed in [9]. Obviously, effort has been invested in optimizing the resonator's sensitivity to the mass accumulated on its surface by analyzing relevant characteristics (for instance, length-to-width ratio) of the resonator's initial shape [10], or even suggesting particularly shaped (as hourglass [11], or rhomboid [12, 13]) resonators.

Another class of resonators of particular interest (also reviewed in [9]) involves those that are parts of a mechanical system (structure, machinery or other) and receive vibration from the system's operation or an external mechanical excitation, rather than an external varying magnetic field [14–18]. These are referred to as passive setups, with the vibrating magnetoelastic part producing variable magnetic flux (of lower intensity compared with active setups) in accordance with the external vibration. Again, voltage is induced in a reception coil in a contact-less manner, with its frequency analysis allowing for contact-less sensing/measuring the dynamics of the magnetoelastic part and, hence, of the system. Interestingly, any faults/failures affecting the system's vibration dynamics (resonant shifts, for instance), will show up in the recorded voltage because, as demonstrated in [14, 16, 17], specific resonants of the system dynamics (as estimated via Finite Element Analysis) are present in the voltage. Based on this principle, fault diagnosis results have been achieved in polymer slabs with built-in magnetoelastic ribbons via 3D printing [14], or joints of structures assembled with such slabs [15]. Again, diagnosis of cracks has been achieved for cantilever metal beams with magnetoelastic ribbon attached onto their surface [16] or rotating metal beams [18]. The possibility of using such passive setups as sensing devices has been demonstrated for short, thin polymer beams (fixed as cantilevers) in [19], and for metal beams also fixed as cantilevers in [17].

This work aims at demonstrating the feasibility of a two-fold extension of previous fault detection and severity estimation results [14, 15] obtained for single (or structures assembled by) slabs with magnetoelastic elements: First, by additionally achieving localization of faults of different magnitudes; and second by using long, highly flexible beams as structures upon which the magnetoelastic elements are attached, instead of being 3D-printed into relatively short rigid slabs (or slab assemblies) [14, 15]. Such flexible beams may also serve as sensing elements when connected to existing (metallic or polymer) vibrating structures, rather than being used as load-bearing structural parts (i.e. slabs). Faults are simulated as loads placed on different positions on the beam surface. Two different loads have been tested, with each one alternatively placed at one of three available positions. The beam retains the cantilever arrangement, but unlike [16–18], the magnetoelastic film attachment position is at the clamped end, with excitation provided via a mini-exciter at its free end. As in [19], voltage is induced in a contact-less manner in the reception coil and is recorded via a conventional oscilloscope. The additional absence of excitation coils (as found in traditional applications [2, 3]) means that complexity and costs are both minimal. For enhanced cost-effectiveness (and unlike [16–18]), effort is invested in optimizing the algorithmic framework used for fault diagnosis, rather than optimizing the hardware (coils, associated circuitry, amplifiers) itself. Hence, faults (loads) are detected-localized by using stochastic model-based procedures optimized for fault detectability via a novel criterion. Even though the excitation is always at the free-end, structural changes (loads) of different intensity and positions on the beam surface are detectable and localizable, meaning that no special excitation position is necessary for obtaining global fault diagnosis results. Hence, long and highly flexible beam arrangements suitably receiving excitation from vibrating (mechanical) systems may, indeed, be used for sensing and diagnostic purposes.

2 Materials and Methods

A 425 mm long, thin and highly flexible beam is formed by tandem connection of two shorter beams with glue. The beams are 3D-printed in Fused Deposition Modeling (FDM) mode with PET-G filament. The final beam is clamped on one end, with the opposite end fixed onto the vibrating rod of a mini-exciter. The Metglas® 2826 MB magnetoelastic film is attached onto the beam surface near the clamp with cyanoacrylate glue. A low-cost Vishay IWAS coil, normally used for wireless charging, is placed 5 mm above the film and receives the emitted magnetic flux (Fig. 1). The SMARTSHAKER™ K2004E01 mini-exciter vibrates the beam as defined by the external waveform generator (SIGLENT, SDG 5122). The reception coil is connected to a conventional oscilloscope, used for voltage data acquisition. The latter is the only signal used for detecting and localizing faults. This *output-only* approach is realistic because in similar real-life applications (bridges, or highly flexible structures of similar shape), the excitation is often unavailable or hard to measure. Moreover, such flexible beam setups may also serve as low-cost sensing elements connected to mechanical systems and driven by their vibrations (see Sect. 3.1).

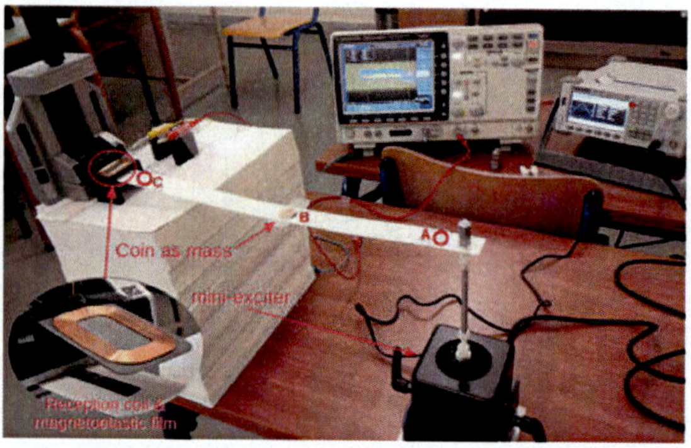

Fig. 1 Experimental setup with a load (coin) at B, and the contact-less sensing concept (zoom)

Table 1 Test scenarios (fault cases), their designations and positions

Test scenario (fault case)	Load object	Mass (g)	Position (from free end)
N-1C	No load (healthy)	0	n/a
A-1C	1 Euro cent	2.3	A (at 35 mm—nearby excitation point)
B-1C	1 Euro cent	2.3	B (at 185 mm—middle)
C-1C	1 Euro cent	2.3	C (at 360 mm—nearby clamp)
N-BN	No load (healthy)	0	n/a
A-BN	Bolt+nuts	6	A (at 35 mm—nearby excitation point)
B-BN	Bolt+nuts	6	B (at 185 mm—middle)
C-BN	Bolt+nuts	6	C (at 360 mm—nearby clamp)

Faults are simulated as loads at three key positions on the beam (Fig. 1 and Table 1), namely near the excitation (A), in the middle (B) and near the clamp (C). The testing scenarios (Table 1) involve cases without load (prefix N) or with a load at position A, B or C. Suffix -1C indicates a "small" load of 1 Euro-cent coin, which is a compact unit of precise weight easily fixable at A, B or C (see prefix) with tape. Suffix -BN indicates a "bigger" load of a bolt fixed with nuts, again at the designated (by prefix) position. Bolts are 2.6 times heavier than 1 Euro-cent coins and easier to fix than the 20 Euro-cent coins of equivalent weight, which are too thick for being fixed with tape on the beam. The aim is to investigate the feasibility of distinguishing faults of different magnitudes and at various locations of a vibrating structure (for instance, a flexible bridge), with magnetoelastic material in only one location.

2.1 Feasibility of Detecting and Localizing Different Faults

As concluded in [14, 15, 19], in passive setups the time history of recorded voltage signals rarely exhibit visible patterns related to faults affecting the (data-generating) associated system. But frequency domain characteristics (natural frequencies ω_n and damping ζ) do change with fault occurrence, since they are physically linked to (and capable of quantifying) fault-induced alterations of the system's structure. Thus, the recorded voltage from a vibrating system is modeled via discrete-time stochastic output-only AutoRegressive (AR) representations, capable (among others) of accurately estimating values for ω_n and ζ, associated to the system's current health state (Table 1). Model-free (see Sect. 3.2) or nonlinear (e.g. Neural Network) fault diagnosis approaches, although highly effective, cannot easily provide results for both ω_n and ζ. In an AR modeled signal, the current value of a process depends on a set of past values up to n previous lags and a series of random[1] values $e[t]$:

$$y[t] = a_1 y[t-1] + a_2 y[t-2] + \cdots + a_n y[t-n] + e[t] \tag{1}$$

with n the AR model order and $e[t]$ a sequence of independent (i.e. uncorrelated) and identically distributed random values (white noise). Estimating n along with parameters $a_1, a_2, \ldots, a_n$ (via Least-Squares criteria) is a task referred to as model identification. The feasibility of detecting and localizing system faults of different magnitudes is demonstrated by comparing characteristics of such AR models, each identified on voltage data from a case in Table 1. The steps for demonstrating the feasibility of detecting and localizing faults are as follows:

1. A recorded voltage dataset from a system in some state (see Table 1) is filtered and subsampled, as shown in Sect. 3.2.
2. Stochastic AR time-series representations are identified on the processed dataset, thus capturing properties of the signal (and the underlying system) dynamics.
3. Natural frequencies ω_n and damping ratios ζ are computed for specific frequency ranges (see Sect. 3.2), and the associated poles are plotted on the s-plane.

The procedure is executed for all available voltage sets. Fault detectability is achieved when pole groups from AR models identified on N-1C data are distinguishable from pole groups corresponding to A-1C, B-1C and C-1C data, inside a given frequency range. Feasibility of localizing faults is demonstrated when pole groups from models identified on A-1C, B-1C and C-1C data are mutually distinguishable in the given frequency range. The same reasoning applies to cases with -BN suffix. Finally, if pole groups from -1C cases may be distinguished from their counterparts resulting from -BN cases, then faults of different magnitudes are detectable and localizable.

[1] $x[t]$ denotes values of x in discrete time with t being the normalized discrete time, $t \times \Delta t$ the absolute time and Δt the sampling period; $x(t)$ denotes values of x at continuous time instant t.

2.1.1 Optimizing Model Order for Enhanced Fault Detectability

As stated in Sect. 2.1, faults are detectable by means of delimiting suitable areas on the complex plane, so that poles from AR models identified on healthy data may only be found inside these areas. Then, enhancing fault detectability means that these areas must be as small/compact as possible. This task may be carried out by suitably optimizing the AR model identification process as follows.

Let an AR model (1) of order n be available for a dataset, with residuals $r[t]$:

$$r[t] = y[t] - \hat{y}[t] \tag{2}$$

where $\hat{y}[t]$ is the sequence of predicted output values from the given AR model and $y[t]$ the measured output $\forall\ t$ of the experiment. The AR model of order n is validated if the sequence $r[t]$ is uncorrelated, as statistically checked via the sample autocorrelation function. With larger values for n yielding less correlated sequences $r[t]$, there exist several criteria (Akaike or Bayesian Information Criterion- AIC or BIC, respectively [20]) which propose an optimal value for n, so that $r[t]$ is quite uncorrelated and the AR model does not become too long. In the current fault detection context, however, once a value for n leading to relatively uncorrelated sequence $r[t]$ is found, one has to make sure that using that n for identifying AR models for *all* healthy system datasets leads to all resulting poles (for the frequency range of interest) lying inside the smallest possible area in the complex plane. This, in turn, means that the risk of this area potentially including poles from AR models identified on faulty system data is minimized.

Consider a set of poles from AR models of a specific order n (identified on a given number of healthy datasets) for the frequency range of interest on the complex-plane. Let the quantity $r_{sum}|_n$ be calculated by summing the absolute distance of every pole from the barycenter of the pole set in the specific frequency range

$$r_{sum}|_n = \sum_{i=1}^{p} r_i \tag{3}$$

where p is the number of datasets (or poles in the selected frequency range) and r_i is the L2-norm (absolute distance) of the i-th pole from the barycenter. Then, one has to simply minimize $r_{sum}|_n$ with respect to n, starting with an initial value for n equal to the smallest AR order that yields uncorrelated $r[t]$.

3 Results and Discussion

The current section features the application of the fault detection and localization scheme presented in Sect. 2.1 on data from vibration testing experiments with the setup in Sect. 2. Two issues are addressed: First, the setup sensing abilities are optimized with respect to noise (induced by contact-less data transmission), by suitably defining the coil height (Sect. 3.1). Then, fault detection and localization results from the optimal setup are presented and commented upon (Sect. 3.2).

3.1 Sensing Properties of the Passive Setup

The setup's sensing properties are assessed and optimized via two testing scenarios: The first one considers signals recorded with the beam at rest, which essentially involve the underlying noise. The second scenario involves data from a vibrating beam with the coil placed at either 20 or 5 mm above the film on the beam surface. A triangular force 160 Hz (supplied by the mini-exciter) sets the beam into a low amplitude (invisible) vibration. Tests at frequencies as low 30 Hz revealed that the power line value (50 Hz) was always present in data, even for frequencies under 50 Hz. Data transmission notably improved for frequencies over 80–100 Hz, as had also been noted for passive setups in [15]. Obviously, multiple excitation locations could probably improve data transmission, but at increased setup complexity. Triangular waveforms were preferred to sinusoids or pulses, since they resemble more to real-life vibration sources due, for instance, to machinery reciprocating parts [15].

For both scenarios, three sets of tests were executed. Figure 2a shows time histories of signals with and without excitation (noise), with both signals being, as expected, almost identical (see Sect. 2.1). Figure 2b shows the Fast Fourier Transform (FFT) plot of the signals in Fig. 2a, with the reception coil placed 20 mm above the beam. Focusing on the highest peaks around 1300 Hz, signals from vibrating beam are clearly different from simple noise, meaning that sensing is indeed feasible. Similar tests with the coil at 5 mm above the beam in Fig. 2c, further highlight the difference between noise and signals from the vibrating beam. Moreover, the latter exhibit peaks around 1350 Hz which are more scattered in Fig. 2b than their counterparts in Fig. 2c.

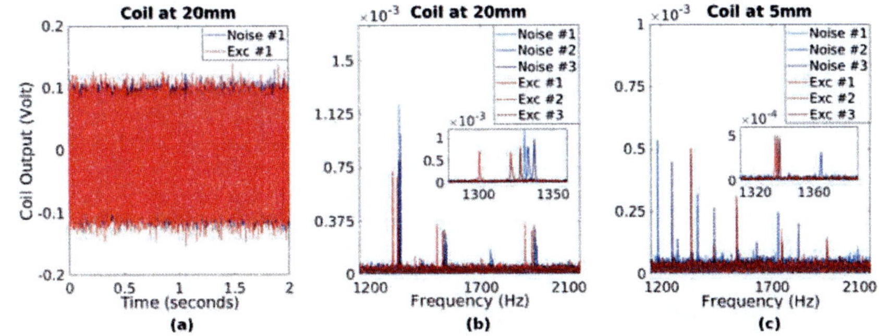

Fig. 2 **a** Time history of signals with and without excitation showing no visible differences **b** Frequency content of signals with and without excitation showing small frequency shifts- coil at 20 mm above the beam. **c** Frequency content as in **b** but with coil at 5 mm above the beam

Thus, sensing results are more consistent when the coil is located as close to the beam surface as possible. With suitable further development, the passive setup could be a cost-effective alternative to pricey sensors (for instance, load cells) used in devices for recording and frequency analysis of (force) signals. A relevant example (that initiated this study) is the Fish Texture Evaluation Tool developed in work package 6 of "FutureEUAqua" project [21].

3.2 Fault Detection and Localization Results

Since the setup sensing properties are validated, testing with fault cases in Table 1 may be carried out. Three tests (each covering two seconds of the beam's vibration) are executed per case, making for twenty-four datasets. Figure 3a shows the signal frequencies from N-1C and B-1C cases, with clear shifting due to load (B-1C) around 1300 Hz. Error bar plots for frequency peaks from all tests in Fig. 3b indicate a clear distinction of fault classes. Signals obtained from passive setups in a contact-less manner are, hence, suitable for detecting loads (faults) and denoting their locations.

Even better, the 3-step algorithm in Sect. 2.1 allows for estimating both frequencies ω_n and damping ratios ζ for fault (load) detection/ localization purposes. The dataset of each experiment is split in half, resulting in six (smaller) sets per case (see Table 1). Hence, transient non-stationary data points are better accounted for during modeling, since they now spread across smaller datasets. Datasets are mildly filtered via 5th order Butterworth filters (with –3 dB cutoff at 3 KHz) and subsampled at 20 kHz (from 1 MHz). The optimal AR model order is estimated (second step in Sect. 2.1.1) along with the resulting discrete-time poles. Figure 4a shows that $r_{sum}|_n$ is minimal for AR orders of 94 and 75 for the N-1C and N-BN cases, respectively. Thus AR(94) and AR(75) models are identified on data from cases with suffix -1C and -BN, respectively. Note that uncorrelated residuals are already obtainable for $n > 45$ for all cases. Pole plots on the z-plane for -1C and -BN cases are presented in Fig. 4b and c, respectively. Around 1300 Hz, poles from N-1C and N-BN cases do

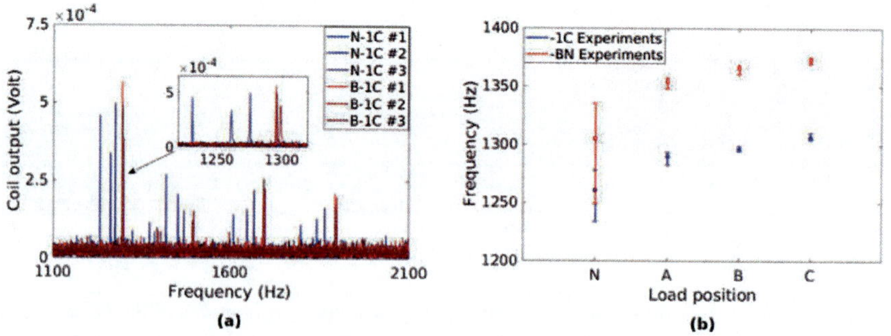

Fig. 3 **a** Frequency content of signals from N-1C versus B-1C tests, showing shifts due to fault (load). **b** Error bars of frequency peaks showing the same shift pattern for -1C and -BN cases

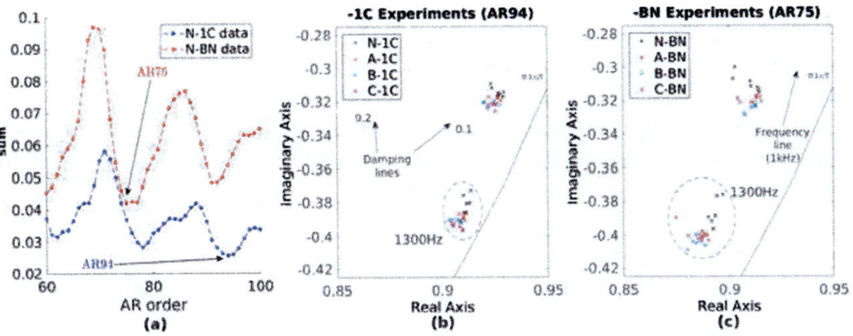

Fig. 4 **a** Values of $r_{sum}|n$ for increasing order n, with optimal values pointed out. **b** Poles corresponding to -1C cases and **c** poles corresponding to -BN cases on the z-plane

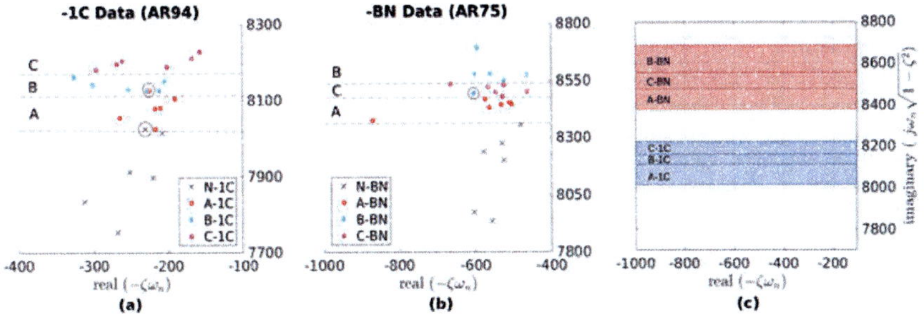

Fig. 5 AR poles corresponding to **a** -1C cases, and **b** -BN cases at 1300 Hz on the s-plane, with horizontal lines indicating that faults (except for those in circles) are correctly localized. **c** Pole areas for -1C and -BN cases showing that localized -1C faults cannot be mistaken for -BN ones

not mix with poles from other -1C/-BN cases, thus ensuring fault (load) detectability. Fault localization is harder, since poles corresponding to data from A-, B- and C-cases seemingly share the same z-plane area. Nonetheless, converting discrete-time poles to their continuous time counterparts and plotting them in Fig. 5, shows one delimited s-plane area for each case, hence ensuring feasibility of fault localization. More importantly, localized -1C faults cannot be mistaken for -BN ones, since their respective s-plane areas do not overlap in Fig. 5c. Thus, the current passive setup may achieve detection and localization of different faults (loads) using magnetoelastic film in only one position of the beam.

4 Conclusions

The feasibility of inducing sensing and fault diagnosis properties to a long flexible cantilever beam by attaching magnetoelastic film on its clamped side was studied. The setup is passive (i.e. without excitation coils), driven from beam vibrations.

The reception coil was optimally placed for contact-less sensing of data produced by beam vibrations and an algorithm for optimal fault detectability was postulated. Testing showed that all faults (loads on the beam) were detected, and almost all of them localized. Most importantly, localized faults of given magnitudes cannot be mistaken for faults of other magnitudes. Results are delivered at minimal cost and complexity, since no special equipment (accelerometers, signal conditioners) is required and the beam is excited at one position for all possible fault locations. Future work will focus on optimizing vibration sensing and fault identifiability by applying excitation to different locations, and assessing performance by extended testing.

Acknowledgements Work supported in part by the project FutureEUAqua-National Matching 2019 Funding. The authors wish to thank Prof. D. Mouzakis for providing the Metglas® film.

References

1. Le Bras, Y., Greneche, J.M.: Magneto-elastic resonance: principles, modeling and applications. In: Awrejcewicz, J. (ed.) Resonance, Chap. 2, IntechOpen (2017). https://doi.org/10.5772/intechopen.70523
2. Grimes, C.A., et al.: Theory, instrumentation and applications of magnetoelastic resonance sensors: a review. Sensors **11**(3), 2809–2844 (2011). https://doi.org/10.3390/s110302809
3. Gonzalez, J.M.: Magnetoelasticity and magnetostriction for implementing biomedical sensors. In: López-Dolado, E., Serrano, M.C. (eds.), Engineering Biomaterials for Neural Applications, pp. 127–147. Springer, Cham (2022). https://doi.org/10.1007/978-3-030-81400-7_6
4. Ren, L., Yu, K., Tan, Y.: Applications and advances of magnetoelastic sensors in biomedical engineering: a review. Materials **12**(7), 1135 (2019). https://doi.org/10.3390/ma12071135
5. Grimes, C.A., et al.: Magnetoelastic microsensors for envinronmental monitoring. In: 14th IEEE International Conference Micro Electro Mechanical Systems (MEMS), pp. 278–281. Interlaken, Switzerland (2001). https://doi.org/10.1109/MEMSYS.2001.906532
6. Baimpos, T., et al.: A polymer-Metglas sensor used to detect volatile organic compounds. Sens. Actuators A: Phys. **158**(2), 249–253 (2010). https://doi.org/10.1016/j.sna.2010.01.020
7. Atalay, S., et al.: Magnetoelastic humidity sensors with TiO_2 nanotube sensing layers. Sensors **20**, 425 (2020). https://doi.org/10.3390/s20020425
8. Samourgkanidis, G., et al.: Hemin-modified SnO_2/Metglas electrodes for the simultaneous electrochemical and magnetoelastic sensing of H_2O_2. Coatings **8**(8), 284 (2018). https://doi.org/10.3390/coatings8080284
9. Dimogianopoulos, D.G.: Sensors and energy harvesters utilizing the magnetoelastic principle: review of characteristic applications and patents. Recent Pat. Electr. Electron. Eng. **5**(2), 103–119 (2012). https://doi.org/10.2174/2213111611205020103
10. Skinner, W.S., et al.: Magnetoelastic sensor optimization for improving mass monitoring. Sensors **22**(3), 827 (2022). https://doi.org/10.3390/s22030827
11. Ren, L., Cong, M., Tan, Y.: An hourglass-shaped wireless and passive magnetoelastic sensor with an improved frequency sensitivity for remote strain measurements. Sensors **20**(2), 359 (2020). https://doi.org/10.3390/s20020359
12. Saiz, P.G., et al.: Enhanced mass sensitivity in novel magnetoelastic resonators geometries for advanced detection systems. Sens. Actuators B: Chem. **296**, 126612 (2019). https://doi.org/10.1016/j.snb.2019.05.089
13. Saiz, P.G., et al.: Influence of the magnetic domain structure in the mass sensitivity of magnetoelastic sensors with different geometries. J. Alloy. Compd. **863**, 158555 (2021). https://doi.org/10.1016/j.jallcom.2020.158555

14. Dimogianopoulos, D.G., Charitidis, P.J., Mouzakis, D.E.: Inducing damage diagnosis capabilities in carbon fiber reinforced polymer composites by magnetoelastic sensor integration via 3D printing. Appl. Sci. **10**(3), 1029 (2020). https://doi.org/10.3390/app10031029
15. Dimogianopoulos, D.G., Mouzakis, D.E.: Nondestructive contactless monitoring of damage in joints between composite structural components incorporating sensing elements via 3D-printing. Appl. Sci. **11**(7), 3230 (2021). https://doi.org/10.3390/app11073230
16. Samourgkanidis, G., Kouzoudis, D.: A pattern matching identification method of cracks on cantilever beams through their bending modes measured by magnetoelastic sensors. Theor. Appl. Fract. Mech. **103**, 102266 (2019). https://doi.org/10.1016/j.tafmec.2019.102266
17. Samourgkanidis, G., Kouzoudis, D.: Characterization of magnetoelastic ribbons as vibration sensors based on the measured natural frequencies of a cantilever beam. Sens. Actuators A: Phys. **301**, 111711 (2020). https://doi.org/10.1016/j.sna.2019.111711
18. Samourgkanidis, G., Kouzoudis, D.: Magnetoelastic ribbons as vibration sensors for real-time health monitoring of rotating metal beams. Sensors **21**(23), 8122 (2021). https://doi.org/10.3390/s21238122
19. Dimogianopoulos, D.G., Mouzakis, D.E.: A versatile interrogation-free magnetoelastic resonator design for detecting deterioration-inducing agents. Lecture Notes in Civil Engineering, vol. 110. Springer, Singapore (2020). https://doi.org/10.1007/978-981-15-9121-1_9
20. Ljung, L.: System Identification: Theory for the User, 2nd edn. Prentice Hall PTR, Upper Saddle River, NJ (1999)
21. FutureEUAqua project (WP6), Horizon2020 grant 817737. https://futureeuaqua.eu/

A Comparative Simulation Study of Localization Error Models for Autonomous Navigation

Anis Koliai, Stephane Bazeille, Michel Basset, and Rodolfo Orjuela

Abstract The navigation and guidance of autonomous vehicles require precise and accurate localization. However, in some situations the use of dead reckoning localization is unavoidable, despite the fact that it introduces an accumulation of errors known as drift. This paper focuses on a comparison in simulation between three different localization drift models based on position measurement, including a proposed model. A guidance and control scheme is implemented, including the proposed drift model. Finally, results from closed loop simulation are exploited to highlight the discontinuity phenomenon caused by erroneous position measurements.

1 Introduction

Over the past few decades, the development of self-driving vehicles has proven to be a challenging research topic for many laboratories and industries. Several researches and applications have been done, however achieving full autonomy still remains a significant task. Autonomous vehicles architecture is structured on three functional levels: perception, planning and control [11]. The perception level aims to localize the vehicle using various sensors as well as to collect information about the surrounding environment (e.g. obstacles, traffic signals, etc.). From the perception information, the planning system computes the desired reference trajectory. The last level is the

A. Koliai (✉) · S. Bazeille · M. Basset · R. Orjuela
Institut de Recherche en Informatique, Mathématiques, Automatique et Signal (IRIMAS), 12 rue des Frères Lumière, 68 093 Mulhouse, France
e-mail: Anis.Koliai@uha.fr

S. Bazeille
e-mail: Stephane.Bazeille@uha.fr

M. Basset
e-mail: Michel.Basset@uha.fr

R. Orjuela
e-mail: Rodolfo.Orjuela@uha.fr

D. Theilliol et al. (eds.), *Recent Developments in Model-Based and Data-Driven Methods for Advanced Control and Diagnosis*, Studies in Systems, Decision and Control 467,
https://doi.org/10.1007/978-3-031-27540-1_17

control, which produces suitable actuator signals (e.g. steering angle, throttle and brake) ensuring that the planned actions are carried out safely.

Knowing the precise and accurate localization of the vehicle is critical for safe autonomous navigation. The localization system must reach 95% of accuracy standards and maintain safety integrity in all situations where operations are expected [13]. Several localization techniques are proposed in the literature. The most well known methods are dependent on the infrastructure like global navigation satellites system (GNSS). Most of the time, GNSS is widely used for precise vehicle localization. Nevertheless, the availability and accuracy are not always guaranteed due to the natural blockage of the signal or malicious spoofing and jamming. Another method to localize the vehicle is to use onboard sensors (e.g. camera, lidar, inertial measurement unit (IMU), wheel encoders, etc.). There are two cases to be considered, when onboard sensors are used: with prior knowledge and without prior knowledge. With prior knowledge, usually predefined maps are used and the methods use sensor data to localize the data in the map. Without prior knowledge, the localization is estimated via a method known as dead reckoning [4].

Each case has advantages and drawbacks. In this research project, we chose to use the dead reckoning method because it does not need any knowledge on the environment and it is easier to implement. This latter exhibits an unbounded drift induced by the accumulating of small errors. The source of the errors differs with the method used.

In this context, several studies are done to propose mathematical models describing the unbounded drift. Among them, Quinchia et al. [12] propose a model to compensate IMU bias drift by considering as input the IMU acceleration and attitude angles rate. Consequently, a mechanization stage must be added to obtain vehicle position, as recently experimented in [14]. New neural network techniques are also used to compensate bias-drift [3, 9] but a training step is unavoidable and the drift model equations cannot be simply used. An empirical formula to model the unbounded cumulative error is proposed in [15]. This model is based on the Ackermann geometry and consider as input the side slip angle, the heading angle and the velocity. A drift model related to visual odometry (VO) drift is developed by Jiang et al. [8]. The model parameters are identified based on the camera estimated position. In the same context using VO, a recent work is done in our lab to propose a drift model which uses the GNSS positions to estimate the drifted position of the vehicle [1]. More advanced techniques can be used to improve visual odometry estimation [5, 6], but here the emphasis is on drift models.

There are many localization techniques integrated in the vehicle to localize it. However, when the localization performances change over time, some undesirable phenomena arise (e.g. position drift, discontinuity). As an example of localization performances changes, let us consider that the localization uses GNSS fused with IMU. When the GNSS is lost, the localization is only based on the IMU integration, which degrades the localization accuracy.

One contribution of this paper is to highlight these kind of phenomena through numerical simulations. For that purpose, two drift models ([8] and [1]) are compared. Based on the obtained results, some improvements to the model [1] are proposed to reduce the added intrinsic disturbance during turn maneuvers. It should be noted that the comparison is given by considering the vehicle in open loop. However, the variation in the localization performances affect the closed loop performances. To this aim, the closed loop behavior in the presence of erroneous localization is further investigated through a simulation scenario. The contributions of this work can be summarized as follows:

1. Proposition of improvements to the drift model developed in our lab [1].
2. Comparison of three different models: the model presented in [8], in [1] and the proposed one.
3. Closed loop simulation of varying localization performances to highlight the effect of discontinuity phenomenon.

The rest of the paper is organized as follows: In Sect. 2 the retained models are exposed and compared by performing open loop simulations. The closed loop scenarios, with simulation and discussion, are presented in Sect. 3. Finally, Sect. 4 concludes the paper.

2 Localization Error Modeling

Many different drift models are proposed in the literature. Generally, a drift model can be constructed by combining a set of equations and a type of noise. An important aspect of the drift modeling concerns the inputs used. Some models use directly the positions as input, while others use other quantities like velocity, acceleration, side slip angle, etc.

In this paper, the choice is made to study the models with positional input. Therefore, the input of all the models is the vehicle true pose $P_T(t) = [X_T(t),\ Y_T(t),\ \Psi_T(t)]^T$, where $X_T(t),\ Y_T(t)$ are the coordinates of the vehicle in the global frame and $\Psi_T(t)$ is the yaw angle. The model's output pose $P_M(t) = [X_M(t),\ Y_M(t),\ \Psi_M(t)]^T$ is calculated such that it drifts over time.

In the following sections, three models are presented and compared. The first model combines a white noise and a first-order Gauss-Markov (GM) process. The second and third models rely on the vehicle displacement equations. For the sake of comparison, the Gaussian noise is applied to the three models. The considered drift models are schematized as in Fig. 1.

2.1 *First Drift Model: Jiang et al. [8]*

The authors in [8] have established a model to represent the drift behavior of visual odometry position estimation. The model has a deterministic part and a stochastic part. The model input is the camera estimated pose. In this work, the estimated pose (drifted pose) is unknown and needed to be found, thus the model can not be used as

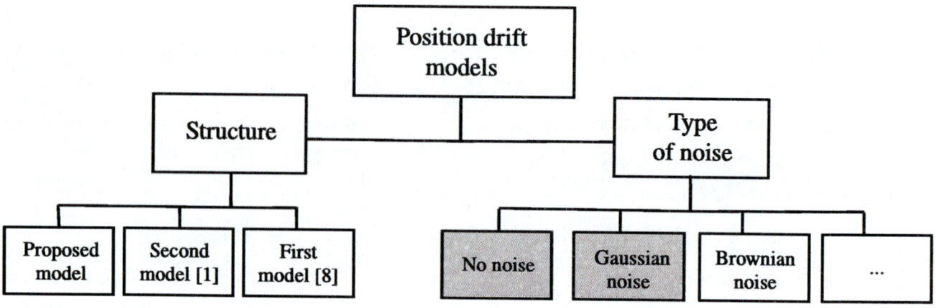

Fig. 1 General scheme for the building of the drift model scheme. In gray, the type of noise used in simulations

it is. Meanwhile, this model is modified slightly and used in the navigation toolbox of Matlab® to illustrate an example of visual-inertial odometry (VIO) [10]. In the example, just the zero mean stochastic part is used to model the uncertainty in the scaling between consecutive frames of the monocular camera. The equations are given as follows:

$$b(t) = \left(1 - \frac{1}{\tau}\right) b(t-1) + W_a(t), \quad \text{(1a)} \quad \text{drift}(t) = b(t) + W_n(t), \quad \text{(1b)}$$

$$\begin{bmatrix} X_M(t) \\ Y_M(t) \end{bmatrix} = scale \begin{bmatrix} X_T(t) \\ Y_T(t) \end{bmatrix} + \text{drift}(t), \quad \text{(1c)} \qquad \Psi_M(t) = \Psi_T(t), \quad \text{(1d)}$$

where $W_a(t) \in \mathbb{R}^{2\times1}$ is a white noise of the GM process with zero mean and $\sigma_a = \sigma_b\sqrt{\frac{2}{\tau} + \frac{1}{\tau^2}}$. The process noise $W_n(t) \in \mathbb{R}^{2\times1}$ is white noise with zero mean and variance σ_n. The model numerical parameters are given as follows: $\tau = 232$, $\sigma_b = \sqrt{1.34}$, $\sigma_n = 0.139$, $scale = 1.2$. As shown in the equation (1c), the actual position is used to estimate the drifted position which is ideal in our context. Note that the yaw angle is not affected by the noise.

Remark: As in [10], the sampling frequency of the model 25 Hz. However, in the simulations (see Sect. 2.4) the frequency is increased 100 Hz.

2.2 *Second Drift Model: Bazeille et al. [1]*

A simple VO localization model is proposed in [1]. This model is able to characterize the erroneous localization given by the VO. This model is based on vehicle position integration, where a reliable system is used to calculate the real current position $(X_T(t),\ Y_T(t))$. The displacement $(\Delta_x(t),\ \Delta_y(t))$ between the current and previous positions is then deduced, using $[m(t),\ o(t)]^T$ as model input.

The algorithm uses the last position added with the displacement $m(t)$ to retrieve the current position. However, the displacement is changed by adding noise. For the

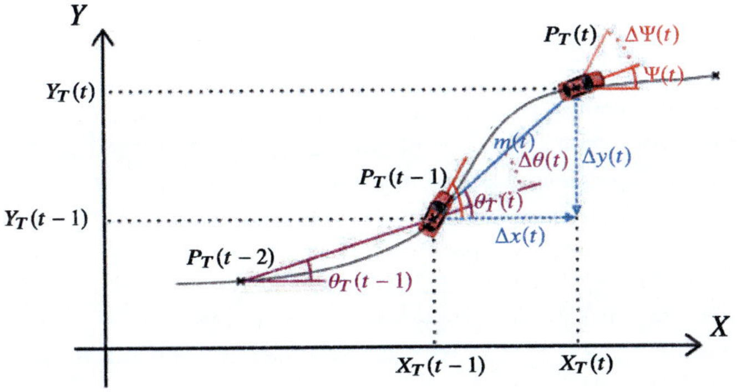

Fig. 2 Vehicle displacement

yaw angle also a noise is added. This noise is integrated while the vehicle is moving and generates an unbounded drift. Variables nomenclature are summarized in Fig. 2 and model equations are presented below:

$$X_M(t) = X_M(t-1) + (m(t) + Kn_m)\cos(\Psi_M(t-1)), \tag{2a}$$
$$Y_M(t) = Y_M(t-1) + (m(t) + Kn_m)\sin(\Psi_M(t-1)), \tag{2b}$$
$$\Psi_M(t) = \Psi_M(t-1) + o(t) + Kn_o, \tag{2c}$$

with

$$\Delta_x(t) = X_T(t) - X_T(t-1), \quad (2d) \qquad m(t) = \sqrt{\Delta_x(t)^2 + \Delta_y(t)^2}, \quad (2g)$$
$$\Delta_y(t) = Y_T(t) - Y_T(t-1), \quad (2e) \qquad o(t) = \text{atan2}(\sin(\Delta_\Psi(t)), \cos(\Delta_\Psi(t))), \quad (2h)$$
$$\Delta_\Psi(t) = \Psi_T(t) - \Psi_T(t-1), \quad (2f)$$

where $m(t)$ is the Euclidean distance between the poses $P_T(t)$ and $P_T(t-1)$ in the 2D plan. However, $o(t)$ is a recalculated $\Delta_\Psi(t)$, for most of the cases $o(t) = \Delta_\Psi(t)$. The equation (2h) is added to avoid trigonometrical orientation problem when the vehicle turns more than 360 °C. The gain factor K adjusts noise amplitude. When $K = 0$, the pose $P_M(t)$ does not accumulate errors. Finally, n_m et n_o are two random variables where, n_m stands for noise on displacement (movement), n_o for noise on orientation:

$$n_m : \alpha \in \Omega \mapsto n_m(\alpha) \in \mathbb{R}, \qquad n_o : \alpha \in \Omega \mapsto n_o(\alpha) \in \mathbb{R}, \quad \text{with } \Omega \in \mathbb{R}. \tag{2i}$$

Remark: The model parameters must be adapted according to the chosen frequency. In the comparison Sect. 2.4 the frequency is fixed 100 Hz.

2.3 Third Drift Model: Proposed Model

In this paper, a third model is proposed based on the algorithm stated in Sect. 2.2. The main change deals with the direction of movement angle employed in (2a) and (2b). The angle $\Psi_T(t)$ is considered as the direction of movement and it is used to calculate $X_M(t)$ and $Y_M(t)$. However, as shown in Fig. 2 the yaw angle $\Psi_T(t)$ is not always the direction of movement $\Psi(t) \neq \theta_T(t)$. Therefore, small errors are induced without the interference of any noise. These small errors are considered as model intrinsic disturbances. For this reason, the angle $\theta_T(t)$ is introduced in the third model, and consequently the model input is $[\Delta_\theta(t),\ m(t),\ o(t)]^T$. The equations become as follows:

$$\theta_M(t) = \theta_M(t-1) + \Delta_\theta(t) + Kn_o, \tag{3a}$$
$$X_M(t) = X_M(t-1) + (m(t) + Kn_m)\cos(\theta_M(t)), \tag{3b}$$
$$Y_M(t) = Y_M(t-1) + (m(t) + Kn_m)\sin(\theta_M(t)), \tag{3c}$$

with

$$\theta_T(t) = \text{atan2}(\Delta_y(t), \Delta_x(t)), \tag{3d}$$
$$\Delta_\theta(t) = \theta_T(t) - \theta_T(t-1). \tag{3e}$$

It is worth to mention that the calculation of $\theta_T(t)$, given by (3d) is feasible, because the $P_T(t)$ is known.

Remark: The model parameters must be adapted according to the chosen frequency. In the comparison Sect. 2.4 the frequency is fixed 100 Hz.

2.4 Open Loop Comparison

Three open loop tests are performed in order to compare the outputs of the three drift models presented in Sect. 2. In the first test no noise is taken into consideration. In the second one, a gaussian noise is considered in the models. In the two tests, the same steering angle $\delta_f(t) = 5°$ and the same velocity ($v_x = 10$ m/s) are transmitted to the CarMaker® [7] (a realistic car simulator) vehicle model. The applied $\delta_f(t)$ provides a turning maneuver. In the third test, the same gaussian noise is applied with a constant steering angle $\delta_f(t) = 0$ deg. The applied $\delta_f(t)$ provides a straight line maneuver.

In all simulations, the vehicle actual pose $P_T(t)$ is assumed well known and it is directly given by the CarMaker vehicle dynamics equations. The pose $P_T(t)$ is then employed as input signals for the three drift models previously presented.

In all simulations, the Euclidean position error $e_p(t) = \sqrt{(X_T(t) - X_M(t))^2 + (Y_T(t) - Y_M(t))^2}$ and the yaw angle error $e_\psi(t) = \Psi_M(t) - \Psi_T(t)$ are shown.

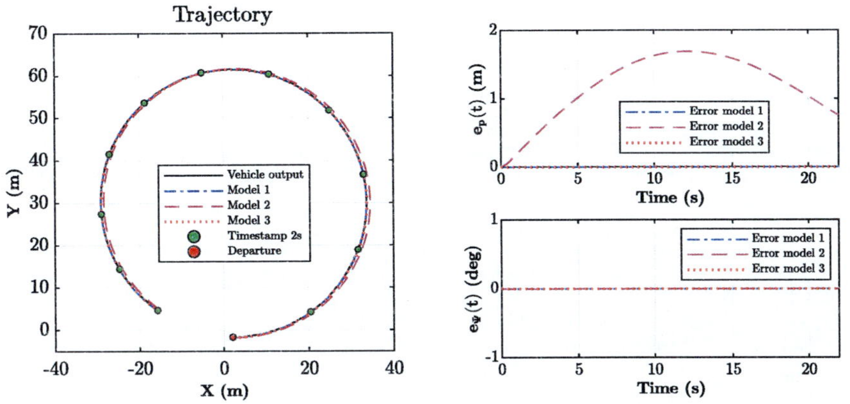

Fig. 3 Vehicle position and drift models outputs (without added noise) in open loop at $v_x = 10$ m/s

Remark: The sampling frequency is chosen 100 Hz which is suitable for a control loop. While increasing the frequency of the model 1 (25 Hz 100 Hz), the amplitude of the added noise becomes higher than the displacement of the vehicle. For this reason, the parameters have been modified as follows: $\sigma_b = \frac{\sqrt{1.34}}{100}$, $\sigma_n = \frac{0.139}{10}$.

First test: without adding noise. For the first model, the $scale = 1$ and $drift(t) = 0$ are considered. For the second and third models the gain $K = 0$ is considered.

The trajectories outputted by the drift models are depicted in Fig. 3. The Euclidean position error $e_p(t)$ and the yaw angle error $e_\psi(t)$ are also plotted. The noise is normally set to zero for all models and the expected position error must be zero. This last behavior is verified for model 1 and 3. However, the model 2 presents a non null position error $e_p(t)$, that is due to the intrinsic disturbances added by the model.

Second test: with gaussian noise. Parameters of the first model are taken as described in Sect. 2.1 with the added modifications. The second and third model parameters are considered as follows: $K = 0.1$, $n_m \sim \mathcal{N}(\mu_1, \sigma_1^2)$ where $\sigma_1 = 0.14$, $\mu_1 = 0.025$ and $n_o \sim \mathcal{N}(\mu_2, \sigma_2^2)$ where $\sigma_2 = 0.003$, $\mu_2 = -0.001$. The choice of the this parameters is justified in [1].

The trajectories of the three models exhibit a position drift with respect to the vehicle true position, as illustrated in Fig. 4. The position error $e_p(t)$ shows that this drift increases over time. Note that the error $e_p(t)$ of model 1 is decreasing towards the end. This is due to the fact that the trajectory is a circle and the cumulative error is compensated. For the same reason, models 2 and 3 show a slowing down of the error growth between 8 s and 15 s. The model 1 shows no yaw drift $e_\psi(t) = 0$, unlike the other models which show an increasing yaw drift. As a result, it is important to note that the proposed model has no intrinsic error.

Third test: with gaussian noise. The parameters of the three models are the same as the second test. The steering angle is different witch gives an other trajectory. As

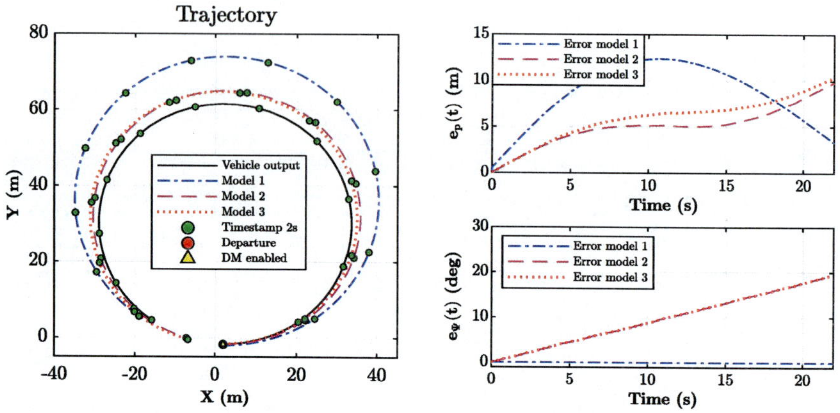

Fig. 4 Vehicle position and drift models outputs (with white noise) in open loop at $v_x = 10\,\text{m/s}$

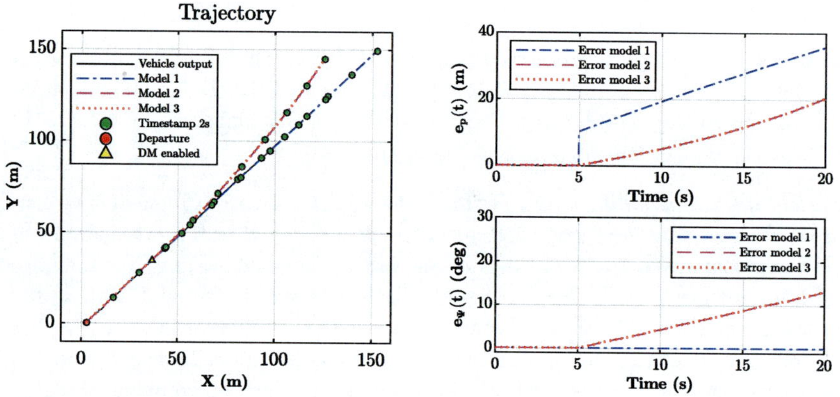

Fig. 5 Vehicle position and drift models outputs (with white noise) in open loop at $v_x = 10\,\text{m/s}$. The three drift models are enabled at 5 s

depicted in Fig. 5, the first 5 s no drift is noticed because the models are disabled. At 5 s, the 3 models are enabled. A large discontinuity is noticed in the results of model 1. In fact, in Eq. (1c) the position is directly multiplied to a scale constant, witch gives such discontinuity. Meanwhile, the model 2 and 3 start to accumulate errors as expected. For this trajectory, the model 2 presents unnoticeable intrinsic errors.

3 Closed Loop Simulation and Results

While the first model presents undesirable discontinuity when its enabled, the second model suffers from undesirable intrinsic error. For this reasons, only the proposed model is chosen to perform closed loop tests.

A simulation scenario is built with the following conditions: at first, the measured pose of the vehicle is the true position. This latter is used to follow a trajectory at constant speed of $v_x = 10$ m/s. After some time, the drift model is enabled. In this way the localization system is deteriorated and the measured pose drifts. After a while, the measured and true pose becomes identical. The guidance and control diagram is shown in Fig. 6.

The controller main objective is to produce the right steering wheel angle to follow the desired path. The lateral controller is provided by a robust state feedback coupled to a feed-forward based on the bicycle model as proposed in [2]. The longitudinal controller is provided by CarMaker. In general, the front steering angle $\delta_f(t)$ is calculated to reduce both lateral error $\epsilon_y(t)$ and orientation error $\epsilon_\Psi(t)$, defined as follows:

$$\dot{\epsilon}_y(t) = v_y(t) + v_x(t)\epsilon_\Psi(t)\ , \quad \epsilon_\Psi(t) = \Psi(t) - \Psi_{ref}(t) \tag{4}$$

where v_x and v_y represent longitudinal and lateral velocities at the center of gravity (CoG) and Ψ_{ref} is the desired yaw angle.

Three path following simulations are performed using the dataset of the track of the Mulhouse Automobile Museum. The initial vehicle position includes a lateral offset of ≈2.1 m. This lateral offset is added to see the ability of the guidance and control system to generate the appropriate steering angle to join the desired path.

For the simulation 1, the switch in Fig. 6 is at position 1, which means that the measured $P_M(t)$ and true $P_T(t)$ positions are identical. The same experiment is repeated in simulation 2. However, during the time 9–15 s the switch is in position 2, which means that the measured position is now affected by the drift and is no longer

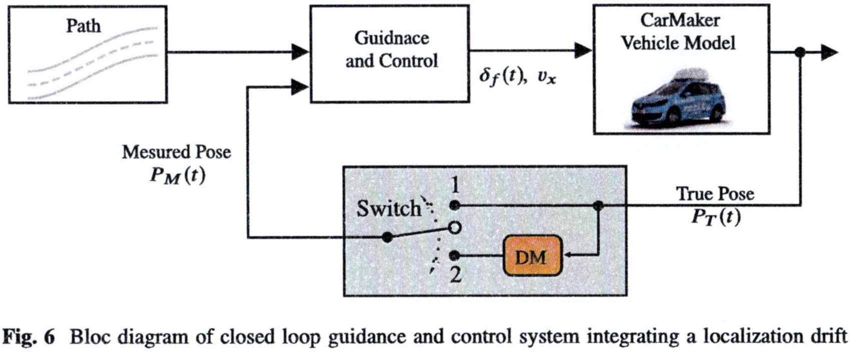

Fig. 6 Bloc diagram of closed loop guidance and control system integrating a localization drift model (DM)

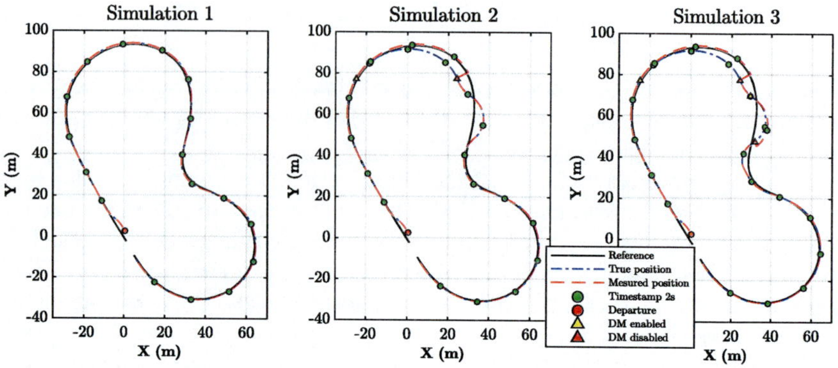

Fig. 7 Trajectory of the vehicle while following the track of the Automobile Museum of Mulhouse. $v_x = 10$ m/s

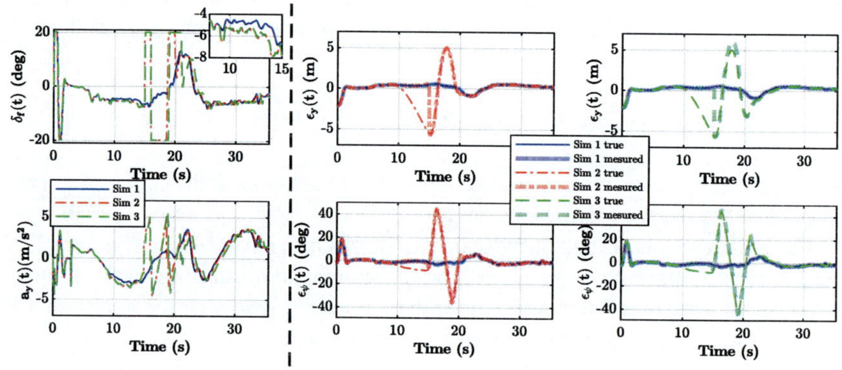

Fig. 8 Control loop signals: steering angle $\delta_f(t)$, lateral error $\epsilon_y(t)$, orientation error $\epsilon_\psi(t)$ and lateral acceleration $a_y(t)$

the actual position. The simulation 3, two switching actions are applied, during the time 9–15 s and from 16 s to 19 s the switch is in position 2. The simulations results are depicted in Figs. 7 and 8.

The simulation 1 results show that the measured and true signals are identical and the path following is carried put as expected. In simulation 2, the results are the same as in simulation 1 except when the drift model is activated. In this last case, the controller generates a steering angle to ensure that the vehicle is on the trajectory, with wrong position measurements. As a consequence, the vehicle deviates from the desired trajectory. Finally, an undesirable measurement discontinuity happens when the true position is recovered. As shown in Fig. 8, the steering angle is saturated due to the large lateral and orientation error caused by the discontinuity. Also, high lateral accelerations are measured. In the simulation 3, these discontinuities happen twice, witch cause longer saturation and larger following errors. Note that, the performances of the controller are deteriorating under this discontinuity.

4 Conclusion

This paper presents a comparison between three different localization drift models based on position measurements. The first model combines a Gauss-Markov process with a white noise. The second one is a reformulation of vehicle displacement equations combined with a white noise. Finally, a third model is proposed here based on the improvements made to the second model to eliminate its intrinsic errors. A guidance and control scheme is implemented, including the third drift model. Then, the closed loop simulation results highlight the discontinuity phenomenon caused by erroneous position measurements when varying the performance of the localization. In future work, the proposed model will be used to test and validate control strategies that take into account measurement discontinuity.

Acknowledgements The authors gratefully acknowledge the National Automobile Museum of Mulhouse (www.musee-automobile.fr) for the access to its test track.

References

1. Bazeille, S., Josso-Laurain, T., Ledy, J., Rebert, M., Al Assaad, M., Orjuela, R.: Characterization of the impact of visual odometry drift on the control of an autonomous vehicle. In: IEEE Intelligent Vehicles Symposium (IV), p. 2020. Las Vegas, NV, USA, October (2020)
2. Boudali, M., Orjuela, R., Basset, M.: A comparison of two guidance strategies for autonomous vehicles. IFAC-PapersOnLine **50**, 1 (2017)
3. Brossard, M., Bonnabel, S.: Learning wheel odometry and IMU errors for localization. In: International Conference on Robotics and Automation (ICRA), Montreal, Canada (2019)
4. Chirca, M.: Perception pour la navigation et le contrôle des robots mobiles. Application à un système de voiturier autonome. Ph.D. thesis, Université Blaise Pascal - Clermont II (2017)
5. Cioffi, G., Scaramuzza, D.: Tightly-coupled fusion of global positional measurements in optimization-based visual-inertial odometry. In: IEEE/RSJ International Conference on Intelligent Robots and Systems (IROS), Las Vegas, NV, USA (2020)
6. Forster, C., Carlone, L., Dellaert, F., Scaramuzza, D.: On-manifold preintegration for real-time visual-inertial odometry. IEEE Trans. Robot. **33**(1), 1–21 (2017). February
7. IPG Automotive GmbH. Reference Manual Version 10.1 (www.ipg-automotive.com)
8. Jiang, R., Klette, R., Wang, S.: Modeling of unbounded long-range drift in visual odometry. In: Fourth Pacific-Rim Symposium on Image and Video Technology (PSIVT), p. 2010. Singapore, Singapore (2010)
9. Mao, N., Jiangning, X., Li, J., He, H.: A LSTM-RNN-based fiber optic gyroscope drift compensation. Math. Prob. Eng. **2021**, 1636001 (2021). July
10. Matlab MathWorks. Visual-Inertial Odometry Using Synthetic Data - MATLAB & Simulink - MathWorks

11. Pendleton, S., Andersen, H., Xinxin, D., Shen, X., Meghjani, M., Eng, Y., Rus, D., Ang, M.: Perception, planning, control, and coordination for autonomous vehicles. Machines **5**(1), 6 (2017). February
12. Quinchia, AG., Falco, G., Falletti, E., Dovis, F., Ferrer, C.: A Comparison between different error modeling of MEMS applied to GPS/INS integrated systems. Sensors **13**(8) (2013)
13. Reid, T.G.R., Houts, S.E., Cammarata, R., Mills, G., Agarwal, S., Vora, A., Pandey, G.: Localization requirements for autonomous vehicles. SAE Int. J. Connect. Autom. Veh. **2**(3) (2019)
14. Vieira, D., Orjuela, R., Spisser, M., Basset, M.: Positioning and Attitude determination for Precision Agriculture Robots based on IMU and Two RTK GPSs Sensor Fusion. Munich, Germany (2022)
15. Zhang, F., Wang, Z., Zhong, Y., Chen, L.: Localization error modeling for autonomous driving in GPS denied environment. Electronics **11**(4), 647 (2022)

Study of Testing Strategy for Performance Analysis of Actuator Layer in Safety Instrumented System

Walid Mechri and Christophe Simon

Abstract This paper analyzes the degradation of actuator layer in Safety Instrumented System (SIS) by considering an intermediate degraded state between the working-and failed states. Sometimes, the current system states are not distinguished perfectly during proof tests. The developed approach consists in using the knowledge of both the system's functioning and, proof testing nature and its parameters. These parameters increase the complexity of the model when considering the unavailability due to the proof test. This approach has been applied to a study case in order to illustrate the effect of proof testing parameters on the SIS performance.

Keywords Dynamic Bayesian network · Unavailability · Safety instrumented system · Proof testing

1 Introduction

Safety Instrumented System (SIS) are usually installed in process industries to protect either people or the environment from real risks involved by dangerous situations such as overpressure, toxic gas, fire, etc.) [2]. The aim of SIS is to bring suitable functions that allow to maintain a safe state of an equipment under control (EUC), in case of a dangerous event occurrence [2]. A SIS is a three layers-system which is made of a sensor layer, a logic solver layer and an actuator layer. This structure aims to take the EUC to a safe state when predetermined conditions are violated.

The requirements of safety function exhibited in [2] introduces a probabilistic measure for the quantitative assessment of the SIS performance. In this context, IEC

W. Mechri
Laboratoire de Recherche MACS, LR16ES22, Université de Gabès, Ecole Nationale d'Ingénieurs de Gabès, 6029 Gabès, Tunisie
e-mail: walid.mechri@isim.rnu.tn

C. Simon (✉)
Université de Lorraine, CRAN CNRS UMR 7039, 54000 Nancy, France
e-mail: christophe.simon@univ-lorraine.fr

D. Theilliol et al. (eds.), *Recent Developments in Model-Based and Data-Driven Methods for Advanced Control and Diagnosis*, Studies in Systems, Decision and Control 467,
https://doi.org/10.1007/978-3-031-27540-1_18

61508 [2] recommends the average probability of failure on demand (PFD_{avg}) as a performance index in the low demand mode. In this demand mode, some failures remain hidden until a proof test is executed or an undesired event occurs on the EUC [7]. These hidden failures can lead to dangerous events with severe consequences. To improve the availability and to increase the failure tolerance, redundant structures are often considered. For example, N redundant shutdown valves are installed on the same pipeline to stop the flow and relieve pressure in case the downstream pressure is too high. When at least one of the N valves works, the EUC is still safe because the SIS works. This redundancy is named 1-out of-N (1ooN) [10].

Modeling a SIS unavailability can be carried out using suitable methods. These methods include fault trees, Markov chains [5], Petri Nets [9], and Dynamic Bayesian Network (DBN)[1]. Indeed, the parameters related to multi-states components in a SIS and the proof test strategy, can be cumbersome when dealing with Markov chain. In this context, the DBN method is an interesting formal tool to model dynamic structure of SIS by considering the events met (failure, degradation, repair) as well as their reliability parameters [7]. The DBN has the ability to also model the dynamic effects associated with the proof test strategy which sometimes induces a modeling complexity and impacts the performance. The DBN model can thus be a powerful tool for multi-state SIS unavailability assessment considering test as exogenous variables.

The remainder of this paper is organized as follows: Sect. 2 illustrates the characteristics of actuator layer in a SIS, as well as the proof test strategy. Section 3 investigates the computation of instantaneous unavailability of a 1oo3 layer structure of a SIS with the proposed test strategy. Section 4 conducts a numerical example to present the performance variation of actuator layer in safety system according to the proof test parameters.

2 SIS Structure and EUC

2.1 SIS Structure

Figure 1 shows the whole structure of a typical SIS that monitors an EUC (chemical reactor). Several architectures of SIS sub-layers are considered. Actuators (FC) and Logic Solvers (LS) are in 1oo3 voting architectures. The sensor layer is divided in two parts (temperature and pressure) each in 1oo2 voting architecture.

For the sake of illustration, this paper considers only the actuator part as a 1oo3 architecture. It is constituted of three identical valves whose goal is to stop the flow and relieve pressure in case the downstream pressure is too high. The process (EUC) is safe when at least one of the three valves works on demand.

Usually, the actuators are subject to degradation phenomena (wear out) given their mechanical property which are the research objective of degradation in SISs and deserve more attentions [10]. The analysis in this paper refers the test function

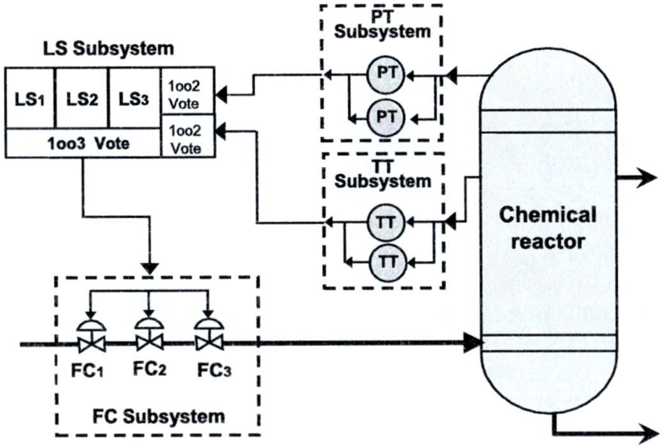

Fig. 1 Example of a typical SIS

for the redundant actuator layer. Therefore, only the actuator subsystem has been considered in this study without loss of generality.

The PFD_{avg} concept is used to qualify SIS operated in low demand mode. In the PFD_{avg} computation to take into account several characteristics like : failure rates, diagnostic coverage (DC), common cause failure (CCF) factor. In addition, testing strategy must be included in the unavailability assessment process of SIS. Proof testing can be executed through many test strategies [6].

2.2 *Proof Test Strategies*

A test is a repetitive activity that can be classified as partial or complete, i.e. applied to all or part of the elements of a SIS. A partial test allows to test some of the components while keeping the availability of the SIF, but with a temporarily degraded performance [3]. For the verification of SIS several proof tests strategies have been defined, in this context Torres-Echeveria et al. [8] proposes a classification of test strategies:

- *Simultaneous test*: All the components are tested together. This requires having a sufficient number of repairmen to test all components. The SIS is made unavailable during simultaneous test [4].
- *Sequential test*: All the components are tested consecutively one after the other. Just after a component is tested and put into service, the next component is tested and so on. This strategy considers that other components are always working [7].
- *Staggered test*: All the components of SIS are tested with their own period of time. This is the most common staggering test. This strategy of test is called uniform staggered test. It is assumed that the other components are working.

- *Random test*: In this strategy, the time interval of test of a component does not follow a specific program, the time interval between two tests of components is randomly chosen [8] or computed according to the current state.

Considering all elements or a part of them for a test defines the strategy. For a redundant layer's elements, testing all or testing one of them are the opposite solutions and allows or not the SIF availability [10]. Moreover, a test has some effectiveness, harmless and duration that impacts the performance index and the modeling effort.

The test is not always able to detect all failures, its ineffectiveness is modeled by ξ. This parameter represents the inability of the test to reveal undetected failures. Other characteristic parameters can be considered like the harmless: the probability of failure due to the test γ [7] and the test duration π.

Considering a non null test duration π and also component redundancy, alternate test strategies can be proposed. Better than testing the layer which renders the layer unavailable, one or more component but less than all of them can be tested. The components under test are taken out of service but the layer remains available. So, the layer structure changes during the test and the unavailability modeling should take it into account to compute the performance index.

3 Performance Analysis Model

Dynamic Bayesian Network (DBN) is an interesting approach that can be used to assess the performance of a SIS [7]. When using DBN, it is possible to make a dynamic analysis of the SIS in each of its functioning phases i.e. by considering the structure modifications. The state of the tested layers are determined given the functioning phase (Normal, Under test) and the test strategy chosen. The probability distribution on the layer states should be reallocated given the structure modification. It represents the main modeling effort.

3.1 Assumptions

For the unavailability modeling the following assumptions are considered:

- All failures follow the exponential distribution.
- After repair, all the SIS components are considered in a state "As Good As New".
- After proof test, the SIS components are "As Good As New" or in undetected failure consequently to the test.
- The β-factor model is used to express CCFs.
- The safe failures are not treated. All failures are dangerous.

These assumptions have for goal to develop a detailed model for having an accurate assessment of the SIS performance (PFD_{avg}).

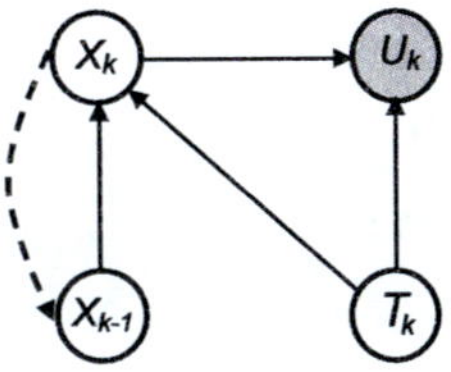

Fig. 2 DBN for unavailability modelling

3.2 *Performance Analysis Modelling*

Considering the system architecture, the proof test state and its characteristics, it is possible to define a DBN model which can be easily simulated, as described in Fig. 2. The gray node U_k represents the unavailability of the modeled layer according to the state probability of X_k representing the layer state at time $k\Delta T$ and the state of the proof test T_k.

Depending on the state of the test there may be several ways to redistribute the probability distribution between states of two consecutive phases due to state of the proof test (T_k=0, $T_k = 0 \rightarrow 1$, $T_k = 1$, $T_k = 0 \rightarrow 1$).

The instantaneous unavailability is computed by summing the probabilities on the states where the layer cannot fulfill its mission when the EUC demand occurs [5]. The performance index (PFD_{avg}) is determined by a discrete time integration of the instantaneous unavailability U_k.

3.3 *Study of a 1oo3 Structure*

The 1oo3 structure is composed of three channels functionally connected in parallel, operating in active redundancy. This structure remains operational until at least one channels is operational [3]. In this study, an alternate testing strategy is adopted, at a fixed frequency. In fact, during test, only one component out of three is tested. The component under test is taken out of service.

In this case, the 1oo3 structure goes to a 1oo2 structure and goes back to a 1oo3 structure after the proof test. In a DBN model, the stochastic process represented by the relation between X_{k-1} and X_k is a Markov model given by Fig. 3a. Considering 3 components and 3 states by components (OK, DD, DU), 27 states should be considered. To reduce the complexity, states grouping can be operated. By the way, the number of states is reduced to 12 but none of the components can then be identified. When testing a particular component, it becomes impossible to allocate specifically the state of the tested component. Thus, three 1oo2 structures that share the previous probability distribution should be considered (cf. Fig. 3b).

The change of the structure from 1oo3 to three 1oo2 through state of proof testing T_k, is realized as follows:

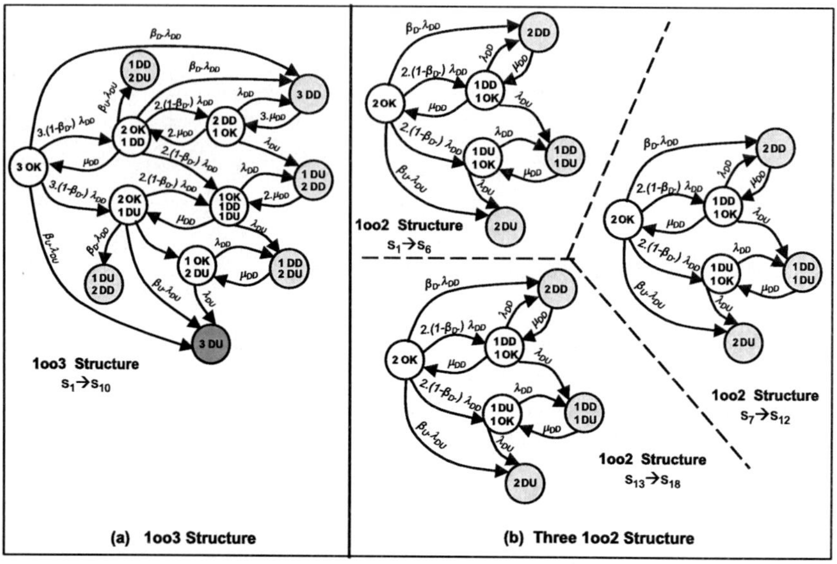

Fig. 3 Cycle of model structures from 1oo3 to a set of three 1oo2

Table 1 CPT of 1oo3 system when ($T_k = 0 \rightarrow 1$)

T_k	X_{k-1}	X_k					
		{2OK}	{1OK,1DD}	{1OK,1DU}	{2DD}	{1DD,1DU}	{2DU}
$0 \rightarrow 1$	{3OK}	1/3	0	0	0	0	0
	{2OK,1DD}	1/6	1/6	0	0	0	0
	{2OK,1DU}	1/6	0	1/6	0	0	0
	{1OK,2DD}	0	1/6	0	1/6	0	0
	{1OK,1DD,1DU}	0	1/9	1/9	0	1/9	0
	{1OK, 2DU}	0	0	1/6	0	0	1/6
	{3DD}	0	0	0	1/3	0	0
	{2DD,1DU}	0	0	0	1/6	1/6	0
	{1DD,2DU}	0	0	0	0	1/6	1/6
	{3DU}	0	0	0	0	0	1/3

- If the test state is not effective the 1oo3 structure and the associated Conditional Probability Table is determined by using the Markov model of Fig. 3a.
- When the test is launched ($T_k = 0 \rightarrow 1$), the state probabilities before the test are reallocated to three anonymous 1oo2 structures where one component is under the test and the others not. The relative CPT is given by Table 1.
- When the test is effective ($T_k = 1$), the 1oo3 is divided in three 1oo2 architectures. The CPT where the 1oo3 structure is under test as described in Table 2 using the Markov model of Fig. 3b.

Table 2 CPT of 1oo3 structure when ($T_k = 1$)

T_k	X_{k-1}	X_k								
		s_1	…	s_6	s_7	…	s_{12}	s_{13}	…	s_{18}
0	s_1	–	–	–	0	…	0	0	…	0
	⋮	–	1oo2	–	⋮	⋱	⋮	⋮	⋱	⋮
	s_6	–	–	–	0	…	0	0	…	0
	s_7	0	…	0	–	–	–	0	…	0
	⋮	⋮	⋱	0	–	1oo2	–	⋮	⋱	⋮
	s_{12}	0	…	0	–	–	–	0	…	0
	s_{13}	0	…	0	0	…	0	–	–	–
	⋮	⋮	⋱	⋮	⋮	⋱	⋮	-	1oo2	–
	s_{18}	0	…	0	0	…	0	–	–	–

- When the test ends ($T_k = 1 \rightarrow 0$), the probabilities are reallocated from three 1oo2 to one 1oo3 architecture by applying the CPT of Table 3. The CPT to apply at time k is given in Table 3.

Once the CPTs are defined (cf. Tables 1, 2 and 3) by using the testing strategy adopted, the DBN model of Fig. 2 is used to compute the instantaneous unavailability of 1oo3 structure according the states proof testing T_k.

4 Numerical Example

To illustrate the proposed model and testing strategy, a numerical example is proposed. The actuator part of the system presented in Fig. 1 is treated. For the simulations, the numerical values of the example are given on Table 4. The reader should remind that the paper focuses on the actuator layer for the sake of illustration without loss of generality. The actuator layer is made up of three valves and structured in 1oo3 architecture.

The proposed alternate testing strategy i.e.. one of three redundant valves is compared with the simple test strategy where the 3 components are tested at the same time. In both cases the same fixed test period is considered.

Figure 4 shows the variation of actuator layer unavailability and its average value PFD_{avg}, for different values of proof testing characteristics ξ, γ and π.

As shown in Fig. 4a, the PFD of layer increases to 1 for all test periods, this is explained by the fact that the 1oo3 architecture is unavailable along the test duration π where the all components of layer are tested simultaneously. As presented in Fig. 4b, the unavailability is less than in the previous case because of the non-simultaneous tests in the layer. This decrease is mainly due to the change of structure of the actuator

Table 3 CPT of 1oo3 structure when ($T_k = 1 \to 0$)

T_k	X_{k-1}	X_k									
		s_1	s_2	s_3	s_4	s_5	s_6	s_7	s_8	s_9	s_{10}
$1 \to 0$	s_1	$(1-\gamma)$	$\gamma.(1-\xi)$	$\gamma.\xi$	0	0	0	0	0	0	0
	s_2	0	$(1-\gamma)$	0	$\gamma.(1-\xi)$	$\gamma.\xi$	0	0	0	0	0
	s_3	0	0	$(1-\gamma)$	0	$\gamma.(1-\xi)$	$\gamma.\xi$	0	0	0	0
	s_4	0	0	0	$(1-\gamma)$	0	0	$\gamma.(1-\xi)$	$\gamma.\xi$	0	0
	s_5	0	0	0	0	$(1-\gamma)$	0	0	$\gamma.(1-\xi)$	$\gamma.\xi$	0
	s_6	0	0	0	0	0	$(1-\gamma)$	0	0	$\gamma.(1-\xi)$	$\gamma.\xi$

Table 4 Reliability parameters values

$\lambda_D(\times 10^{-6}/h) = 7$;	$DC = 0.4$;	$\text{MTTR} = 10\text{h}$;	$\text{T}_i(h) = 2190$;
$\xi = 0.4$;	γ=0.06;	$\pi = 20\text{h}$	

$\mu_{DD} = \frac{1}{MTTR}$

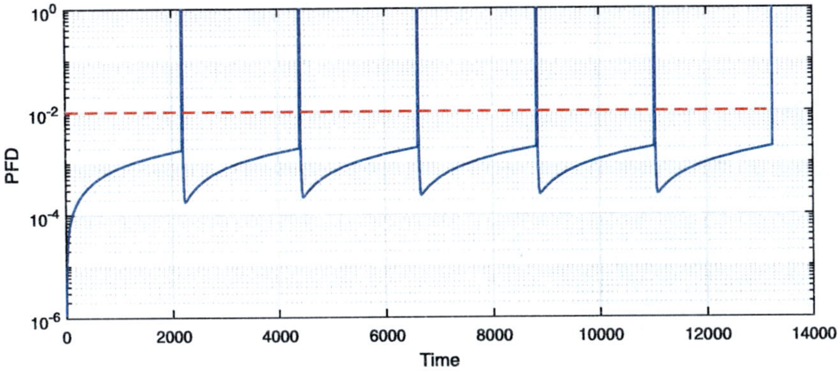

(a) PFD variation with simultaneous test

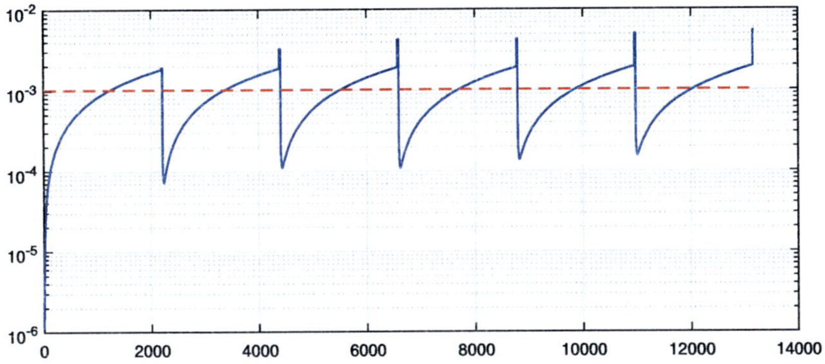

(b) PFD variation with sequential test

Fig. 4 PFD and PFD_{avg} of actuator layer with proposed testing strategy

layer. Indeed, by applying the proposed test strategy, the actuator layer goes from a 1oo3 structure to a 1oo2 structure. The variation of the PFD_{avg} due to a variation of test strategy is clearly visible. The value of PFD_{avg} varies from 1.0134×10^{-2} (Fig. 4a) to 0.10276×10^{-2} (Fig. 4b).

Here, discussions center on the effects of different proof test parameters on system performance and the strategy testing with parameters in Table 4. The variation PFD_{avg} is computed based on developed model, as presented in Fig. 5. This figure depicts

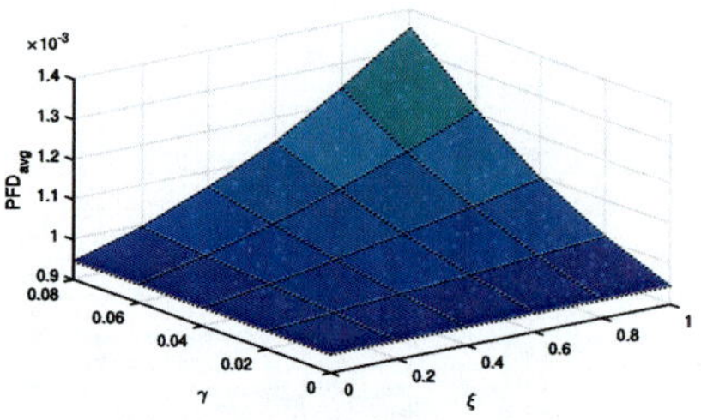

(a) PFD_{avg}f(γ, ξ) with π=20h

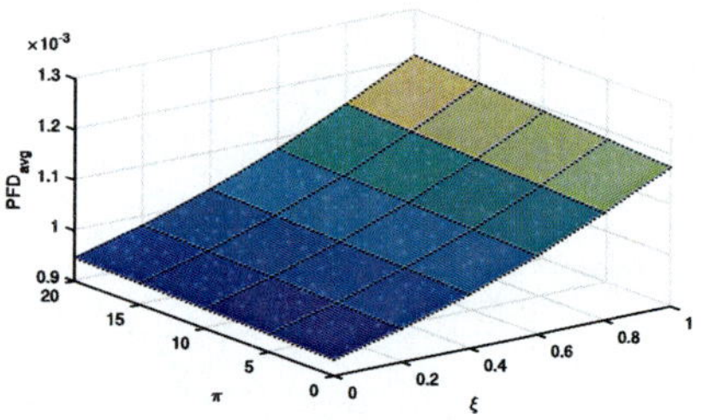

(b) PFD_{avg}f(ξ, π) with γ=0.06

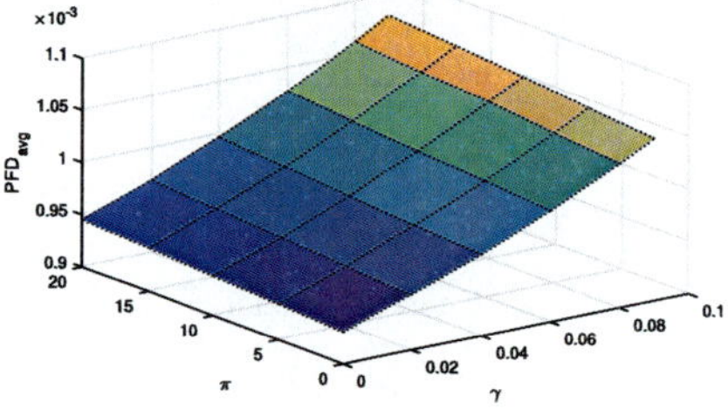

(c) PFD_{avg}f(γ, π) with ξ=0.4

Fig. 5 PFD_{avg} variation of actuator layer according to proof test parameters

3D-representations of the PFD_{avg} according to the parameters γ, ξ and π. Three cases of simulation are considered.

Obviously, the layer performance is sensitive to the characteristic values of the proof test and also the testing strategy. On one hand, γ and ξ each have a slight effect on the performance. On the other hand, their combined effect is particularly deleterious (cf. Fig. 5a). According to Fig. 5b the test duration π have less effect than the test inefficiency ξ or test harmless γ at the fixed test frequency. In third case (cf. Fig. 5c) the variation of PFD_{avg}, is presented according to γ and π with ξ constant.

Considering the previous remarks, it seems clear that the test strategy should focus more on detecting faults and not deteriorating the tested components better than working on reducing the test duration. Nevertheless, the test periods are critical because the structure modification is less fault tolerant and the most influent on the unavailability.

5 Conclusion

This paper studied the sensitivity of a SIS layer unavailability to proof test strategy and characteristics. For this purpose a DBN is proposed for its ability to model the impact of structure modification and considering test test as an exogenous variable.

A numerical example has been provided to illustrate the effect of the proposed testing strategy on the SIS performance. Based on the assumption, for a 1oo3 structure, we found that proposed strategy can contribute to a better system performance compared to simultaneous tests.

The DBN offers a simplified construction of the overall model for the assessment of the SIS performance. The integration of the proof tests parameters allowed the ability to formulate the behavior of the component block during the test and therefore configure the effect of the test on the SIS components.

The results obtained dependent on the assumptions made for the construction of DBN model. Currently, the failure rate were considered constant which is a usual assumption. However, it is necessary to develop a more precise DBN with the ability to model ageing effect on unavailability which is usually modeled by Weibull distributions. DBN remains a satisfying tools for this purpose.

References

1. Cai, B., Liu, Y., Fan, Q.: A multiphase dynamic bayesian networks methodology for the determination of safety integrity levels. Reliab. Eng. Syst. Saf. **150**(Supplement C), 105 – 115 (2016). https://doi.org/10.1016/j.ress.2016.01.018
2. IEC61508: Functional safety of Electrical/Electronic/Programmable Electronic Safety related systems. part 1-7 (2010)
3. Innal, F., Lundteigen, M.A., Liu, Y., Barros, A.: Pfdavg generalized formulas for sis subject to partial and full periodic tests based on multi-phase Markov models. Reliab. Eng. Syst. Saf. **150**, 160–170 (2016)
4. Mechri, W.: Evaluation de la performance des systèmes instrumentés de sécurité à paramètres imprécis. University of Tunis El-Manar, Tunisie, Thèse (2011)
5. Mechri, W., Simon, C., Ben Othman, K.: Switching Markov chains for a holistic modeling of sis unavailability. Reliab. Eng. Syst. Saf. **133**, 212–222 (2015)
6. Mechri, W., Simon, C., Bicking, F., Ben Othman, K.: Probability of failure on demand of safety systems by multiphase Markov chains. In: Conference on Control and Fault-Tolerant Systems (SysTol), pp. 98–103 (2013). https://doi.org/10.1109/SysTol.2013.6693839
7. Simon, C., Mechri, W., Capizzi, G.: Assessment of safety integrity level by simulation of dynamic bayesian networks considering test duration. J. Loss Prevent. Process Ind. **57**, 101–113 (2019)
8. Torres-Echeverria, A., Martorell, S., Thompson, H.: Multi-objective optimization of design and testing of safety instrumented systems with moon voting architectures using a genetic algorithm. Reliab. Eng. Syst. Saf. **106**, 45–60 (2012)
9. Wu, S., Zhang, L., Lundteigen, M.A., Liu, Y., Zheng, W.: Reliability assessment for final elements of siss with time dependent failures. J. Loss Prevent. Process Ind. **51**, 186–199 (2018)
10. Zhang, A., Wu, S., Fan, D., Xie, M., Cai, B., Liu, Y.: Adaptive testing policy for multi-state systems with application to the degrading final elements in safety-instrumented systems. Reliab. Eng. Syst. Saf. **221**, 108360 (2022)

Localization and Navigation of an Autonomous Vehicle in Case of GPS Signal Loss

S. N. Oubouabdellah, S. Bazeille, B. Mourllion, and J. Ledy

Abstract In recent years, autonomous vehicles have become an axis of academic and industrial research. Localizing these vehicles without a GPS signal represents a challenge for researchers, because the other sensors are usually less accurate, less fast and require more computation. Among localization methods, dead reckoning ones do not need prior knowledge as they are easier to implement for real time purposes. However, their biggest flaw is the accumulation of errors over time. In this work, we present an onboard localization method dedicated to autonomous vehicles for short time navigation without GPS. We developed a method with a high rate inertial-visual data fusion module that allows locating the vehicle in real-time. This method has been validated offline and tested online in a path following control loop on an experimental vehicle.

1 Introduction

The Global Positioning System (GPS) is one of the main localization methods when it comes to autonomous vehicles. However, there are situations such as tunnel, forest, dense city, interference problem, satellite incidents where GPS is no longer reliable or available.

S. N. Oubouabdellah (✉) · S. Bazeille · B. Mourllion · J. Ledy
IRIMAS, Université de Haute-Alsace, Brunstatt-Didenheim, France
e-mail: selma.oubouabdellah@uha.fr

S. Bazeille
e-mail: stephane.bazeille@uha.fr

B. Mourllion
e-mail: benjamin.mourllion@uha.fr

J. Ledy
e-mail: jonathan.ledy@uha.fr

D. Theilliol et al. (eds.), *Recent Developments in Model-Based and Data-Driven Methods for Advanced Control and Diagnosis*, Studies in Systems, Decision and Control 467,
https://doi.org/10.1007/978-3-031-27540-1_19

1.1 Context

Several researches have been conducted in the field of onboard sensors based localization for autonomous vehicles [1]. The most used technique is the Visual-Inertial-Odometry (VIO) [2, 3] because it relies on two sensors, camera and Inertial Measurement Unit (IMU), that have complementary properties. VIO can be applied with monocular cameras [4, 5] or stereo cameras [6, 7]. There are two types of data fusion: tight coupling [8] and loose coupling [9]. The tight coupling exploits raw data measured by sensors and is known for its robustness while the loose coupling works with processed data and is known for its simplicity to implement. Data fusion using the Kalman filter is widely used in the VIO context [10]. Mainly, the Extended Kalman Filter (EKF) [11, 12] but also the Multi State Constraint Kalman Filter (MSCKF) [13, 14].

1.2 Motivation

The IRIMAS institute works since many years on many topics linked to autonomous vehicles. One of its contributions is an efficient lateral controller based on the GPS. It was developed and validated on the experimental platform of the institute [15, 16]. Then, a Visual Odometry (VO) algorithm based on a monocular camera dedicated to smalls robots was proposed [17] to self-localize without any dependence on the infrastructure. This algorithm did not require unrealistic computation time and did not require prior knowledge on the environment. Considering these two working modules we decided to try to apply onboard localization to a bigger vehicle in a high rate control loop. The compatibility of the lateral controller linked to the drifting algorithm was studied [18] and tested in simulation. In this work, we show the results of the implementation of the onboard localization system in our experimental platform in real time in a control loop. We choose to implement a loosely coupled data fusion method based on an IMU, a monocular camera, and wheel velocity measurements. A Kalman Filter has been implemented to allow localization at the same rate of GPS. It does not need any prior knowledge and can replace the GPS for a short time when it is not available. This method has been tested offline on real data and on the experimental autonomous vehicle of the IRIMAS called ARTEMIPS.

1.3 Summary

The structure of the paper is the following: in the second section, the localization methods allowing the exploitation of IMU and camera measurements are presented as well as the proposed Kalman Fusion. In the third section, the offline evaluation of the method is presented and in the fourth section, the real-time experimentation results are displayed and discussed.

2 Localization Method

Dead reckoning is the process of calculating current position of moving vehicles by using a previously determined position, and then incorporating estimates of speed, heading direction over elapsed time. Dead reckoning is subject to cumulative errors so it is not accurate for long time localization compared with a GPS. To be able to calculate the vehicle's position using an IMU and a camera, we implemented inertial and visual odometry algorithms, and then we combined them with a Kalman filter. In this work, the rate of inertial measurement 100 Hz, the rate of the camera (1–10 frames per second), and the rate of the wheel 25 Hz.

2.1 Inertial Odometry (IO)

This technique [19, 20] allowed us to estimate the 3D position and orientation of a vehicle on North-East-Down (NED) frame by using IMU measurements which are the linear accelerations and the angle rates. Its principle consists first in calibrating the sensor, then projecting the angle rates from body to NED frame in order to obtain the vehicle's orientation. This orientation permits to project the linear accelerations to NED frame then to reduce the gravity components. Finally, the position of the vehicle is obtained by double integration.

2.2 Visual Odometry (VO)

The VO algorithm used in this work has been developed by Rebert et al. [17]. Its working principle consists first in detecting features from the pre-processed images using an improved version of Harris corners detector [21]. Then these features are matched between each two successive images and homography or a fundamental matrix is searched from the matched features. This matrix is then used to estimate the relative movement of the vehicle that allows obtaining its absolute position by adding to the previous position. In this work, we use a monocular camera with an image size of 2464×1300 pixels. As we are in a monocular case, we have to estimate our position up to a scale factor. In order to obtain this scale factor, we use the wheel velocity from the vehicle.

2.3 Visual-Inertial Data Fusion Using a Kalman Filter

A discrete Kalman filter has been implemented in order to combine the inertial and visual data. This filter allows us to estimate the state of a system and its variance thanks to its linear model by taking control and measurement as inputs. It works in two steps, the prediction step and the correction step:

- The prediction step uses the estimated state at the previous time and the control signal at the current time to estimate the current state. In our case, we take the inertial data as the control signal.
- The correction step uses the measurement at the current time to correct the predicted state in order to obtain a more accurate estimation. In our case, we take the visual data as the measurement.

As a reminder, in this work, the rate of inertial measurement (100 Hz) is faster than the rate of the camera (1–10 frames per second). Since the acquisition rates of inertial and visual data are different, the filter will repeat the prediction step using the inertial data until a visual data is available to perform the correction step.

First, a linear system of $n = 4$ order has been chosen to describe the vehicle dynamics. It is defined by the following state representation:

$$Z_{k+1} = \underbrace{\begin{bmatrix} 1 & 0 & dt & 0 \\ 0 & 1 & 0 & dt \\ 0 & 0 & 1 & 0 \\ 0 & 0 & 0 & 1 \end{bmatrix}}_{A_k} Z_k + \underbrace{\begin{bmatrix} 0 & 0 \\ 0 & 0 \\ dt & 0 \\ 0 & dt \end{bmatrix}}_{B_k} u_k$$

$$S_k = \underbrace{\begin{bmatrix} 0 & 0 & 1 & 0 \\ 0 & 0 & 0 & 1 \end{bmatrix}}_{C_k} Z_k$$

where:

- Z represents the state of the system:

$$Z = \left[X \ Y \ V_X \ V_Y\right]^T$$

  With:

  - (X, Y) is the position of the vehicle on NED frame.
  - (V_X, V_Y) are the linear velocities of the vehicle on NED frame.

- u represents the control which is defined as follows:

$$u = \left[a_X \ a_Y\right]^T$$

  - a_X et a_Y are the linear accelerations on NED frame obtained by IO.

- S represents the measured output. It is the linear speed of the vehicle on NED deduced from VO results.
- dt is the acquisition period of the inertial data.

In order to implement a Kalman filter on the system, we proceed as follows:

Initialization Step Initialization of the state and covariance matrix of the estimated state P and definition of the covariance matrices Q and R that express respectively the confidence in the model and the confidence in the measurement.
Prediction Step

- State prediction: $\hat{Z}_{k|k-1} = A_k\ \hat{Z}_{k-1|k-1} + B_k\ u_k$
- Predicted state's covariance estimation: $P_{k|k-1} = A_k\ P_{k-1|k-1}\ A_k^T + Q$

Correction step

- State error computation: $\varepsilon = S_k - C_k\ \hat{Z}_{k|k-1}$
- Innovation covariance matrix computation: $s = C_k\ P_{k|k-1}\ C_k^T + R$
- Kalman gain computation: $K_{f_k} = P_{k|k-1}\ C_k^T\ s^{-1}$
- Estimated state update: $\hat{Z}_{k|k} = \hat{Z}_{k|k-1} + K_{f_k}\ \varepsilon$
- Estimated covariance matrix update: $P_{k|k} = (I - K_{f_k}\ C_k)\ P_{k|k-1}$.

The proposed Kalman fusion allows estimating the position and the linear velocity of the vehicle in the NED frame but it does not give any information about the heading. For the real-time implementation, we used the heading estimated by inertial odometry. It was chosen because it is a bit more accurate than the one estimated by the visual odometry algorithm.

3 Offline Evaluation

The proposed localization methods have been evaluated using recorded inertial, visual and GPS data from the experimental vehicle of the IRIMAS institute. These measurements were recorded at the circuit of the automobile museum of Mulhouse.

3.1 *Experimental Platform*

ARTEMIPS, depicted in Fig. 1, is the autonomous vehicle of the IRIMAS institute. It is a Renault Grand Scenic 3 equipped with multiple sensors and actuators. The sensors used in this work are:

- High precision Oxford IMU RT-3002 with DGPS technology. This sensor will be considered as the reference position (ground truth).
- Manta G507C camera. Visual data are provided at the rate 10 Hz at best.
- Xsens Mti IMU. Inertial data are provided at the rate 100 Hz.
- Wheels velocity provided 25 Hz.

Fig. 1 Experimental vehicle of the IRIMAS institute: ARTEMIPS

ARTEMIPS is also equipped with 3 actuators: 2 integrated servo motors MAC141 to pilot the steering wheel and the brake of the vehicle, as well as a NI multi-function DAQ system to pilot the engine. In this work, only the steering wheel actuator is used; the engine and brake are manually controlled.

3.2 *Validation of the Localization Methods*

Figure 2 presents the results of the different localization methods tested on data recorded with ARTEMIPS. The reference trajectory of the vehicle is represented in blue, inertial odometry is depicted in red, visual odometry is shown in green and the Kalman localization is displayed in yellow. The trajectory begins at the Start point (in green), then makes a whole turn of the track to end at the Finish point (in red). We notice that the IO accumulates drift from the beginning to the end of the tests, while

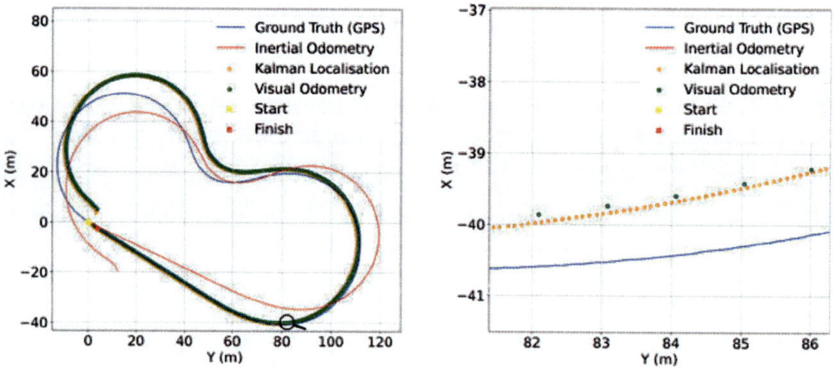

Fig. 2 Comparison of the localisation algorithms on a trajectory

Table 1 Representation of the localization errors

Errors	Inertial odometry	Visual odometry	Kalman fusion
Mean error (m)	8.49	4.58	4.31
Mean heading error (deg)	1.60	4.75	1.60
Maximal error (m)	19.21	10.49	10.54
Maximal heading error (deg)	5.15	11.57	5.15
Final error (m)	19.21	7.47	7.36
Final heading error (deg)	1.26	11.57	1.26

the VO accumulates more errors in turns than in lines. Let us also remark that IO works 100 Hz while VO works at a lower frequency (10 Hz at best). Finally, results presented in Fig. 2 shows that the Kalman fusion was able to estimate the vehicle trajectory with an accuracy that can be compared to the VO; the yellow and green layouts are almost overlaid. It is important to note that this precision is fulfilled at the same rate of the IO (100 Hz).

Results presented in Table 1 confirm the accuracy of the Kalman fusion. We notice that the Kalman fusion allows reducing the estimation errors of the visual and inertial odometries. Thus, in case of not favorable environment (bad weather conditions for example), the inertial estimations prevent high trajectory drift. Therefore, the proposed method provides 100 Hz estimation of the vehicle's localization without GPS while combining the accuracy obtained by VO and IO. Although this accuracy is not sufficient for long time autonomous car navigation (i.e. drift increases over time), it can be used in the context of an emergency stop when the GPS signal is lost.

4 Real-Time Implementation

After performing the validation process described in Sect. 3.2, the Kalman localization method has been implemented specifically for ARTEMIPS vehicle to be linked to the autonomous lateral controller [15], instead of the GPS one. The reference trajectory for real-time experimentation has been chosen as shown in Fig. 3 (the proposed method has been tested with a different reference trajectory and a different start point). Concerning the camera setup, an acquisition frequency of 3 images per second was adjusted. This configuration was required to allow the visual odometry to perform its task, because the average computation time of the real-time visual odometry is 0.193 s and higher frame rate would cause delays for image processing. The vehicle was forced to drive at lower speed than usual to reduce the distance between consecutive images. This is particularly critical during turns. To sum up, the speed profile has been adapted to the situation so we had a speed of 15 km/h in the straight

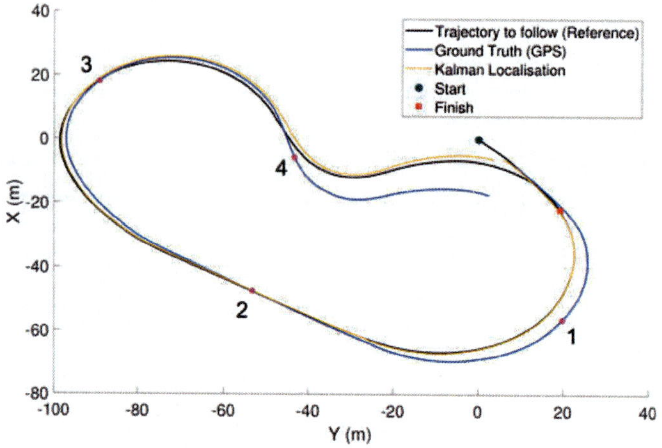

Fig. 3 Localisation results while following a recorded trajectory

Fig. 4 Images from the camera

line and 7 km/h in turns. Finally, in term of weather, the sky was partially cloudy which means that the conditions were favorable for the VO algorithm. Examples of images are shown in Fig. 4, they correspond to the magenta dots in Fig. 3.

Figure 3 presents the experimental results of the real-time implementation. We observe that the trajectory obtained thanks to Kalman fusion localization is accurate considering the real position of the vehicle (GPS) until moment 4. At that time, the accumulated drift was too important for the vehicle to stay on the road and the localization error was no longer acceptable. However, we let the vehicle continue its navigation after moment 4 in order to see the effect of the drift accumulation. This

Table 2 Representation of the localization errors

Errors	Inertial odometry	Visual odometry	Kalman fusion
Mean error (m)	370.00	3.87	3.72
Mean heading error (deg)	1.89	3.26	1.89
Maximal error (m)	726.15	11.68	11.37
Maximal heading error (deg)	5.84	22.51	5.84
Final error (m)	726.15	11.37	11.34
Final heading error (deg)	0.97	7.61	0.97

resulted an important final error shown in Table 2. We also remark that the precision is better in the straight line than in turns which is due to the VO performances. It should also be noted that the IO accumulates a significant drift when the speed of the vehicle is low because its acquisition frequency is high (100 Hz) so the vehicle appears stationary and the information will be hard to distinguish from the noise. Above all, this result demonstrates that the ARTEMIPS vehicle was able to drive the whole circuit of the automobile museum of Mulhouse with the use of the proposed localization and of the lateral controller. ARTEMIPS left the road during the 'S' maneuver of the circuit (image 4 of Fig. 4), at $t = 171$ s after travelling more than 500 m without GPS localization. To summarize, the proposed fusion method allowed the ARTEMIPS vehicle to drive for about 3 min in autonomous mode without using the GPS signal to locate itself. It has to be noted that this result was obtained with a not optimized version of the VO algorithm, so it will be improved in terms of frequency and accuracy in the close future.

In Table 2 we can see that the proposed Kalman fusion allowed us to obtain less estimation errors than VO and IO. These results support previously discussed ones in Sect. 3.2. Table 2 as they confirm that the IO position error is extremely important. Nevertheless, the heading obtained by IO is accurate with respect to the GPS. However, given the amount of error accumulated over time, this method is not accurate enough for long-time autonomous navigation (over 171 s in this case). It can be used for short term applications such as emergency stop when the GPS signal is lost or to replace this signal if it is disturbed and not available 100 Hz.

5 Conclusion

In this work, we have proposed a loosely coupled visual-inertial data fusion that has been evaluated offline and then validated with a real-time implementation. This method allowed improving the estimation accuracy obtained by visual odometry and inertial odometry as it permitted to get a localization frequency 100 Hz for the autonomous control mode of the ARTEMIPS vehicle. During this work, we also reached a total independence from the GPS signal. To conclude, this work constitutes a new step toward autonomy for the autonomous vehicle ARTEMIPS. In future works, the fusion method will be improved by using, for example, an Extended Kalman Filter on a non-linear system that describes better the vehicle dynamics. We will also optimize the VO algorithm in order to be able to increase the images acquisition frequency thus the vehicle speed.

Acknowledgements Thanks to ANR for financing the Evi-Deep support project. Thanks to Sébastien Jung who worked on the implementation of IO and VO during his internship in 2021. Thanks to the automobile museum that made the experimentation's circuit available.

References

1. Marden, S., Whitty, M.A.: Gps-free localisation and navigation of an unmanned ground vehicle for yield forecasting in a vineyard (2014)
2. Gui, J., Dongbing, G., Wang, S., Huosheng, H.: A review of visual inertial odometry from filtering and optimisation perspectives. Adv. Robot. **29**(20), 1289–1301 (2015)
3. Leutenegger, S., Lynen, S., Bosse, M., Siegwart, R., Furgale, P.: Keyframe-based visual-inertial odometry using nonlinear optimization. Int. J. Robot. Res. **34**(3), 314–334 (2015)
4. Qin, T., Li, P., Shen, S.: Vins-mono: a robust and versatile monocular visual-inertial state estimator. IEEE Trans. Robot. **34**(4), 1004–1020 (2018)
5. Tanskanen, P., Naegeli, T., Pollefeys, M., Hilliges. Semi-direct ekf-based monocular visual-inertial odometry. In: 2015 IEEE/RSJ International Conference on Intelligent Robots and Systems (IROS), pp. 6073–6078 (2015)
6. Usenko, V., Engel, J., Stückler, J., Cremers, D.: Direct visual-inertial odometry with stereo cameras. In: 2016 IEEE International Conference on Robotics and Automation (ICRA), pp. 1885–1892 (2016)
7. Sun, K., Mohta, K., Pfrommer, B., Watterson, Mi., Liu, S., Mulgaonkar, Y., Taylor, C.J., Kumar, V.: Robust stereo visual inertial odometry for fast autonomous flight. IEEE Robot. Autom. Lett. **3**(2), 965–972 (2018)
8. He, Y., Zhao, J., Guo, Y., He, W., Yuan, K.: Pl-vio: tightly-coupled monocular visual-inertial odometry using point and line features. Sensors **18**(4) (2018)
9. Al Bitar, N., Gavrilov, A.I.: Comparative analysis of fusion algorithms in a loosely-coupled integrated navigation system on the basis of real data processing. Gyroscopy and Navigat. **10**(4), 231–244 (2019)
10. Delmerico, J., Scaramuzza, D.: A benchmark comparison of monocular visual-inertial odometry algorithms for flying robots. In: 2018 IEEE International Conference on Robotics and Automation (ICRA), pp. 2502–2509 (2018)
11. Forster, C., Carlone, L., Dellaert, F., Scaramuzza, D.: On-manifold preintegration for real-time visual-inertial odometry. IEEE TRO **33**(1), 1–21 (2017)
12. Huai, Z., Huang, G.: Robocentric visual-inertial odometry. In: 2018 IEEE/RSJ International Conference on Intelligent Robots and Systems (IROS), pp. 6319–6326 (2018)
13. Li, Mingyang, Mourikis, Anastasios I.: High-precision, consistent EKF-based visual-inertial odometry. Int. J. Robot. Res. **32**(6), 690–711 (2013)
14. Ma, F., Shi, J., Yang, Y., Li, J., Dai, K.: Ack-msckf: Tightly-coupled ackermann multi-state constraint kalman filter for autonomous vehicle localization. Sensors **19**(21) (2019)
15. Boudali, M.T., Orjuela, R., Basset, M.: A comparison of two guidance strategies for autonomous vehicles. IFAC-PapersOnLine **50**, 12539–12544 (2017)
16. Laghmara, H., Boudali, M.T., Laurain, T., Ledy, J., Orjuela, R., Lauffenburger, J.P., Basset, M.: Obstacle avoidance, path planning and control for autonomous vehicles. In: 2019 IEEE Intelligent Vehicles Symposium (2019)

17. Rebert, M., Monnin, D., Bazeille, S., Cudel, C: Parallax beam: a vision-based motion estimation method robust to nearly planar scenes. JEI **28**(2) (2019)
18. Bazeille, S., Josso-Laurain, T., Ledy, J., Rebert, M., Al Assaad, M., Orjuela, R.: Characterization of the impact of visual odometry drift on the control of an autonomous vehicle. In: 2020 IEEE IV, pp. 2037–2043 (2020)
19. Titterton, D., Weston, J.L., Weston, J.: Strapdown Inertial Navigation Technology, vol. 17. IET (2004)
20. Vieira, D., Orjuela, R., Spisser, M., Basset, M.: Positioning and attitude determination for precision agriculture robots based on IMU and two RTK GPSs sensor fusion. In: 7th IFAC Conference Sensing, Control and Automation for Agriculture (2022)
21. Sánchez, J., Monzón, N., Salgado De La Nuez, A.: An Analysis and Implementation of the Harris Corner Detector, vol. 8 (2018)

Nonlinear Analysis of SRF-PLL: Hold-In and Pull-In Ranges

Tatyana A. Alexeeva, Nikolay V. Kuznetsov, Mikhail Y. Lobachev, Marat V. Yuldashev, and Renat V. Yuldashev

Abstract A Synchronous Reference Frame Phase-Locked Loop (SRF-PLL) is a nonlinear control circuit widely used in power engineering for the synchronization and control of three-phase grid-connected converters. New applications of power converters in sustainable energy generators have led to the problem of nonlinear analysis of SRF-PLL stability. In this paper, a continuous nonlinear model of SRF-PLL with a first-order loop filter is studied and the hold-in range of the model is analysed. Using the qualitative theory of dynamical systems and classical methods of control theory, we conduct a nonlinear analysis of the SRF-PLL model and suggest an analytical estimate for the pull-in range of the considered model. The obtained estimate is compared with known engineering estimates of the pull-in range. MATLAB Simulink is used to perform computer simulation which confirms theoretical results.

1 Introduction

Control problems related to synchronization in electrical networks are important for developing green energy and sustainable future. One of these problems is connection of a solar or wind generator, which produces direct current (DC) to a grid [1, 2]. The grid is usually an alternating current (AC) network, therefore it is necessary to use DC-AC power converter as part of the generator. It is important to measure grid AC voltage and synchronize the output voltage of the power converter to the main grid before connection to avoid power surges and related damage to the equipment. After

T. A. Alexeeva
HSE University, Kantemirovskaya st., 3A, Saint Petersburg, Russia

N. V. Kuznetsov (✉)
Institute for Problems in Mechanical Engineering RAS, V.O., Bolshoy pr., 61, Saint Petersburg, Russia
e-mail: nkuznetsov239@gmail.com

N. V. Kuznetsov · M. Y. Lobachev · M. V. Yuldashev · R. V. Yuldashev
St. Petersburg State University, University Embankment, 7/9, Saint Petersburg, Russia

D. Theilliol et al. (eds.), *Recent Developments in Model-Based and Data-Driven Methods for Advanced Control and Diagnosis*, Studies in Systems, Decision and Control 467,
https://doi.org/10.1007/978-3-031-27540-1_20

connection, monitoring of the frequency and phase of the grid allows controlling the balance of active and reactive power supplied to the network by the generator [3]. The most frequently used solution is Synchronous Reference Frame Phase-Locked Loop (SRF-PLL). This nonlinear control device monitors the grid's phase and frequency to control the inverter.

In the present paper, we study SRF-PLL model with a first-order loop filter. There are also several modifications of the SRF-PLL, which are based on some pre-filtering (adding filter before the loop) and in-loop filtering (adding filter inside the loop), such as the DDSRF-PLL [4], moving average filter (MAF) PLLs [5, 6], delayed signal cancellation [7], and numerous other methods (see, e.g., [8–10]). Adding pre-loop filter does not affect the performance (speed, noise, and stability) of the SRF-PLL itself, therefore it is omitted for simplicity. However, additional in-loop filtering requires to redo the whole study for every modification. Nonlinear stability analysis is especially difficult to carry over to SRF-PLLs with additional in-loop filtering and is beyond the scope of this article.

The paper is organized in the following way. In Sect. 2, a continuous time mathematical model of SRF-PLL is derived. Synchronization properties and stability of the corresponding system is studied in Sect. 3. An analytical estimate of global stability is obtained and plotted on a two-dimensional diagram for all SRF-PLL parameters. In Sect. 4, the analytical formula is compared with engineering estimates.

2 Mathematical Model of the SRF-PLL

Consider a simplified model of an inverter (Fig. 1) connected to a three-phase electrical grid [2, 3, 11] and corresponding phase (line-to-neutral) voltages

$$\begin{aligned} u_{\mathrm{a}}(t) &= u\sin(\omega_{\mathrm{ref}}t),\\ u_{\mathrm{b}}(t) &= u\sin(\omega_{\mathrm{ref}}t-\frac{2}{3}\pi),\\ u_{\mathrm{c}}(t) &= u\sin(\omega_{\mathrm{ref}}t+\frac{2}{3}\pi). \end{aligned} \tag{1}$$

Here ω_{ref} is the grid (reference) frequency and u is the Root-Mean-Squared voltage.[1] Denote the reference phase $\omega_{\mathrm{ref}}t$ by $\theta_{\mathrm{ref}}(t)=\omega_{\mathrm{ref}}t$. The measured line voltages $u_{\mathrm{abc}}=\left(u_{\mathrm{a}}\ u_{\mathrm{b}}\ u_{\mathrm{c}}\right)^T$ are converted into dq0 reference frame $u_{\mathrm{dq0}}=\left(u_{\mathrm{d}}\ u_{\mathrm{q}}\ u_{0}\right)^T$ by the Park transformation $u_{\mathrm{dq0}}=P(\theta_{\mathrm{PLL}})u_{\mathrm{abc}}$ where

[1] In the present work, we consider a SRF-PLL model, which assumes that the reference signal has constant frequency and amplitude. When the reference signal is unbalanced or has DC component or is distorted, some modifications of SRF-PLL can be considered, e.g., a three-phase EPLL (3EPLL) [2].

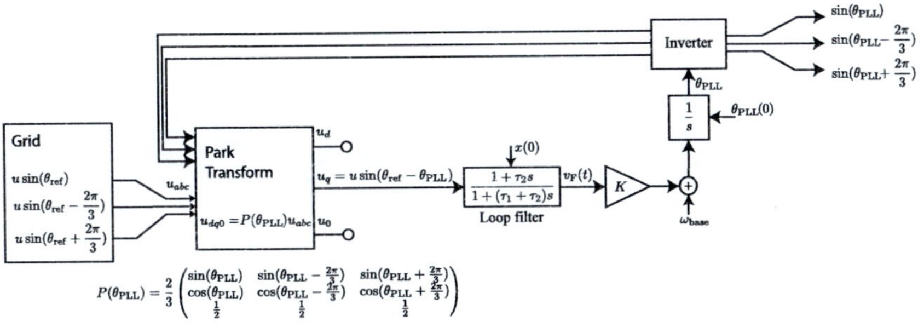

Fig. 1 Nonlinear model of SRF-PLL [2]

$$P(\theta_{\text{PLL}}) = \frac{2}{3}\begin{pmatrix} \sin(\theta_{\text{PLL}}) & \sin(\theta_{\text{PLL}} - \frac{2\pi}{3}) & \sin(\theta_{\text{PLL}} + \frac{2\pi}{3}) \\ \cos(\theta_{\text{PLL}}) & \cos(\theta_{\text{PLL}} - \frac{2\pi}{3}) & \cos(\theta_{\text{PLL}} + \frac{2\pi}{3}) \\ \frac{1}{2} & \frac{1}{2} & \frac{1}{2} \end{pmatrix}$$

and $\theta_{\text{PLL}} = \theta_{\text{PLL}}(t)$ is the phase of the inverter output voltage.

The q component of the transformed set of signals is the error signal

$$\begin{aligned} u_{\text{q}}(t) = &\frac{2u}{3}\Big(\sin(\theta_{\text{ref}})\cos(\theta_{\text{PLL}}) + \sin\left(\theta_{\text{ref}} - \frac{2}{3}\pi\right)\cos\left(\theta_{\text{PLL}} - \frac{2}{3}\pi\right) \\ &+ \sin\left(\theta_{\text{ref}} + \frac{2}{3}\pi\right)\cos\left(\theta_{\text{PLL}} + \frac{2}{3}\pi\right)\Big) \\ &= u\sin(\theta_{\text{ref}} - \theta_{\text{PLL}}). \end{aligned} \tag{2}$$

Denote by $\theta_e(t)$ the phase error:

$$\theta_e(t) = \theta_{\text{ref}}(t) - \theta_{\text{PLL}}(t). \tag{3}$$

The error signal $u_{\text{q}}(t)$ is an input of the first-order loop filter which transfer function has the form [12–14]

$$F(s) = \frac{1 + \tau_2 s}{1 + (\tau_1 + \tau_2)s}, \quad \tau_1 > 0, \ \tau_2 > 0. \tag{4}$$

Denote by $x(t) \in \mathbb{R}$ the state of the loop filter. The output of the loop filter $v_{\text{F}}(t) = \frac{1}{\tau_1+\tau_2}x + \frac{\tau_2}{\tau_1+\tau_2}u\sin\theta_e$ is used to control the frequency $\omega_{\text{PLL}}(t)$ of the inverter output voltage, which is proportional to the control voltage:

$$\omega_{\text{PLL}}(t) = \dot{\theta}_{\text{PLL}}(t) = \omega_{\text{base}} + K v_{\text{F}}(t), \tag{5}$$

where $K > 0$ is a gain and ω_{base} is a base frequency.

Combining (2)–(5) we get equations of the SRF-PLL:

$$\begin{aligned} \dot{x} &= -\frac{1}{\tau_1+\tau_2}x + \frac{\tau_1}{\tau_1+\tau_2}u\sin\theta_e, \\ \dot{\theta}_e &= \omega_e - K\Big(\frac{1}{\tau_1+\tau_2}x + \frac{\tau_2}{\tau_1+\tau_2}u\sin\theta_e\Big), \end{aligned} \tag{6}$$

where $\omega_e = \omega_{\text{ref}} - \omega_{\text{base}}$ is a frequency error.

3 SRF-PLL Stability Analysis

3.1 *Local Stability (Small-Signal Analysis)*

Observe that system (6) is 2π-periodic in θ_e. If $\omega_e < uK$, then it has an infinite number of equilibria $(x^{\text{eq}},\ \theta_e^{\text{eq}})$ which satisfy

$$x^{\text{eq}} = \frac{\tau_1\omega_e}{K}, \quad \sin\theta_e^{\text{eq}} = \frac{\omega_e}{uK}.$$

Equilibria of system (6) correspond to locked states of SRF-PLL (the phase error $\theta_e(t)$ is constant when the PLL is locked). In engineering literature, a *hold-in range* concept is widely used in order to characterize the ability of the loop to maintain phase-locked conditions when the frequency error ω_e varies slowly. Strict mathematical definition of the hold-in range can be found in [15–17].

System (6) has asymptotically stable equilibria $\big(\frac{\tau_1\omega_e}{K},\ \arcsin\frac{\omega_e}{uK} + 2\pi m\big)$ and unstable equilibria $\big(\frac{\tau_1\omega_e}{K},\ \pi - \arcsin\frac{\omega_e}{uK} + 2\pi m\big)$, and the hold-in range is $[0,\ \omega_h) = [0,\ uK)$.

3.2 *Global Stability (Large-Signal Analysis)*

To study the hold-in range it is sufficient to analyse the PLL system in vicinities of equilibria (local analysis). In PLL literature, concept of pull-in range is widely used in order to characterize global stability properties of the corresponding system. A *pull-in range* is the largest interval of frequency errors $|\omega_e| \in [0, \omega_p)$ from the hold-in range for which any trajectory of system (6) tends to an equilibrium (for brevity, we shall call such systems *globally stable*).

Notice that analysis of global stability requires consideration of the whole phase space and nonlinearity cannot be neglected. For global analysis of PLL models and estimation of the pull-in range various nonlinear methods can be used such as the direct Lyapunov method (see, e.g., [18, 19]), the method of two-dimensional comparison systems (see, e.g., [20]), phase portrait analysis (see, e.g., [21]), the harmonic balance method (see, e.g., [22, 23]), and others [24–26].

Theorem 1 *Pull-in range of model* (6) *satisfies the inequality* $\omega_p \geq \omega_p^{\text{est}}$ *where* $\omega_p^{\text{est}} > 0$ *is the unique solution of equation*

$$\arcsin \frac{\omega_p^{\text{est}}}{uK} + \sqrt{\left(\frac{uK}{\omega_p^{\text{est}}}\right)^2 - 1} = \frac{\pi \tau_1}{4(\sqrt{\tau_2(\tau_1 + \tau_2)} - \tau_2)}. \tag{7}$$

Proof To analyse the pull-in range of system (6), we apply the direct Lyapunov method and the corresponding theorem on global stability for the cylindrical phase space (see, e.g., [27, 28]). If there is a continuous function $V(x, \theta_e) : \mathbb{R}^2 \to \mathbb{R}$ such that
(i) $V(x, \theta_e + 2\pi) = V(x, \theta_e) \quad \forall x \in \mathbb{R}, \ \forall \theta_e \in \mathbb{R}$,
(ii) for any solution $(x(t), \theta_e(t))$ of system (6) the function $V(x(t), \theta_e(t))$ is nonincreasing,
(iii) if $V(x(t),\ \theta_e(t)) \equiv V(x(0),\ \theta_e(0))$, then $(x(t), \theta_e(t)) \equiv (x(0), \theta_e(0))$,
(iv) $V(x,\ \theta_e) + \theta_e^2 \to +\infty$ as $|x| + |\theta_e| \to +\infty$,
then any trajectory of system (6) tends to an equilibrium.

For system (6) we consider the following Lyapunov function satisfying condition (iv):

$$V(x,\ \theta_e) = \frac{1}{2}(x - \frac{\tau_1 \omega_e}{K})^2 + \frac{\tau_1}{K} \int_0^{\theta_e} \left(\sin\sigma - \frac{\omega_e}{uK} + \beta_0 |\sin\sigma - \frac{\omega_e}{uK}|\right) d\sigma$$

where

$$\beta_0 = -\frac{\int_0^{2\pi} (\sin\sigma - \frac{\omega_e}{uK})\, d\sigma}{\int_0^{2\pi} |\sin\sigma - \frac{\omega_e}{uK}|\, d\sigma} > 0.$$

The special form of coefficient β_0 allows the Lyapunov function to be 2π-periodic and, hence, to satisfy the first condition of the theorem. Computation of the Lyapunov function derivative along the trajectories of system (6), considering it as a quadratic form, and providing its negative-definiteness lead to fulfilling conditions (ii), (iii) and, hence, to global stability of the system. □

Notice that the left-hand side of (7) is a monotonous function and ω_p^{est} can be evaluated numerically. Moreover, Eq. (7) can be considered as an equation in two variables: $\frac{\tau_2}{\tau_1}$ and $\frac{\omega_p^{\text{est}}}{uK}$. Figure 2 shows stability regions observed in the second-order SRF-PLL. The shaded area bounded by curve (7) corresponds to global stability estimate by the Lyapunov function approach. Above the line $\frac{\omega_e}{uK} = 1$ no equilibria exist in system (6). For SRF-PLL parameters from the gap between the line $\frac{\omega_e}{uK} = 1$ and curve (7), system (6) has asymptotically stable equilibria, but the global stability is not guaranteed.

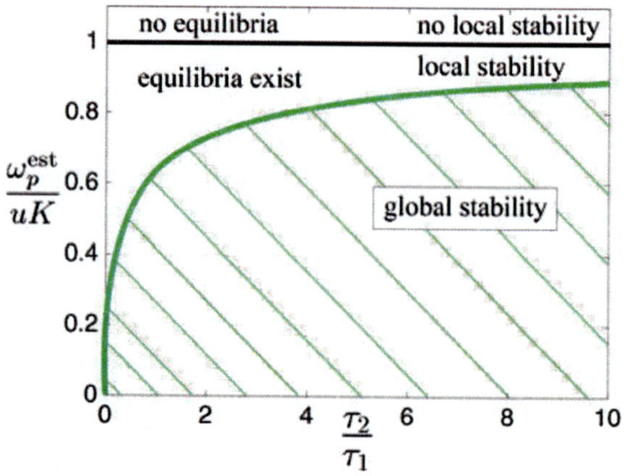

Fig. 2 Stability regions observed in the second-order SRF-PLL. The solid line $\frac{\omega_e}{uK} = 1$ corresponds to equilibria existence, the curve which bounds the global stability domain is Eq. (7) from Theorem 1

Remark 1 Shaded area in Fig. 2 is a domain in the parameters space such that system (6) with parameters from this domain is globally stable, i.e., a trajectory with initial data from any point of the phase space tends to an equilibrium. Notice that many works studying PLL-based systems conduct local analysis only and estimate equilibria's domains of attraction, not analyzing the whole phase space (see, e.g., recent papers [29–33]). For global analysis of systems in the cylindrical phase space the corresponding theorem from [27, 28] should be used.

4 Comparison with Known Results

In this section, we compare estimate (7) for the pull-in range of SRF-PLL with engineering results from [34–38]. Whereas conservative estimate (7) from Theorem 1 is strictly mathematically justified, the rest rely on approximations.

One of the first equations used in engineering design was derived by D.Richman [34]:

$$\omega_p^{\text{Richman}} \approx uK\sqrt{\frac{2\tau_2}{\tau_1+\tau_2} - \left(\frac{\tau_2}{\tau_1+\tau_2}\right)^2}. \tag{8}$$

In [34], Richman uses phase plane descriptions and derives an approximate formula for the pull-in time T_F assuming $uK\tau_2 \gg 1$. Equation (8) for the pull-in frequency was introduced in [38], [13, p. 168] by setting $T_F \to +\infty$.

Another formula based on estimations was derived by Viterbi [35, 38, 39] for $uK\tau_2 \gg 1$:

$$\omega_p^{\text{Viterbi}} \approx uK\sqrt{\frac{2\tau_2}{\tau_1+\tau_2}} > \omega_p^{\text{Richman}}. \tag{9}$$

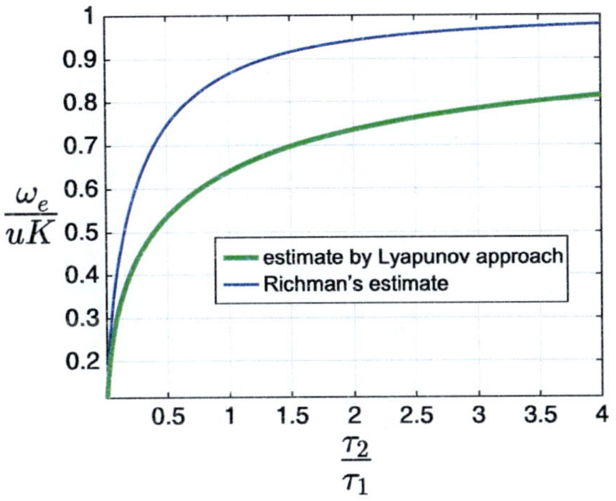

Fig. 3 Comparison of the pull-in range conservative estimate ω_p^{est} from Theorem 1 with the Richman's estimate (8) $\omega_p^{\text{Richman}} \approx uK\sqrt{\frac{2\tau_2}{\tau_1+\tau_2} - \left(\frac{\tau_2}{\tau_1+\tau_2}\right)^2}$

Later Lindsey and Mengali derived approximate formulae for the case of arbitrary periodic nonlinearity in equation (6) [36, 37]. Their formulae for sinusoidal nonlinearity are identical to each other and to the Viterbi's formula (9).

Notice that formula (9) is not valid for $\tau_2 > \tau_1$: this case leads to $\frac{\omega_e}{uK} > 1$, whereas equilibria do not exist in this case and the system is not globally stable. In contrast, $\sqrt{\frac{2\tau_2}{\tau_1+\tau_2} - \left(\frac{\tau_2}{\tau_1+\tau_2}\right)^2} \in (0,\ 1)$ and the pull-in range estimate (8) is less than the hold-in range.

The global stability estimate ω_p^{est} provided by Theorem 1 is conservative, i.e., $\omega_p^{\text{est}} \leq \omega_p$. Although the global stability of SRF-PLL model (6) is guaranteed for any $\omega_e < \omega_p^{\text{est}}$, the exact global stability domain can be wider. Figure 3 shows that the approximate formula (8) provides the wider global stability domain for system (6) than estimate (7). Consequently, the following inequality is valid for the considered estimates:

$$\omega_p^{\text{est}} < \omega_p^{\text{Richman}} < \omega_p^{\text{Viterbi}}.$$

To analyse estimates (8) and (9), we simulate SRF-PLL model (6) in MATLAB Simulink, which is widely used for the study of PLL-based circuits [14, 40]. Simulation of PLL models in SPICE can be found in [41] and [42].

Let us consider SRF-PLL model (6) in MATLAB Simulink with the following parameters:

$$\tau_1 = 0.0448,\ \tau_2 = 0.4,\ K = 2500,\ u = 1.$$

Theorem 1 guarantees that the pull-in frequency $\omega_p \geq \omega_p^{\text{est}} \approx 2208$, whereas equation (8) provides the following pull-in range estimate $\omega_p^{\text{Richman}} \approx 2487.3$.

In Fig. 4, simulation of trajectory of system (6) with initial data $x(0) = -\tau_1$, $\theta_e(0) = 0$ is shown. In the left subfigure, the trajectory tends to an equilibrium point, what

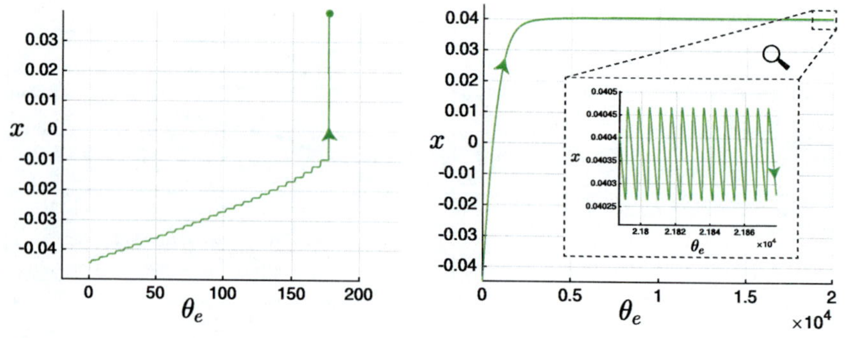

Fig. 4 Simulation of system (6). Initial data: $x(0) = -\tau_1,\ \ \theta_e(0) = 0$. Parameters: $\tau_1 = 0.0448,\ \tau_2 = 0.4,\ K = 2500,\ u = 1$. In the left subfigure, $\omega_e = 2208 \approx \omega_p^{\text{est}} \le \omega_p$ and the trajectory tends to an equilibria, what corresponds to theoretical results (the system is globally stable). In the right subfigure, $\omega_e = 2487.3 \approx \omega_p^{\text{Richman}}$ and the system experiences a persistent oscillation (see the magnified domain), which indicate that the Richman's estimate may yield an unstable behavior

fits the theoretical results. However, oscillations appear in the right subfigure (see the magnified domain), hence, there is no global stability. The simulation shows that estimate (8) should be used carefully.

5 Conclusion

In this work, global analysis of SRF-PLL stability was conducted. As a result, the analytical formula for the pull-in range for SRF-PLL was derived and compared with estimates of Viterbi, Richman, Lindsey, and Mengali, which rely on approximations. Simulation results show that those estimates known from the literature may lead to an unstable behaviour and should be used carefully.

Acknowledgements The work is supported by the Leading Scientific Schools of Russia project NSh-4196.2022.1.1 (Sects. 1, 3) and the Russian Science Foundation project 22-11-00172 (Sects. 2, 4).

References

1. Burton, T., Jenkins, N., Sharpe, D., Bossanyi, E.: Wind Energy Handbook. Wiley (2011)
2. Karimi-Ghartemani, M.: Enhanced Phase-Locked Loop Structures for Power and Energy Applications. Wiley (2014)
3. Teodorescu, R., Liserre, M., Rodriguez, P.: Grid Converters for Photovoltaic and Wind Power Systems, vol. 29. Wiley (2011)

4. Rodríguez, P., Pou, J., Bergas, J., Candela, J.I., Burgos, R.P., Boroyevich, D.: Decoupled double synchronous reference frame PLL for power converters control. IEEE Trans. Power Electron. **22**(2), 584–592 (2007)
5. Golestan, S., Ramezani, M., Guerrero, J.M., Monfared, M.: dq-frame cascaded delayed signal cancellation- based PLL: analysis, design, and comparison with moving average filter-based PLL. IEEE Trans. Power Electron. **30**(3), 1618–1632 (2015)
6. Ghoshal, A., Vinod, J.: Performance evaluation of three phase SRF-PLL and MAF-SRF-PLL. Turk. J. Electr. Eng. Comput. Sci. **23**(6), 1781–1804 (2015)
7. Messo, T., Sihvo, J., Yang, D., Wang, X., Blaabjerg, F.: Improved delayed signal cancellation-based SRF-PLL for unbalanced grid. In: 2017 IEEE Energy Conversion Congress and Exposition (ECCE), pp. 3103–3110 (2017)
8. Young, K., Dougal, R.A.: SRF-PLL with dynamic center frequency for improved phase detection. In: 2009 International Conference on Clean Electrical Power, pp. 212–216 (2009)
9. Reza, S., Ciobotaru, M., Agelidis, V.G.: Single-phase grid voltage frequency estimation using teager energy operator-based technique. IEEE J. Emerg. Sel. Top. Power Electron. **3**(4), 1218–1227 (2015)
10. Shakerighadi, B., Ebrahimzadeh, E., Taul, M.G., Blaabjerg, F., Bak, C.L.: Modeling and adaptive design of the SRF-PLL: nonlinear time-varying framework. IEEE Access **8**, 28 635-28 645 (2020)
11. Zhong, Q.-C., Hornik, T.: Control of Power Inverters in Renewable Energy and Smart Grid Integration, vol. 97. Wiley (2012)
12. Gardner, F.: Phaselock Techniques, 3rd edn. Wiley, New York (2005)
13. Egan, W.: Phase-Lock Basics, 2nd edn. Wiley, New York (2007)
14. Best, R.: Phase-Locked Loops: Design, Simulation and Application, 6th ed. McGraw-Hill (2007)
15. Kuznetsov, N., Leonov, G., Yuldashev, M., Yuldashev, R.: Rigorous mathematical definitions of the hold-in and pull-in ranges for phase-locked loops. IFAC-PapersOnLine **48**(11), 710–713 (2015)
16. Leonov, G., Kuznetsov, N., Yuldashev, M., Yuldashev, R.: Hold-in, pull-in, and lock-in ranges of PLL circuits: rigorous mathematical definitions and limitations of classical theory. IEEE Trans. Circuits Syst.-I: Regular Pap. **62**(10), 2454–2464 (2015)
17. Best, R., Kuznetsov, N., Leonov, G., Yuldashev, M., Yuldashev, R.: Tutorial on dynamic analysis of the Costas loop. IFAC Ann. Rev. Control **42**, 27–49 (2016)
18. Kuznetsov, N., Lobachev, M., Yuldashev, M., Yuldashev, R.: The Egan problem on the pull-in range of type 2 PLLs. Trans. Circuits Syst. II: Express Briefs **68**(4), 1467–1471 (2021)
19. Kuznetsov, N., Lobachev, M., Yuldashev, M., Yuldashev, R., Volskiy, S., Sorokin, D.: On the generalized Gardner problem for phase-locked loops in electrical grids. Dokl. Math. **103**(3), 157–161 (2021)
20. Kuznetsov, N., Lobachev, M., Yuldashev, M., Yuldashev, R., Tavazoei, M.: The Gardner problem on the lock-in range of second-order type 2 phase-locked loops. IEEE Trans. Autom. Control (2023), (accepted)
21. Kuznetsov, N., Arseniev, D., Blagov, M., Lobachev, M., Wei, Z., Yuldashev, M., Yuldashev, R.: The Gardner problem and cycle slipping bifurcation for type-2 phase-locked loops. Int. J. Bifurcat. Chaos **32**(9) (2022). art. num. 2250138
22. Homayoun, A., Razavi, B.: On the stability of charge-pump phase-locked loops. IEEE Trans. Circuits Syst. I: Regular Pap. **63**(6), 741–750 (2016)
23. Kuznetsov, N., Lobachev, M., Yuldashev, M., Yuldashev, R., Kolumbán, G.: Harmonic balance analysis of pull-in range and oscillatory behavior of third-order type 2 analog PLLs. IFAC-PapersOnLine **53**(2), 6378–6383 (2020)
24. Kuznetsov, N.: Theory of hidden oscillations and stability of control systems. J. Comput. Syst. Sci. Int. **59**(5), 647–668 (2020)
25. Kuznetsov, N., Matveev, A., Yuldashev, M., Yuldashev, R., Bianchi, G.: Stability of charge-pump phase-locked loops: the hold-in and pull-in ranges. IFAC-PapersOnLine **53**, 2022–2026 (2020)

26. Kuznetsov, N., Matveev, A., Yuldashev, M., Yuldashev, R.: Nonlinear analysis of charge-pump phase-locked loop: the hold-in and pull-in ranges. IEEE Trans. Circuits Syst. I: Regular Pap. **68**(10), 4049–4061 (2021)
27. Leonov, G., Kuznetsov, N.: Nonlinear Mathematical Models of Phase-Locked Loops. Stability and Oscillations, Cambridge Scientific Publishers (2014)
28. Kuznetsov, N., Lobachev, M., Yuldashev, M., Yuldashev, R., Kudryashova, E., Kuznetsova, O., Rosenwasser, E., Abramovich, S.: The birth of the global stability theory and the theory of hidden oscillations. In: 2020 European Control Conference Proceedings, pp. 769–774 (2020)
29. Dai, Z., Li, G., Fan, M., Huang, J., Yang, Y., Hang, W.: Global stability analysis for synchronous reference frame phase-locked loops. IEEE Trans. Ind. Electron. (2021)
30. Chen, Y., de Oliveira, M.C.: Revisiting stability analysis of phase-locked loops with the Popov stability criterion. IFAC J. Syst. Control (2021). art. num. 100177
31. Mansour, M.Z., Me, S.P., Hadavi, S., Badrazadeh, B., Karimi, A., Bahrani, B.: Nonlinear transient stability analysis of phase-locked loop based grid-following voltage source converters using Lyapunov's direct method. IEEE J. Emerg. Sel. Top. Power Electron. (2021)
32. Zhang, Y., Zhang, C., Cai, X.: Large-signal grid-synchronization stability analysis of PLL-based VSCs using Lyapunov's direct method. IEEE Trans. Power Syst. **37**(1), 788–791 (2021)
33. Peng, Q., Buticchi, G., Tan, N.M.L., Guenter, S., Yang, J., Wheeler, P.: Modeling techniques and stability analysis tools for grid-connected converters. IEEE Open J. Power Electron. **3**, 450–467 (2022)
34. Richman, D.: Color-carrier reference phase synchronization accuracy in NTSC color television. Proc. IRE **42**(1), 106–133 (1954)
35. Viterbi, A.: Principles of Coherent Communications. McGraw-Hill, New York (1966)
36. Lindsey, W.: Synchronization Systems in Communication and Control. Prentice-Hall, New Jersey (1972)
37. Mengali, U.: Acquisition behavior of generalized tracking systems in the absence of noise. IEEE Trans. Commun. **21**(7), 820–826 (1973)
38. Greenstein, L.: Phase-locked loop pull-in frequency. IEEE Trans. Commun. **22**(8), 1005–1013 (1974)
39. Viterbi, A.: Acquisition and tracking behavior of phase-locked loops. Jet Propulsion Laboratory, California Institute of Technology, Pasadena, External Publ, vol. 673 (1959)
40. Tranter, W., Bose, T., Thamvichai, R.: Basic Simulation Models of Phase Tracking Devices Using MATLAB. Synthesis lectures on communications, Morgan & Claypool (2010)
41. Bianchi, G., Kuznetsov, N., Leonov, G., Seledzhi, S., Yuldashev, M., Yuldashev, R.: Hidden oscillations in SPICE simulation of two-phase Costas loop with non-linear VCO. IFAC-PapersOnLine **49**(14), 45–50 (2016)
42. Kuznetsov, N., Leonov, G., Yuldashev, M., Yuldashev, R.: Hidden attractors in dynamical models of phase-locked loop circuits: limitations of simulation in MATLAB and SPICE. Commun. Nonlinear Sci. Numer. Simul. **51**, 39–49 (2017)

Machine Learning, Artificial Intelligence and Data-Driven Methods

Complementary Reward Function Based Learning Enhancement for Deep Reinforcement Learning

George Claudiu Andrei, Mayank Shekhar Jha, and Didier Theillol

Abstract Several complex sequential decision-making problems have been successfully implemented using reinforcement learning (RL) for continuous optimal control. However, the sample efficiency of data collection process during the learning phase is still not well addressed. The convergence rate to the optimal policy as well as the time of the learning process are strongly influenced by the efficiency of the data collected by the agent during the learning phase. This paper proposes a method to generate efficient sample data which allows the agent to collect high reward trajectories more frequently, decreasing the learning phase time. The proposed method consists of Complementary reward (CR) function augmented to the traditional reward function. The CR tends to infinity when the control input leads to the system performance that meets the given requirements very accurately. Consequently, the control policy which maximizes the reward function can render the system to optimal performance. The main contributions of this study include the following aspects: (1) a new proposed Complementary reward that is augmented to the reward function which improves performance of the reinforcement learning based controller in terms of system requirements; (2) speed-up of training phase via generation of more efficient data resulting in a better learned policy.

1 Introduction

Reinforcement Learning (RL) has become one of the most important and useful approach in control engineering. RL uses a trial-and-error learning process to maximize a decision-making agent's total reward observed from the environment. Com-

G. C. Andrei · M. S. Jha (✉) · D. Theillol
Centre de Recherche en Automatique de Nancy (CRAN), UMR 7039, CNRS, Université de Lorraine, 54506 Vandoeuvre-lès-Nancy Cedex, France
e-mail: mayank-shekhar.jha@univ-lorraine.fr

G. C. Andrei
e-mail: george.andrei@etu.unilim.fr

D. Theillol
e-mail: mayank-shekhar.jha@univ-lorraine.fr

D. Theilliol et al. (eds.), *Recent Developments in Model-Based and Data-Driven Methods for Advanced Control and Diagnosis*, Studies in Systems, Decision and Control 467,
https://doi.org/10.1007/978-3-031-27540-1_21

pared to optimal control theory, this maximization problem can be viewed as minimizing the cost function since the reward function is designed based on the control problem and system requirements. Recently, significant progress has been made by combining advances in deep learning (DL) with RL methods for solving optimal control problems resulting in the "Deep Q Network" (DQN) algorithm [1, 2]. DQN solves RL problems with high-dimensional observation spaces but it can deal only with discrete action spaces. Deep Deterministic Policy Gradient (DDPG) method was introduced in 2015 in order to overcome the curse of dimensionality problem as well as to deal with continuous spaces [1]. DDPG requires an actor-critic framework making the algorithm easy to implement and applicable to optimal control problems characterized by continuous spaces. A key mechanism used in the DDPG algorithm is the use of replay buffers to store trajectories of experience in order to extract a batch of samples during training phase as well as break the correlation between data. A prioritized experience replay buffer was introduced as improvement of the previous technique in order to make the samples characterized by high rewards be selected more frequently [1]. The efficiency of the stored trajectories in terms of reward defines training phase duration and convergence of the policy: more efficient collected trajectories in terms of high reward are and the faster the agent will learn the optimal policy. This process could make the training phase very slow and increases the agent's ability to learn a rather sub-optimal policy. This paper proposes an approach to generate efficient sampled data during the learning phase which consists of a Complementary reward (CR) function augmented to the traditional reward function in order to guide control learning for speeding it up and favouring the convergence to the optimal policy. Similar to the Barrier function concept [3], the proposed CR tends to infinity when the control input leads to the system performance meet its requirements very accurately. The main contributions of this study include the following aspects: (1) a new proposed Complementary reward that is augmented to the reward function which improves performance of the reinforcement learning based controller in terms of system requirements; (2) speed-up of training phase via generation of more efficient data resulting in a better learned policy.

The rest of the paper is organized as follows. Section 2 provides a general state-of-the-art of reinforcement learning including DDPG algorithm. Section 3 presents problem formulation. Section 4 provides more details on RL-based control for balancing a rotary inverted pendulum formulation problem. The proposed method using Complementary function-based reinforcement learning will be discussed in Sect. 4. Finally, the results of simulation and its analysis in comparison with traditional reward function method is discussed in conclusions are outlined in Sect. 5.

2 Preliminaries

RL involves an agent exploring an unknown environment to achieve a goal: the agent builds up its knowledge of the environment by gaining experience. Beyond the agent and the environment, it is required to identify four main sub-elements that need to be defined in a RL problem: a *policy*, a *reward signal*, a *value function* and optionally a *model* of the environment [2]. A *policy* $\pi(s_t)$ defines the learning agent's way of behaving at a given time t. A *reward signal* $r_{t+1} = r(s_t, a_t)$ is sent to the agent at any time t by the environment. It defines the goal in a RL setting and it is used for measuring the performance of the agent based on the same concept as cost function in control theory. The objective of the agent is to learn an optimal policy by maximizing the accumulated discounted reward $\sum_{i=t}^{T} \gamma^{i-t} r(s_t, u_t)$ from current time step t to a future time step T, where $\gamma \in (0\ 1)$ is the discount factor which discounts the value of future rewards. A *value* of a state usually denoted as $V_\pi(s_t)$ is the expected accumulated reward over the future following the policy π and starting from a particular system state s_t: $V_\pi(s_t) = E_\pi[r_{t+1} + \gamma r_{t+2} + \gamma^2 r_{t+3} + ...|s_t]$, where E_π is the expected value under policy π. By considering how good it is to be in a state by taking into account the action taken by the agent, it is necessary to refer to the *action-value function*, also known as *Q-function* or *Q-value function* as $Q_\pi(s_t, a_t)$.

$Q_\pi(s_t, a_t)$ gives the expected return for performing action a_t in state s_t at time step t, under the policy π: $Q_\pi(s_t, a_t) = E_\pi[r_{t+1} + \gamma r_{t+2} + \gamma^2 r_{t+3} + ...|s_t, a_t]$. The fourth and final element is a *model* of the system to be controlled. Methods for solving reinforcement learning problems that use models are called model-based methods, as opposed to simpler model-free methods that are explicitly trial-and-error learners. Even in model-free methods, the system model may be required in order to simulate the system making the agent able to collect data if the hardware is not available.

Generally, RL algorithms are based on Markov decision process (MDP) framework for modeling the environment, the policy and the agent. It is defined by the tuple (S, A, r, P):

1. A set of states, S, which contains all possible states of the environment. In other terms, all the possible discrete or continuous measurable outputs.
2. A set of possible actions, $A(s)$, which contains all possible actions or control inputs which can be applied to the system in each state.
3. A reward function $r_{t+1} = r(x_t, u_t)$, which is the immediate reward perceived after transition from state s_t to s_{t+1} as consequences of taking an action a_t.
4. A transition model, $P(s_{t+1}|s_t, a_t)$, which denotes the probability of reaching state s_{t+1} when performing action a_t in state s_t in case of stochastic environments. It can be assumed equal to one in deterministic environments.

2.1 Deep Deterministic Policy Gradient Method

DDPG is a *model-free*, *off-policy* and *actor-critic* framework based algorithm with continuous action spaces, proposed by Dr. Lillicrap et al. in 2015. This method is an extension of two other algorithms, Deep Q-Networks (DNQ) and Deterministic Policy Gradient (DPG) which uses the experience replay and target network as main techniques. DDPG uses two neural networks in the actor-critic and it employs the use of off-policy data and the Bellman equation to learn the Q-function $Q(s_t, a_t)$ which is then used to derive and learn the policy π: $Q_\pi(s_t, a_t) = E_{s_{t+1}\sim E}[r_{t+1} + \gamma E_{a_{t+1}\sim\pi}[Q_\pi(s_{t+1}, a_{t+1})]$, where r_{t+1} is the reward observed from the environment after the action a_t at time step t, $s_{t+1} \sim E$ means that the transition is sampled from the environment defined as E, and $a_{t+1} \sim \pi$ means that the action is sampled from policy π. If the policy is deterministic, it is usually denoted as μ, and the inner expectation of the Bellman equation can be avoided: $Q_\mu(s_t, a_t) = E_{s_{t+1}\sim E}[r_{t+1} + \gamma Q_\phi(s_{t+1}, \mu(s_{t+1}))]$. Q_μ can be learned off-policy, by using transitions generated by a different policy β since the expectation only depends on the environment. The function approximator is represented by the critic neural network parameterized by θ_ϕ and it is used to approximate the Q-function. The mean-squared error can be used as loss function to be minimized in the optimization process: $L(\theta_Q) = E_{s_t,\rho^\beta,a_t\sim\beta,r_{t+1}\sim E}[(y_t - Q(s_t, a_t)|\theta_\phi)^2]$. y_t is the *target value*: $y_t = r_{t+1} + \gamma Q(s_{t+1}, \mu(s_{t+1})|\theta_\phi)$, where ρ^β is the discounted state transition for the policy β, and $\mu(s_t) = \underset{a}{\operatorname{argmax}} Q(s_t, a_t)$ is the policy defined by acting greedily. The deterministic policy $\mu(s_t)$ is represented by the actor neural network parameterized by θ_μ. The Bellman equation is then used as in a conventional RL method to update the critic network. The actor neural network is updated using the following sampled policy gradient to maximize the expected discounted reward: $\nabla_{\theta_\mu} J \approx E[\nabla_{\theta_\mu} Q(s_t, \mu(s_t|\theta_\mu)|\theta_\phi)]$. The off-policy data defined as experience tuple $e_t = (s_t, a_t, r_{t+1}, s_{t+1})$ is generated during the training phase and then is stored in a replay buffer in order to break the inter-correlations between experiences which are sampled from the memory during the updating phase of the networks. This procedure makes the training phase easier to converge by leading the target value as well as Q-function prediction be independent from each other resulting in a more stable learning. The idea is that the weights of the target networks are initialized as copies of the actor and critic networks weights, but updated more slowly to keep them fixed for some time steps. Finally, the parameters of the target network are updated using moving averages for both actors and critics:

$$\theta_{\phi'} \leftarrow \tau\theta_\phi + (1 - \tau)\theta_{\phi'}$$

$$\theta_{\mu'} \leftarrow \tau\theta_\mu + (1 - \tau)\theta_{\mu'}$$

where $\tau << 1$ is the soft target update rate which makes the target networks change slowly improving learning stability.

An important problem in RL is the trade-off between *exploration* and *exploitation*. To achieve high rewards, the agent has to choose actions by applying a control input it has tried before and knows to be a good one. *Exploration* refers to the RL agent exploring the environment to collect unseen trajectories with high rewards. *Exploitation* means simply following the current better policy from the experiences collected in the past to gain as much reward as possible. Thus, DDPG uses a noise sampled from a noise process N is added to the actor policy when selecting an action during training as exploration law: $\pi(s_t) = \mu(s_t|\theta_\mu) + N$. This is called action noise. Different noise processes can be used. The original DDPG paper suggest a time correlated one known as *Ornstein-Uglenbeck process* [1]. It is common practice to employ a variance of the action noise between 1% and 10% of the maximum control input action to reduce the training failure rate.

3 Problem Formulation

In this paper, the balance control of a rotary inverted pendulum system based on the DDPG algorithm has been implemented. The control of a pendulum has been one of extensively applied problems in control field and its choice is justified by the fact that balancing an inverted pendulum in up-right position from the its stable position is not a trivial task. The rotary inverted pendulum consists of a motor arm, which is actuated by a DC servo motor, with a swinging pendulum arm attached to its end. The measured output coming from the system is defined by the motor angle $\theta \in [-\frac{\pi}{2} \;\; \frac{\pi}{2}]$ and by the inverted pendulum angle $\alpha \in [-2\pi \;\; 2\pi]$. The set of all possible values of the two angles among with their rate of change defines the continuous observation space: $S \in [\theta \;\; \alpha \;\; \dot{\theta} \;\; \dot{\alpha}]$. The unstable equilibrium point defines the reference point positioned at the origin of the reference system: $\alpha_0 = 0$ and $\theta_0 = 0$. The control input u_t is the applied voltage by the DC Motor and the set of all possible values defines the continuous action space: $A \in [-10 \;\; 10]$ V. The reward function indicates whether the task is successful or not during the training phase and it is one of the most critical part of the RL design.

The reward function is then designed: $r_{t+1} = -\left(q_{11}\theta_t^2 + q_{22}\alpha_t^2 + q_{33}\dot{\theta}_t^2 + q_{44}\,\dot{\alpha}_t^2\right)$.

The quadratic reward function indicates the agent how far or close his performance is from the reference position. It has been designed in such a way as to explicitly emphasizes the importance of minimizing the control error with respect to the up-right position of the inverted pendulum ($\alpha_{ref} = 0$ and $\theta_{ref} = 0$) positioned at reference system origin. Consequently, the learned policy has to make the control error e_t converge to zero as close as possible at each time step t as shown in Eq. 1.

$$e_t = |e_\alpha| \simeq 0 \text{ with } e_\alpha = |\alpha_{ref} - \alpha| = |0 - \alpha|. \tag{1}$$

The absolute value is used to not differentiate between positive (counter clockwise) or negative rotations (clockwise) of the inverted pendulum.

Additionally, the motor angle's position needs to be controlled as addition system requirement: $\theta \in [-30 \; 30]$ [deg] when balancing the inverted pendulum.

4 Complementary Reward Function Method

In RL problems, the agent learns the optimal control policy which maximizes the reward function by exploring the range of control strategies and selecting the optimal control strategy which maximizes the reward function. In this context, a new method is proposed to speed-up the control learning process and favouring the convergence to the optimal policy by generating more trajectories with high rewards. Such trajectories are generated by a CR included in the reward function which tends to infinity when the system response leads to optimal performance making the agent collect a higher reward during the learning phase. In this sense, in the case of balancing control problem defined in Sect. 3, the CR defined in Eq. 2 is proposed:

$$C(s_t, e_\alpha, \dot{\alpha}) : \begin{cases} q_{55} e^{\left(\frac{1}{\gamma|\alpha ref-\alpha|}\right)} + q_{66} e^{\left(\frac{1}{\gamma|\dot{\alpha}|}\right)} \text{ if } \theta < \pm A \text{and } \alpha < \pm B \text{ and} \dot{\alpha} < \pm D \\ 0 \quad \text{otherwise} \end{cases} \tag{2}$$

where γ is the damping coefficient defining the speed divergence of the function with respect to the control error e_t, q_{55} and q_{66} are weighted parameters while A and B and D are CR conditions to be fixed based on the control problem or system requirements.

As shown in Fig. 1, the proposed CR is inversely proportional control error (see Eq. 1) tending to infinity when e_α is very close to zero. The rate of convergence of the CR is dictated by the damping coefficient as shown in the zoom sub-figure. The choice of γ is very important and must be chosen according to system requirements and control problem.

The CR is included in the DDPG agent reward function defined in Sect. 3 as shown in Eq. 3:

$$R_{t+1} = r_{t+1} + C(s_t, e_\alpha, \dot{\alpha}) \tag{3}$$

where r_{t+1} is the reward function defined in Eq. 3, $C(s_t, \alpha, \dot{\alpha})$ in Eq. 2 respectively.

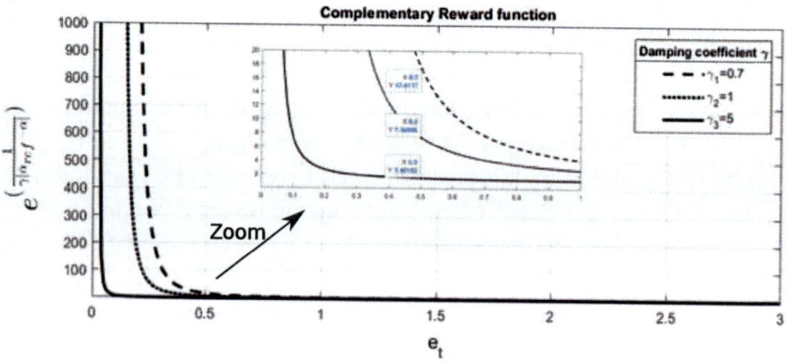

Fig. 1 Complementary function with different damping coefficient

The traditional reward function becomes negligible when at least e_α is close to zero: $R_{t+1} = r_{t+1} + C(s_t, e_\alpha, \dot{\alpha}) \simeq C(s_t, e_\alpha, \dot{\alpha})$ if at least $e_\alpha \simeq 0$ and $\dot{\alpha} \simeq 0$.

The corresponding pseudo-algorithm concerning the proposed method implemented with DDPG algorithm is given below Algorithm 1 with M, T and n being the total number of episodes, episode duration and number of samples respectively.

Algorithm 1 DDPG using Complementary reward function

Step 1. Initialization
Randomly initialize actor network $\mu(s_t|\theta_\mu)$ and critic network $Q(s_t, a_t|\theta_\phi)$
with weights θ_μ and θ_ϕ respectively
Copy the weights to target network μ' and Q', $\theta_{\mu'} \leftarrow \theta_\mu, \theta_{\phi'} \leftarrow \theta_\phi$
Initialize replay buffer
Step 2. Exploration
for episode = [1:M]
Initialize the noise process N $\sim$ OU for exploration
for t = [1:T]
Obtain the current system state s_t from the environment
Generate control input action $a_t = \mu(s_t|\theta_\mu) + N$ based on the current actor policy exploration action noise
Execute action a_t, then **if**: e_θ < A and e_α < B, receive reward R_{t+1}(3) otherwise r_{t+1}
Obtain the resulting system state s_{t+1}
Store experience (s_t, a_t, R_{t+1} *or* r_{t+1}, s_{t+1}) into the replay buffer
Step 3. Update
Extract a mini-batch of n experiences (s_t, a_t, R_{t+1} or r_{t+1}, s_{t+1}) from replay buffer
Compute the target value: $y_t = r_{t+1} + \gamma Q(s_{t+1}, \mu(s_{t+1})|\theta_\phi)$
Update critic network by minimizing the loss $L = \frac{1}{n}\sum_i(y_i - Q(s_{i+1}, a_{i+1}|\theta_\phi))^2$
Update the actor policy by using the policy gradient
$\nabla_{\theta_\mu} J \approx \frac{1}{n}\nabla_{\theta_{\mu_i}} Q(s_i, \mu(s_i|\theta_{\mu_i})|\theta_\phi)$
Update the target networks
$\theta_{\phi'} \leftarrow \tau\theta_\phi + (1-\tau)\theta_{\phi'}$
$\theta_{\mu'} \leftarrow \tau\theta_\mu + (1-\tau)\theta_{\mu'}$
end
end

To demonstrate the effectiveness of the proposed approach while using DDPG algorithm for optimal control, a comparison between traditional method and the proposed one has been carried out. The training settings, hyper and weighted parameters, neural networks configuration remain the same. The training phase duration is set to 600 episodes whose duration is defined by 2500 number of steps each resulting in 25 s since the sample time is set to 0.01. The learned policy by the agent is then tested on the same simulink model that was used to train the agent for both methods with simulation time of 10 [sec]. The position of the pendulum and motor angles are always initialized in the same way at the beginning of each training episode: $\alpha_0 = \pm\pi$ [rad] and $\theta_0 = 0$ which consist on vertically down position for the pendulum and motor angle positioned at the origin of the reference frame respectively,

During each episode, the agent will generate online trajectories by observing the measured output and reward coming from the environment and will subsequently use them to improve and update their policy.

4.1 Traditional Reward Function Approach

The trained agent performance without using CR function is first analyzed. Figure 2 shows the output response of the inverted pendulum angle α with respect to the unstable equilibrium point $\alpha_{ref} = 0$. As it can be noticed, the trained agent is not able to balance the pendulum in up-right position but leads to oscillations around the stable equilibrium point which is positioned at $\alpha = +180$ [deg] as well as $\alpha = -180$ [deg] depending on the rotation sense. On the other hand, Fig. 3 shows that θ output response is respecting the system requirements of keeping the motor angle in between ± 30 [deg]. These performances demonstrate that the agent learnt a sub-optimal policy and it would need to be trained for more than 600 episodes for converging to a better policy. The sub-optimal convergence can be observed in the episode reward function as shown in Fig. 6 where the steady state response has an offset of -500 with respect to zero episode reward which is the maximized one.

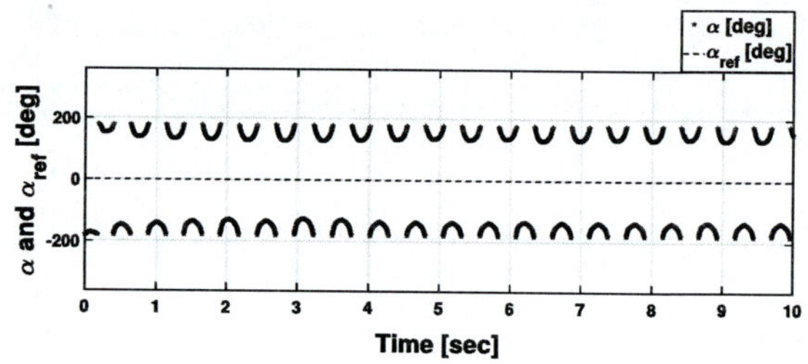

Fig. 2 Inverted Pendulum angle response with no CR

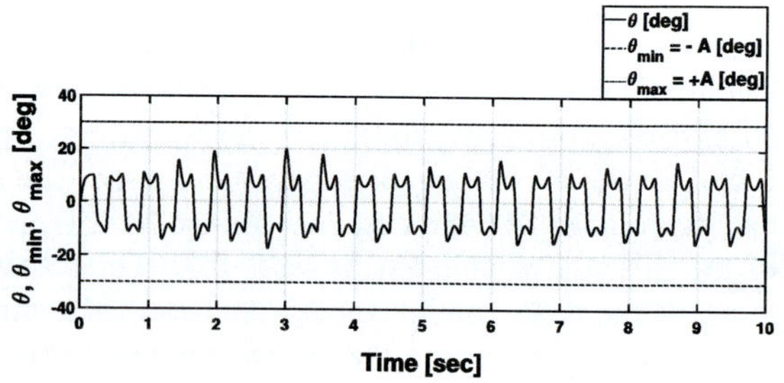

Fig. 3 Motor angle response with no CR

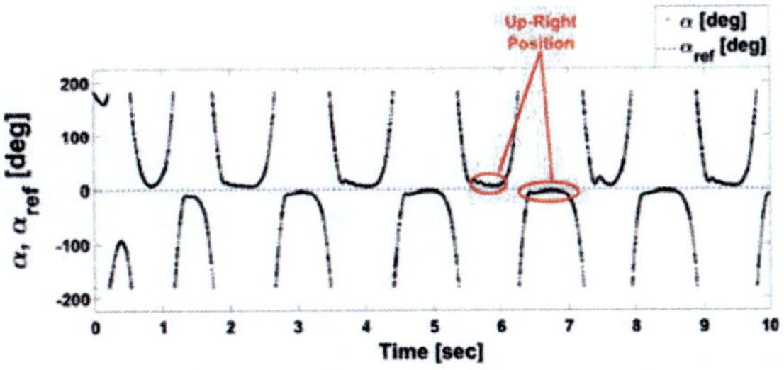

Fig. 4 Inverted Pendulum angle response with CR and $\gamma = 0.7$

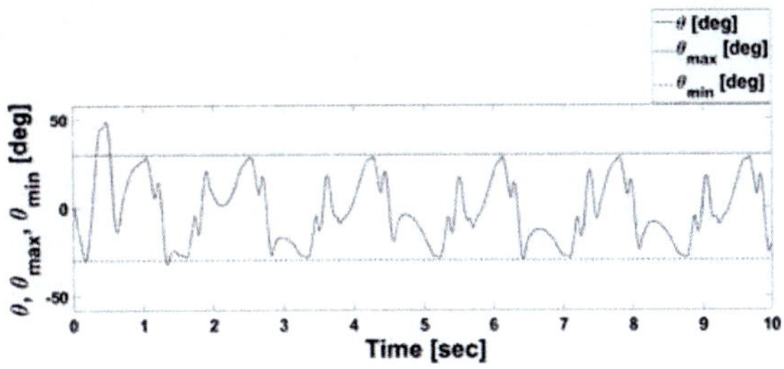

Fig. 5 Motor angle response with CR and $\gamma = 0.7$

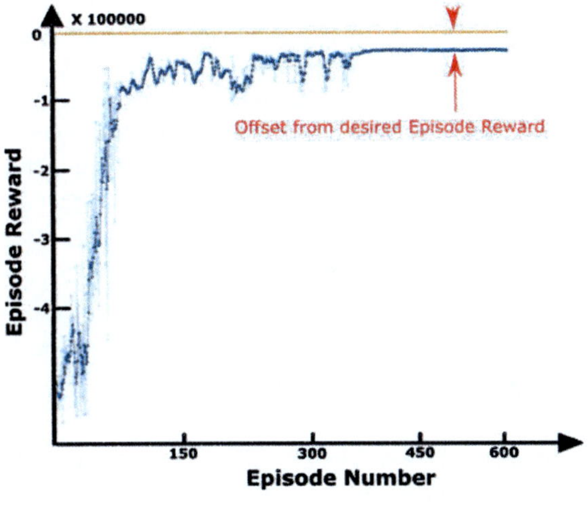

Fig. 6 Episode reward function with no CR

4.2 Complementary Function Approach

The implementation of the proposed method via CR is carried out with the same training and structure of the neural networks and a damping coefficient $\gamma = 0.7$. Figure 4 shows the inverted pendulum angle α with respect to the reference position $\alpha_{ref} = 0$. As it can be noticed, the trained agent is able to keep the inverted pendulum in up-right position in less than 1 [sec]: $\alpha \pm 5$ [deg]. However, the trained agent is not able to balance the pendulum for more than one 1.5 [sec] since the learned policy is not optimal. Figure 5 shows that θ output response is respecting

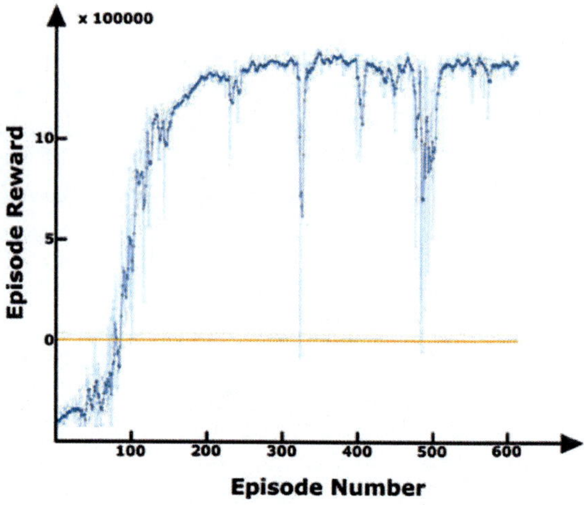

Fig. 7 Episode reward function with CR and $\gamma = 0.7$

the system requirements of keeping the motor angle in between ±30 [deg]. This improvement in terms of performance and convergence to a better policy in the same amount of episodes took place thanks to the introduction of the CR which allowed the agent to collect higher rewards with respect to the traditional approach resulting in a better performance. From episode reward shown in Fig. 7 is possible to notice that at the beginning of the training phase the reward function response is similar to the traditional approach until the agent is within the CR conditions. As soon as the agent collects high trajectories, he focuses on generating others with higher reward by updating the policy. Consequently, CR speed-up the convergence rate to a better policy with respect to the traditional approach.

5 Conclusions and Future Work

The proposed approach improves the policy learning by speeding up its convergence rate via Complementary reward function making the agent able to learn a better policy in less amount episodes thanks to its capability of rewarding the agent with high values when the control task achieves the required performance levels. The proposed approach can be implemented in any kind of RL problems since it acts at the level of reward function design which is a step inherent to all the RL algorithms. Thus, the proposed CR function concept can be viewed as a general contribution to RL based approach. Numerical issues that might occur when the CR tends to infinity represent a mathematical constraint of this method. For this reason, the CR must be up bounded accurately based on the weighted parameter and damping coefficient so the CR function diverges to the up bound which can only be reached when the control error is very close to zero. However, the comparison between the traditional method and the proposed one demonstrates that CR function based approach can provide a better performance under similar conditions. The next step will be to tune the hyper parameters of the CR function, in particular, the damping coefficient in order to speed-up the convergence of the optimal policy in as few episodes as possible.

References

1. Timothy, P., Lillicrap, et al.: Continuous control with deep reinforcement learning (2015). https://doi.org/10.48550/ARXIV.1509.02971
2. Sutton, R.S., Barto, A.G.: Reinforcement Learning: An Introduction. Adaptive Computation And Machine Learning Series, 2nd edn. The MIT Press, Cambridge, Massachusetts (2018). ISBN: 9780262039246
3. Long Vu, T. et al.: Barrier Function-based Safe Reinforcement Learning for Emergency Control of Power Systems (2021). https://doi.org/10.48550/ARXIV.2103.14186

Data-Driven Dissipative Verification of LTI Systems: Multiple Shots of Data, QDF Supply-Rate and Application to a Planar Manipulator

Tábitha E. Rosa and Bayu Jayawardhana

Abstract We present a data-driven dissipative verification method for LTI systems based on using multiple input-output data. We assume that the supply-rate functions have a quadratic difference form corresponding to the general dissipativity notion known in the behavioural framework. We validate our approach in a practical example using a two-degree-of-freedom planar manipulator from Quanser, with which we demonstrate the applicability of multiple datasets over one-shot of data recently proposed in the literature.

1 Introduction

In the design and control of complex high-tech systems, the provision of accurate models of the systems is often a time-consuming and costly process. Correspondingly, the data-driven approach has recently gained attention to complement the model-based methods.

In recent years, several approaches have been presented for analyzing systems properties directly from one-shot of input-output data that is sufficiently rich in information, or typically referred to as the persistently exciting property [1–4]. The systems properties that can be verified directly from data include dissipativity [2, 5–7], the persistence of excitation properties [3], and data informativity [8]. In these cases, it is of utmost importance to have enough and suitable data to verify the information we desire. However, obtaining a single trajectory that provides such information on the system can be problematic in many real-life applications. In many cases, performing new or several experiments to get the proper data shot can

T. E. Rosa (✉) · B. Jayawardhana
Engineering and Technology Institute Groningen, University of Groningen, Nijenborgh 4, 9747 Groningen, The Netherlands
e-mail: t.esteves.rosa@rug.nl

B. Jayawardhana
e-mail: b.jayawardhana@rug.nl

D. Theilliol et al. (eds.), *Recent Developments in Model-Based and Data-Driven Methods for Advanced Control and Diagnosis*, Studies in Systems, Decision and Control 467,
https://doi.org/10.1007/978-3-031-27540-1_22

be costly and time-consuming. In other cases, we can obtain missing or corrupted data in one-shot of data or get one-shot of poorly exciting data.

Rather than obtaining systems properties directly from the data, one can first identify linear time-invariant (LTI) models before analyzing the systems. In this context, several techniques in systems identification that can deal with missing data or multiple datasets have been proposed, among many others, in [9, 10]. These approaches have been applied, for instance, in [11, 12].

Inspired by these results from systems identification literature, there are recent results on data-driven analysis using multiple-shots of input-output data, see for instance [3, 13, 14]. In [14], the authors propose, among others, a method to verify the persistence of excitation of a trajectory that recovers a set of multiple trajectories aligned over at least a certain number of steps. In [3], the datasets can come from different batches of measurements or experiments without the need to overlap.

As our main contribution, we extend the data-driven dissipative analysis using one-shot data to multiple shots of data following the approach pursued in [3]. In particular, we generalize the data-driven verification of dissipativity with quadratic input-output supply-rate function as presented in [1, 2, 5], which is extended recently in [5] to the quadratic differential form (QDF) of supply-rate function. The use of QDFs generalizes the well-known concept of QSR dissipativity and broadens the applicability of the dissipativity notion beyond passive systems and ℓ_2 stable systems. For instance, it encompasses negative-imaginary systems [15, 16] and clockwise systems [17, 18]. Apart from the study and control of this class of systems, QDFs have also found application, for instance, in the stability analysis of mass-spring-damper systems with hysteresis [17] and in fault detection systems [19]. For the rest of this work, we will consider the time-difference form of the QDF.

In recent data-driven verification literature, it is common to consider a finite-time horizon in the summation of the supply-rate function, which can be verified easier than the infinite horizon case employing Willems' fundamental lemma. In this context, when the time horizon of dissipativity is limited to an integer $L > 0$, a data-driven verification method of *L-QSR dissipativity* is proposed in [2] using only one-shot of data which is convex in its formulation and whose data are required to be persistently excited. This method resolves the non-convexity issue in [1], which proposes a method that also incorporates the time-difference form in the QDF. When the largest time-difference in the QDF is given by an integer $N > 0$, the authors in [1] study the data-driven verification of the so-called *(L, N)-QSR dissipativity* via the behavioural framework. One crucial aspect in the verification of L-QSR dissipativity in [2] is that the condition must be checked not only for the time horizon L but also for the time horizon $L - \nu$ for all admissible ν, which we refer to in the work as the *(L, ν)-QSR dissipativity*. Motivated by these results, the concept of (L, ν)-QSR and L-QSR dissipativity are extended to the time-difference formulation of QDF in [5] using the notion of (L, ν, N)-QSR and (L, N)-QSR dissipativity. In Sect. 2, we present a precise definition of these notions. In these results, the need to test all admissible ν and the extension to the infinite-time horizon case is not trivial.

Correspondingly, the contributions of this work are as follows. Firstly, we propose a data-driven verification method employing multiple datasets whose combination is collectively persistently exciting. Secondly, we consider the QSR-dissipativity using the time-difference form of the supply-rate. Thirdly, using a single computable criterion, the proposed method can verify (L, N)-QSR dissipativity without the need for verification at different time horizons $L - \nu$ for all admissible ν.

Notation: The set of vectors (matrices) of order n ($n \times m$) with real entries is represented by $\mathbb{R}^n$ ($\mathbb{R}^{n\times m}$) and correspondingly, that with integer entries is denoted by $\mathbb{Z}^n$ ($\mathbb{Z}^{n\times m}$). Similar notation is applied to denote a vector (matrix) with zeros and ones by 0^n and 1^n (or $0^{n\times m}$ and $1^{n\times m}$), respectively. The $n \times n$ identity matrix is denoted by I^n. The set of positive (or negative) integers is denoted by $\mathbb{Z}_+$ (or $\mathbb{Z}_-$, respectively). The Kronecker product is indicated by $\otimes$. The space of square-summable discrete-time signals is denoted by $\ell_2(\mathbb{R}^\bullet)$. Given $e \in \ell_2(\mathbb{R}^\bullet)$, we denote

$$e_{[i,j]} = \left[e(i)^\top \; e(i+1)^\top \cdots e(j)^\top\right]^\top. \tag{1}$$

For a set of q vectors $e^i_{[0,T_i-1]}$, $i = 1, \ldots, q$, we define

$$\mathbf{e}_{[0,\mathbf{T}-1]} := \left[{e^1_{[0,T_1-1]}}^\top \; {e^2_{[0,T_2-1]}}^\top \cdots {e^q_{[0,T_q-1]}}^\top\right]. \tag{2}$$

A Hankel matrix with $L \in \mathbb{Z}_+$ block rows of a finite sequence $e_{[0,T-1]}$ is given by

$$H_L(e_{[0,T-1]}) = \begin{bmatrix} e(0) & e(1) & \cdots & e(T-L) \\ e(1) & e(2) & \cdots & e(T-L+1) \\ \vdots & \vdots & \ddots & \vdots \\ e(L-1) & e(L-2) & \cdots & e(T-1) \end{bmatrix}. \tag{3}$$

Lemma 1 (Finsler's Lemma [20]) *If there exist $w \in \mathbb{R}^n$, $Q \in \mathbb{R}^{n\times n}$, $B \in \mathbb{R}^{m\times n}$ with* $\text{rank}(B) < n$ *and $B^\perp$ is a basis for the null space of B, that is, $BB^\perp = 0$, then all the following conditions are equivalent: (i). $w^\top Q w < 0$, $\forall w \neq 0 \; : \; Bw = 0$; (ii). ${B^\perp}^\top Q B^\perp < 0$; (iii). $\exists \mu \in \mathbb{R} \; : \; Q - \mu B^\top B < 0$; and (iv). $\exists \mathcal{X} \in \mathbb{R}^{n\times m} \; : \; Q + \mathcal{X}B + B^\top \mathcal{X}^\top < 0$.*

2 Problem Formulation

We consider the following discrete-time linear time-invariant (LTI) system

$$\Sigma : \begin{cases} x(k+1) = Ax(k) + Bu(k), & x(0) = x_0, \\ \quad y(k) = Cx(k) + Du(k), \end{cases} \tag{4}$$

where $x(k) \in \mathbb{R}^n$ corresponds to the state vector, $u(k) \in \mathbb{R}^m$ is the control input and $y(k) \in \mathbb{R}^p$ is the output, with $u \in \ell_2(\mathbb{R}^m)$. We also assume that the state-space matrices are the minimal realization of the system Σ. We consider that we have access to the inputs and outputs for all instants of time $k = 0, \ldots, T_f$, where T_f is any arbitrary given time.

In this work, we address the problem of verifying the (L, N)-dissipativity properties of Σ following [1, 5]. For a given $\nu \geq n$, the system Σ in (4) is said to be (L, ν, N)-*dissipative* with respect to a supply-rate $w(y_{[k,k+N]}, u_{k,k+N]})$ if $\sum_{k=0}^{L-\nu-1} w(u_{[k,k+N]}, y_{[k,k+N]}) \geq 0$ holds for all trajectories $(u_{[0,L+N-1]}, y_{[0,L+N-1]})$ with $x_0 = 0$. Furthermore, it is called (L, N)-*dissipative* if Σ is (L, ν, N)-dissipative for all $n \leq \nu < L$. Following the QDF formulation in [5, 21], we consider a quadratic difference form of supply function w as follows

$$w(y_{[k,k+N]}, u_{k,k+N]}) = \sum_{i,j=0}^{N} \begin{bmatrix} y(k+i) \\ u(k+i) \end{bmatrix}^{\top} \Phi_{ij} \begin{bmatrix} y(k+j) \\ u(k+j) \end{bmatrix} \tag{5}$$

for every $k \geq 0$, where each Φ_{ij} is the usual QSR matrix given by $\begin{bmatrix} Q_{ij} & S_{ij} \\ S_{ij}^{\top} & R_{ij} \end{bmatrix}$ with $Q_{ij} = Q_{ij}^{\top}$ and $R_{ij} = R_{ij}^{\top}$. Note that the N in (L, N)-QSR dissipativity refers to the largest time differences in the QDF of the supply function. The matrix Φ_{ij} above gives the dissipativity relationship between a pair of data $\big(y(k+i), u(k+i)\big)$ and $\big(y(k+j), u(k+j)\big)$ for a given time shift i and j. For passive and ℓ_2 systems, they satisfy the supply-rate (5) with $N = 0$. We note here that in the formulation of quadratic programming, later on, we use the combination of all Φ_{ij} via $\Phi_N = \begin{bmatrix} \Phi_{00} & \cdots & \Phi_{0N} \\ \vdots & \ddots & \vdots \\ \Phi_{N0} & \cdots & \Phi_{NN} \end{bmatrix}$ where $\Phi_{ji} = \Phi_{ij}^{\top}$ for all $i, j = \{0, \ldots, N\}$. Specifically, for every ν, we will refer the (L, ν, N) dissipative property with the supply function w given by (5) as (L, ν, N)-*QSR dissipativity*.

An important assumption used in the literature (as in [1, 2, 5]) for verifying L-QSR-dissipativity is the persistent excitation of the input data. The main idea of having the input measurements persistently exciting is that by using a single shot of data, we can obtain the rest of the admissible trajectories of (4), which is known as Willems' fundamental lemma. In practice, where missing data, corrupted data or an insufficient amount of data can take place, the available one-shot of data $z_{[0,T-1]}$ may not satisfy the central hypothesis in this lemma (namely, the persistence of excitation condition). Consequently, we cannot obtain information on the rest of the admissible trajectories. To deal with such cases, the following *collectively persistently exciting* notion is introduced in [3], which plays an essential role in the present work.

Definition 1 ([3]) Consider a set of q measured trajectories given by $\mathbf{e}_{[0,\mathbf{T}-1]}$ as in (2). This set of trajectories is *collectively persistently exciting* of order L with $0 < L \leq T_i$ for all $i = 1, 2, \ldots, q$, if the following mosaic-Hankel matrix

$$\mathcal{H}_L(\mathbf{e}_{[0,\mathbf{T}-1]}) = \big[\, H_L(e^1_{[0,T_1-1]})\ H_L(e^2_{[0,T_2-1]})\ \cdots\ H_L(e^q_{[0,T_q-1]}) \,\big], \tag{6}$$

where $\mathbf{e}_{[0,\mathbf{T}-1]}$ is the set of q shots of trajectories, and has full row rank.

Using this notion, a set of measured trajectories with possible different lengths can be used together to obtain every possible trajectory of Σ with a time horizon L, as given in the following lemma.

Lemma 2 ([3]) *Suppose that* $(\mathbf{y}_{[0,\mathbf{T}-1]}, \mathbf{u}_{[0,\mathbf{T}-1]})$, *for* q *snapshots, are trajectories of* Σ. *If the set of inputs* u^i *is collectively persistently exciting of order* $L+n$ *with* $0 < L \leq T_i$ *for all* $i = 1, 2, \ldots, q$, *then* $(\bar{y}_{[0,L-1]}, \bar{u}_{[0,L-1]})$ *is an admissible trajectory of* (4) *if and only if there exists a vector* $\alpha \in \mathbb{R}^{q(1-L)+\sum_{i=1}^{q} T_i}$ *such that*

$$\begin{bmatrix} \mathcal{H}_L\left(\mathbf{y}_{[0,\mathbf{T}-1]}\right) \\ \mathcal{H}_L\left(\mathbf{u}_{[0,\mathbf{T}-1]}\right) \end{bmatrix} \alpha = \begin{bmatrix} \bar{y}_{[0,L-1]} \\ \bar{u}_{[0,L-1]} \end{bmatrix}. \tag{7}$$

Given the information mentioned above, in this work, we address the problem of verifying the dissipativity with the QDF supply-rate function of a system Σ based on multiple shots of measured trajectories which are collectively persistently excited.

Multiple shots (L, ν, N)-QSR dissipativity verification problem: Given q multiple shots of trajectories $(\mathbf{y}_{[0,\mathbf{T}+N-1]}\mathbf{u}_{[0,\mathbf{T}+N-1]})$, of Σ, verify whether Σ is (L, ν, N)-QSR dissipative for some $\nu \geq n$ with $0 < L \leq T_i$ for all $i = 1, \ldots q$.

3 Main Result

In the following theorem, we present our main result where a data-driven verification method is given to check the (L, N)-QSR-dissipativity of Σ. For any time $k \geq 0$,

$$Z(k) := \left[y(k)^\top \; u(k)^\top \; \cdots \; y(k+N)^\top \; u(k+N)^\top\right]^\top. \tag{8}$$

Using this windowed data vector of size N, the ith batch data $Z^i_{[0,T_i-1]}$ is understood as in (1) and the expression of Hankel matrix $H_L(Z^i_{[0,T_i-1]})$ follows the composition in (3). Following the expression of mosaic-Hankel matrix in Definition 1, we define the mosaic-Hankel matrix $\mathcal{H}_L(\mathbf{Z}_{[0,\mathbf{T}-1]})$ as in (6) using q snapshots of data $\mathbf{Z}_{[0,\mathbf{T}-1]}$, following . Additionally, we denote

$$U = \left[U_{\text{aux}} \; 0^{(m+p)\nu\times(m+p)(L-\nu)(N+1)}\right], \;\; U_{\text{aux}} = I^\nu \otimes \left(I^{m+p} \; 0^{(m+p)\times(m+p)N}\right). \tag{9}$$

Theorem 1 *Let the integer* $L > 0$ *and* $(\mathbf{y}_{[0,\mathbf{T}-1]}, \mathbf{u}_{[0,\mathbf{T}-1]})$, *be a set of* q *trajectories of* (4) *with* n *be the order of the system,* $\sum_{i=1}^{q} T_i \geq (L-1)(m+p+q)$ *and* $n+1 < L \leq T_i$ *for all* $i = 1, \ldots, q$. *Suppose that the set of* q *snapshots of inputs* $\mathbf{u}_{[0,\mathbf{T}+N-1]}$, *is collectively persistently exciting of order* $L+N+n$ *and there exists* ν *s.t.* $n \leq \nu < L$ *and*

$$U_\perp{}^\top \mathcal{H}_L(\mathbf{Z}_{[0,\mathbf{T}-1]})^\top \Phi_L \mathcal{H}_L(\mathbf{Z}_{[0,\mathbf{T}-1]}) U_\perp \succeq 0, \tag{10}$$

holds, where $\Phi_L = I_L \otimes \Phi_N$, *and* $U_\perp = (U\mathcal{H}_L(\mathbf{Z}_{[0,\mathbf{T}-1]}))^\perp$ *with* U *given in* (9). *Then* (4) *is* (L, λ, N)-QSR *dissipative for all* $\nu \leq \lambda < L$.

Proof First, by the hypotheses of the theorem, we have that the set of q snapshots of inputs $\mathbf{u}_{[0,\mathbf{T}+N-1]}$, is collectively persistently exciting of order $L+N+n$ holds with $L+N \leq T_i$, for all $i = 1, \ldots, q$, such that (10) holds for an integer ν in the interval $[n, L)$. Following the main results in [3], it follows from Definition 1 and Lemma 2 that for a given set of q trajectories $\left(\mathbf{y}_{[0,\mathbf{T}+N-1]}, \mathbf{u}_{[0,\mathbf{T}+N-1]}\right)$, if u is collectively persistently exciting then other admissible trajectory $\left(\bar{y}_{[0,L+N-1]}, \bar{u}_{[0,L+N-1]}\right)$ of Σ satisfies

$$\begin{pmatrix}\mathcal{H}_{L+N}\left(\mathbf{y}_{[0,\mathbf{T}+N-1]}\right)\\ \mathcal{H}_{L+N}\left(\mathbf{u}_{[0,\mathbf{T}+N-1]}\right)\end{pmatrix}\alpha = \begin{bmatrix}\bar{y}_{[0,L+N-1]}\\ \bar{u}_{[0,L+N-1]}\end{bmatrix} \tag{11}$$

for some α. Likewise, if (11) holds then we can rewrite it using $\mathbf{Z}$ such that

$$\mathcal{H}_L(\mathbf{Z}_{[0,\mathbf{T}-1]})\alpha = \bar{Z}_{[0,L-1]} \tag{12}$$

holds with the same α as in (11) and with $\bar{Z}_{[0,L-1]}$ be constructed through the rearrangement and stacking of the elements in $\left(\bar{y}_{[0,L+N-1]}, \bar{u}_{[0,L+N-1]}\right)$ as used in $\mathbf{Z}_{[0,\mathbf{T}+N-1]}$. In other words, verifying (12) is the equivalent of verifying (11). Thus, the matrices multiplying α in (11) and (12) share the same rank, meaning that we can verify if the persistence of excitation conditions hold using any of these matrices.

From the hypotheses of the theorem, we have that $U_\perp$ is a null space of $U\mathcal{H}_L(\mathbf{Z}_{[0,\mathbf{T}-1]}))$, e.g., $U\mathcal{H}_L(\mathbf{Z}_{[0,\mathbf{T}-1]})U_\perp = 0$ holds. Combining this fact with (12) implies that $U\mathcal{H}_L(\mathbf{Z}_{[0,\mathbf{T}-1]})\alpha = 0$, for any α that satisfies (12) with zero initial condition $[\,\bar{y}_{[0,\nu-1]}^\top \; \bar{u}_{[0,\nu-1]}^\top\,]^\top = 0$. Note that this happens due to the definition of U as in (9), as the non-zero elements are in place to ensure that the initial conditions are null. Using conditions 1) and 2) from Finsler's lemma in Lemma 1, and having in mind the aspects above of $U_\perp$, the inequality in (10) is equivalent to $\alpha^\top\mathcal{H}_L(\mathbf{Z}_{[0,\mathbf{T}-1]})^\top\Phi_L\mathcal{H}_L(\mathbf{Z}_{[0,\mathbf{T}-1]})\alpha \succeq 0$, for all α as before (which results in admissible trajectories with zero initial conditions). It follows immediately that $\sum_{k=0}^{L-1}\bar{Z}(k)^\top\Phi_N\bar{Z}(k) = \sum_{k=0}^{L-1}\sum_{i,j=0}^{N}\begin{bmatrix}\bar{y}(k+i)\\ \bar{u}(k+i)\end{bmatrix}^\top\Phi_{ij}\begin{bmatrix}\bar{y}(k+j)\\ \bar{u}(k+j)\end{bmatrix} \geq 0$, which is the equivalent of verifying $\sum_{k=0}^{L-1}\bar{Z}(k)^\top\Phi_N\bar{Z}(k) = \bar{Z}_{[0,L-1]}^\top\Phi_L\bar{Z}_{[0,L-1]} \geq 0$ for all trajectories $\left(\bar{y}_{[0,L+N-1]}, \bar{u}_{[0,L+N-1]}\right)$ with initial conditions $[\,\bar{y}_{[0,\nu-1]}^\top \; \bar{u}_{[0,\nu-1]}^\top\,]^\top = 0$. Consequently, considering the ν initial conditions, the previous results are equivalent to verifying whether

$$\sum_{k=0}^{L-\nu-1}\tilde{Z}(k)^\top\Phi_N\tilde{Z}(k) = \sum_{k=0}^{L-1}\bar{Z}(k)^\top\Phi_N\bar{Z}(k) \geq 0 \tag{13}$$

holds for any trajectory $\left(\tilde{y}_{[0,L+N-\nu-1]}, \tilde{u}_{[0,L+N-\nu-1]}\right)$ with initial conditions $\tilde{x}_0 = 0$, where x is the state of an arbitrary minimal realization of Σ, and for any trajectory $\left(\bar{y}_{[0,L+N-1]}, \bar{u}_{[0,L+N-1]}\right)$ with initial conditions $[\,\bar{y}_{[0,\nu-1]}^\top \; \bar{u}_{[0,\nu-1]}^\top\,]^\top = 0$. Additionally, we have that if (13) holds for some value of ν, than we recover the definition of (L, ν, N)-QSR-dissipativity, as presented in [2, 5].

Now it remains to prove that (L, λ, N)-QSR dissipativity holds for all $\lambda > \nu$. For this purpose, let us take $\nu < \lambda < L$. We will now show that the solvability of (10) also implies that we have (L, λ, N)-QSR dissipativity, i.e., (10) holds with U and U_{aux} defined in (9) to be replaced by $U_\lambda = \left[U_{\text{aux}}\; 0^{(m+p)\lambda\times(m+p)(L-\lambda)(N+1)}\right]$,

$U_{\text{aux},\lambda} = I^{\lambda} \otimes \left(I^{m+p}\ 0^{(m+p)\times(m+p)N}\right)$, respectively. First we note that the null space $U_{\lambda,\perp} := (U_{\lambda}\mathcal{H}_L(\mathbf{Z}_{[0,\mathrm{T}-1]}))^{\perp}$ with the above choice of U_{λ} exists if $\sum_{i=1}^{q} T_i \geq \lambda(m + p) + (L-1)q$. By the hypothesis of the theorem where we assume that $\sum_{i=1}^{q} T_i \geq (L-1)(m+p+q)$ and $\lambda < L$, it follows immediately that $\sum_{i=1}^{q} T_i \geq \lambda(m+p) + (L-1)q$ holds. Thus the null space $U_{\lambda,\perp}$ is non-empty. By the construction of U in (9) and U_{λ} with $\nu < \lambda$, it follows that $\mathrm{Im}(U) \subset \mathrm{Im}(U_{\lambda})$. This implies that $U_{\lambda,\perp} \subset U_{\perp}$. Consequently, if (10) holds for some value of ν then (10) also holds with U and U_{aux} be replaced by U_{λ} and $U_{\text{aux},\lambda}$ for all $\lambda > \nu$. This proves the claim. □

One direct implication from Theorem 1 is that when (10) holds with $\nu = n$ then we obtain the (L, N)-QSR dissipativity of Σ. Consequently, we can use ν to upper-bound the order of the system n as discussed in [2, 5].

Theorem 1 covers existing results in [2, 5], which is restricted to the one-shot data case. On the one hand, by taking $q = 1$, we immediately recover the results in [5] with the main difference on the assumption of T. When we apply Theorem 1 in [5] using the hypotheses in Theorem 1 above, then we immediately verify the (L, N)-QSR dissipativity of the system without the need to check for each ν as posited in [5]. On the other hand, by setting $N = 0$ and taking $q = 1$, we recover the L-QSR-dissipativity, which is also studied in [2]. The idea in Theorem 1 and that in [5] share many commonalities with the approach in [2]. Note, however, that the main difference is in how we construct the main inequality and arrange the data used in the numerical verification.

By modifying the matrices Q_{ij}, S_{ij} and R_{ij} in Φ_{ij}, one can directly verify various systems properties of Σ that involve particular dissipative inequality as discussed in the Introduction. This includes passive systems, ℓ_2-stable systems, negative-imaginary/counter-clockwise systems, and positive-imaginary/clockwise systems. Additionally, we can potentially extend this result to the case where the system is affected by noise by applying, for instance, the results in [22].

4 Example—2 Degree-of-Freedom Planar Manipulator

For validating our technique experimentally, we consider experiments performed in a two-degree-of-freedom (DoF) planar manipulator from Quanser [23], using a rigid joint configuration as detailed in [24]. As presented in [24], the robot is modelled by state equations with four state variables representing the generalized positions and momenta of the two joints, with input variables of electrical current (in Amperes) to the actuation motors and with output variables of the generalized position of the joints. From the property of Euler-Lagrange systems [25], it is well-known that the open-loop behaviour of any robotic manipulator without damping (lossless) is dissipative with respect to the QDF supply-rate function of $w(y(t), u(t)) = \dot{y}(t)^{\top}u(t)$ (in the continuous-time case) or its associated discrete-time version $w(y_{[k,k+N]}, u_{k,k+N]}) = (y(k+1) - y(k))^{\top}u(k+1)$. In our data-driven

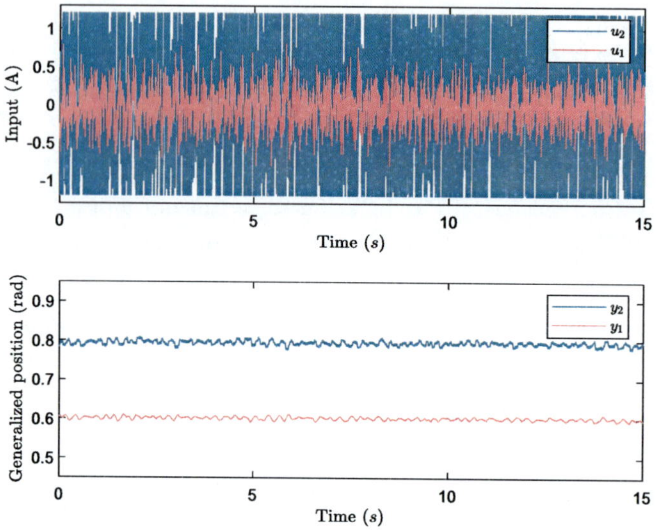

Fig. 1 One set of experimental data of a 2-DoF planar manipulator from Quanser. The closed-loop system (robot + controller) operates at an equilibrium point $(q^*, p^*) = (\left[\begin{smallmatrix}0.6\\0.8\end{smallmatrix}\right], \left[\begin{smallmatrix}0\\0\end{smallmatrix}\right])$ while perturbed by an external signal with a normal distribution, a standard deviation of 0.05 and zero-mean

dissipative verification, we consider a closed-loop system using the controller proposed in Case 2 of Sect. 5.1 from [24], while the robot is operated in the neighbourhood of the generalized positions $q^* = [\,0.6\;0.8\,]^\top$ and generalized momenta $p^* = [\,0\;0\,]^\top$ so that the dynamics can be approximated by an LTI system (4). In this case, the QDF supply-rate function of the closed-loop system will also contain dissipation terms introduced by the friction damping and feedback controller.

For generating the input-output data, we excite the system with an external signal with a normal distribution, a standard deviation of 0.05 and a zero-mean that is calculated and applied separately to each joint. The time-series input-output data is collected with a sampling time of $T_s = 0.005$ s. Figure 1 shows the data from one of the experiments where we can see that the robots operate in the neighbourhood of (q^*, p^*). Note that the input for the second motor is saturated. A video of this experiment can be found at youtu.be/2kg4Tp3qp3Y.

We consider five snapshots ($q = 5$) of data obtained from five different batches of experiments, representing the case where a batch of measurement data is unavailable. The size of the snapshots is given by $T_i \in \{10, 9, 13, 10, 12\}$, where each of them is taken from a different point of the experiment and each snapshot has the form $(\mathbf{u}_{[0,\mathrm{T}+N-1]}, \mathbf{y}_{[0,\mathrm{T}+N-1]})$. Additionally, we take $L = 4$ and $N = 1$, where the choice of $L = 4$ comes from the knowledge of the order of the system and $N = 1$ is from the QDF supply-rate function of open-loop robotic manipulators, as discussed before.

We use the numerical software Matlab (R2020a) in conjunction with the parser YALMIP [26] and the solver Mosek [27] for the optimization procedures of finding feasible solutions $\Phi \succ 0$ in Theorem 1 via linear matrix inequality (LMI) conditions.

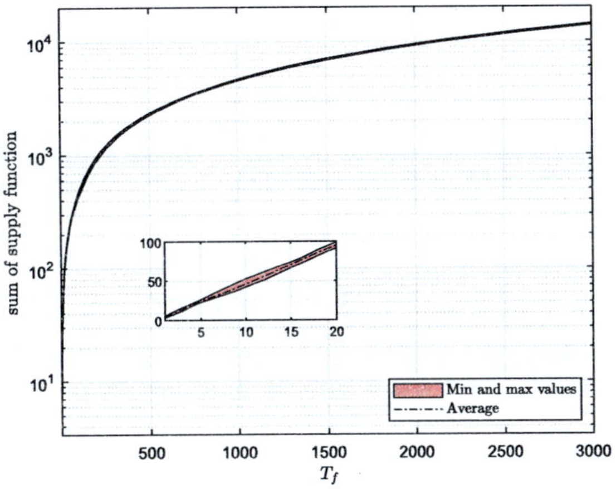

Fig. 2 The plot of $\Upsilon(T_f) := \sum_{k=0}^{T_f} w(u(k), y(k))$ as a function of the time T_f, where the supply-rate function w is given by (5) with the identified Φ_1

Using the aforementioned multiple datasets, the obtained matrix Φ_1 is given by

$$\Phi_1 = \text{diag}\left\{1, 1, 1, 1, 1.999, 2, \left[\begin{smallmatrix} 1.463 & -0.098 \\ -0.098 & 0.674 \end{smallmatrix}\right]\right\}. \tag{14}$$

For comparison purposes, we apply the methods in [2, 5] to check whether they can find a supply function to which the system is dissipative. Both of these methods are used for verifying the dissipativity, however, they consider using one single data shot with the input persistently excited. Thus, we test each of the small snapshots separately, using $N = 1$ and $N = 0$ for the method in [5] and $N = 0$ for [2]. As expected, the searches using all the snapshots of data separately cannot provide results since each dataset individually is not persistently excited.

We can verify numerically whether the identified supply function (14) holds for all batches of data. Particularly, we evaluate whether $\Upsilon(T_f) := \sum_{k=0}^{T_f} w(u(k), y(k)) \geq 0$ holds for arbitrary $T_f \geq 0$ with the identified Φ_1 as in (14). This is presented in Fig. 2, which shows numerically that the system is indeed dissipative with respect to Φ_1 for an extended time horizon beyond the prescribed time horizon of L.

5 Conclusions

This work presents a data-driven method to verify dissipativity properties using a quadratic difference form of the supply-rate function and multiple data sets that collectively satisfy the persistent excitation condition. The method is validated on experimental data from a 2-DoF robotic manipulator. Future work includes a fault detection method using the concept of data-driven dissipativity analysis, in which we identify the dissipativity inequality using a data-driven approach similar to the one in this work.

References

1. Maupong, T., Mayo-Maldonado, J.C., Rapisarda, P.: On Lyapunov functions and data-driven dissipativity. IFAC-PapersOnLine **50**(1), 7783–7788 (2017)
2. Romer, A., Berberich, J., Köhler, J., Allgöwer, F.: One-shot verification of dissipativity properties from input-output data. IEEE Control Syst. Lett. **3**(3), 709–714 (2019)
3. van Waarde, H.J., De Persis, C., Camlibel, M.K., Tesi, P.: Willems' fundamental lemma for state-space systems and its extension to multiple datasets. IEEE Control Syst. Lett. **4**(3), 602–607 (2020)
4. Markovsky, I., Dörfler, F.: Behavioral systems theory in data-driven analysis, signal processing, and control. Annu. Rev. Control **52**, 42–64 (2021)
5. Rosa, T.E., Jayawardhana, B.: On the one-shot data-driven verification of dissipativity of LTI systems with general quadratic supply rate function. In: Proceedings of the 19th European Control Conference, (Rotterdam), pp. 1291–1296. IEEE (2021)
6. van Waarde, H.J., Camlibel, M.K., Rapisarda, P., Trentelman, H.L.: Data-driven dissipativity analysis: application of the matrix S-lemma. IEEE Control Syst. Mag. **42**(3), 140–149 (2022)
7. Koch, A., Berberich, J., Allgower, F.: Provably robust verification of dissipativity properties from data. IEEE Trans. Autom. Control (2021)
8. Van Waarde, H.J., Eising, J., Trentelman, H.L., Camlibel, M.K.: Data informativity: a new perspective on data-driven analysis and control. IEEE Trans. Autom. Control **65**(11), 4753–4768 (2020)
9. Markovsky, I.: Exact system identification with missing data. In: 52nd IEEE Conference on Decision and Control, pp. 151–155. IEEE (2013)
10. Liu, Z., Hansson, A., Vandenberghe, L.: Nuclear norm system identification with missing inputs and outputs. Syst. Control Lett. **62**(8), 605–612 (2013)
11. Ding, T., Sznaier, M., Camps, O.: A rank minimization approach to fast dynamic event detection and track matching in video sequences. In: 2007 46th IEEE Conference on Decision and Control, pp. 4122–4127. IEEE (2007)
12. Noh, S., Zoltowski, M.D., Sung, Y., Love, D.J.: Pilot beam pattern design for channel estimation in massive MIMO systems. IEEE J. Sel. Top. Signal Process. **8**(5), 787–801 (2014)
13. Markovsky, I.: A missing data approach to data-driven filtering and control. IEEE Trans. Autom. Control **62**(4), 1972–1978 (2016)
14. Berberich, J., Allgöwer, F.: A trajectory-based framework for data-driven system analysis and control. In: 2020 European Control Conference (ECC), pp. 1365–1370. IEEE (2020)
15. Lanzon, A., Petersen, I.R.: Stability robustness of a feedback interconnection of systems with negative imaginary frequency response. IEEE Trans. Autom. Control **53**(4), 1042–1046 (2008)
16. Mabrok, M.A., Alyami, M.A., Mahmoud, E.E.: On the dissipativity property of negative imaginary systems. Alex. Eng. J. **60**(1), 1403–1410 (2021)
17. Ouyang, R., Andrieu, V., Jayawardhana, B.: On the characterization of the Duhem hysteresis operator with clockwise input-output dynamics. Syst. Control Lett. **62**(3), 286–293 (2013)
18. Ouyang, R., Jayawardhana, B.: Absolute stability analysis of linear systems with Duhem hysteresis operator. Automatica **50**(7), 1860–1866 (2014)
19. Lei, Q., Munir, M.T., Bao, J., Young, B.: A data-driven fault detection method based on dissipative trajectories. IFAC-PapersOnLine **49**(7), 717–722 (2016)
20. de Oliveira, M.C., Skelton, R.E.: Stability tests for constrained linear systems. In: Perspectives in Robust Control, pp. 241–257. Springer, Berlin (2001)
21. Trentelman, H., Willems, J.: Every storage function is a state function. Syst. Control Lett. **32**(5), 249–259 (1997)
22. De Persis, C., Tesi, P.: Formulas for data-driven control: stabilization, optimality, and robustness. IEEE Trans. Autom. Control **65**(3), 909–924 (2019)
23. Quanser.: 2 DOF Serial Flexible Joint, Reference Manual. Doc. No. 800, Rev 1 (2013)
24. Chan-Zheng, C., Borja, P., Scherpen, J.M.: Passivity-based control of mechanical systems with linear damping identification. IFAC-PapersOnLine **54**(19), 255–260 (2021)

25. Ortega, R., Perez, J.A.L., Nicklasson, P.J., Sira-Ramirez, H.J.: Passivity-Based Control of Euler-Lagrange Systems: Mechanical, Electrical and Electromechanical Applications. Springer Science & Business Media (2013)
26. Löfberg, J.: YALMIP: a toolbox for modeling and optimization in MATLAB, pp. 284–289. Taipei, Taiwan (2004)
27. ApS, M.: The MOSEK optimization toolbox for MATLAB manual. Version 9.0 (2019)

Hybrid Model/Data-Driven Fault Detection and Exclusion for a Decentralized Cooperative Multi-robot System

Zaynab EL Mawas, Cindy Cappelle, and Maan EL Badaoui EL Najjar

Abstract In this paper, a decentralized cooperative localization system is developed, with a hybrid (model/data driven) diagnosis phase. The fault detection and exclusion (FDE) is based on the usage of two approaches: model based, establishing the fault detection indicators based on information theory, followed by a data-driven method, deducing the behaviour of the system by generated indicators thresholding. Based on the indicators, a Multi-Layer Perceptron (MLP) is effectively used for each phase: the detection of the occurrence of fault in one hand and then the isolation and exclusion of the faulty sensor in the system in the other hand. The approach is tested on real data acquired by three Turtlebot3 equipped with wheel encoders (for odometry), a gyroscope (for the yaw angle), a *Marvelmind* localization system (for the position), and an *Optitrack* system (for ground truth).

1 Introduction

The advances in the field of autonomous vehicles requires the design and establishment of a technology panel in which security and communication are key factors. As vehicles become smarter, they ought to validate multiple requirements as their awareness evolves. They also need to gain the trust of the pedestrians and passengers

Z. E. Mawas (✉) · C. Cappelle · M. E. B. E. Najjar
University Lille, CNRS, Centrale Lille, UMR 9189 CRIStAL, Villeneuve D'Ascq, France
e-mail: zaynab.elmawas@univ-lille.fr

C. Cappelle
e-mail: cindy.cappelle@univ-lille.fr

M. E. B. E. Najjar
e-mail: maan.el-badaoui-el-najjar@univ-lille.fr

C. Cappelle
CIAD UMR 7533, University Bourgogne Franche-Comté, UTBM, 90010 Belfort, France

D. Theilliol et al. (eds.), *Recent Developments in Model-Based and Data-Driven Methods for Advanced Control and Diagnosis*, Studies in Systems, Decision and Control 467,
https://doi.org/10.1007/978-3-031-27540-1_23

alike, which increases the focus on the creation of a diagnosis layer leading to much tolerant systems. This is done by studying problematic scenarios where the fault can cause severe effects.

In the vehicle environment, the tasks are carried out cooperatively when it is divided among several agents who carry it out individually. This means that every vehicle navigates in its own trajectory and tends to localize themselves and other vehicles that they come across sharing the same environment. This "Cooperative Positioning System (CPS)" has been introduced by *Kurazume et al.* in [1] to resolve the accumulation problem of the dead reckoning approach in a multi-robot system, by founding the concept of "portable landmarks" for the navigation in an unknown environment. *Roumelioutis et al.* [2] adopted this technique and integrated the Kalman filter to process the available positioning information shared from the members of the team and produce a pose estimate for every one of them.

Along the use of Kalman filter, and in multi-robot multi-sensor system, the natural sensors redundancy makes the fault detection and exclusion (FDE) algorithms more feasible. In [3, 4], this redundancy was used to exclude the faulty sensors from the fusion procedure. It is an observer-based FDE approach that uses information theory indicators to simultaneously localize a group of robots and detect their faulty sensors.

In [5], a decentralized version was introduced, in order to increase the robustness of the system by deploying an architecture where each robot processes its own information and only exchanges (point-to-point) the informational vector and matrix of correction towards an observed neighboring robot [6].

The main motivation behind this work is weakness that lies within this method. They require a prior initializing value of the threshold, which can be far from the actual case of the robot. Moreover, the error's module can vary given the situation, and we cannot always presume having a non-faulty case for the trajectory to base the study on for the calculation of the probabilities of false alarm and detection, as real acquisition errors can occur. This requires merging the technique with a tool that can deduces patterns from data: a Machine Learning (ML) algorithm that learns multiple error cases with various modules to detect abnormal behavior of the system. This hybrid method is feasible, and proved by [7] in which they used multiple model adaptive estimation technique based on Kalman filter to recognize failure patterns.

The residuals between the estimators and the real system result are used as input to a Neural Network (NN) that detects and identifies the faults. ML techniques are used in various other applications and cases. For example, in [8], a single back-propagation (BP) neural network is used to detect malfunctions based on the current and past states of a multi-robot system and for monitoring the behaviour of the system. Moreover, a fault detection framework for autonomous robots through sensory data was presented in [9]. It also uses BP NN to synthesize fault detection components based on the data collected in the training runs and Time-Delay Neural Networks (TDNNs) as classifiers.

This paper presents a decentralized multi-sensor data fusion approach for cooperative multi-vehicle system, with hybrid fault detection and exclusion (H-FDE), using model-based and data-driven methods. It is organized as the following: this introduction is followed in Sect. 2 by the proposed approach presentation, where the various tools are explained, for position estimation as well as the fault tolerant fusion application and the ML use. The experimental platform is described in Sect. 3 as well as the results of the method and the final section deals with conclusion and perspectives.

2 Proposed Approach

The proposed approach includes a hybrid fault detection and isolation step that merges the use of model based with data-driven techniques. It uses information theory and Multi-Layer Perceptron (MLP) to learn and detect the pattern of the fault in order to exclude it. It is presented in Fig. 1.

2.1 Description of the Approach

The approach has 6 main parts, and start by an initialization of the size and value of the state/observation vectors and matrices.

1. **Data Acquisition**: The data is acquired on board of the various types of sensors: encoder for model evolution, "*intra*" measurement on the robot itself and "*inter*" measurements providing pose estimation that is computed and sent by the the neighboring robot's sensors whenever it perceives a robot nearby.
2. **Information Filtering** (2.2): It consists of two parts: prediction and correction. The prediction model is based on encoder data providing the odometric model. The correction's model integrates a Batch Covariance Intersection Informational Filter (B-CIIF) which is obtained by summing the weighted informational contributions

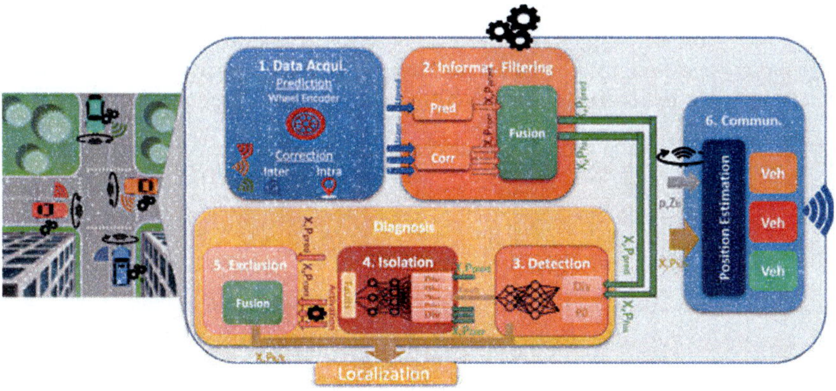

Fig. 1 Hybrid Fault tolerant system

of the sensors in the configuration. The weights are integrated with the dependent measurement whose cross correlation is not negligible.

3. **Fault detection**: The approach then proceeds to the diagnosis phase on each robot: a generalized informational residual gD_{JS}, describing the dissimilarities between joined sensor measurements and the predicted solution is generated (2.3). The divergence used in this work for dissimilarity measurement is the Jensen Shannon divergence. The occurrence of a default is determined by the use of a MLP that establishes the limit above which the value of the divergence implies the passage to an abnormal situation. This MLP takes the generalized residual along with the prior probability of the no fault hypothesis (P_0).
4. **Fault Isolation**: Once a faulty case is detected, an isolation step is applied, by calculating the divergence between the prediction and corrections: joined corrections solution deprived of each of the corrections once then twice (i.e. Generalized Observer Scheme [10]). They are then fed to the isolation's MLP, along with the prior probability of the no fault hypothesis (P_0) for each sensor. The choice of residuals is based on the operation and tolerance requirements, ensuring that all the sensor-faults are detectable.
5. **Fault exclusion**: The localization solution excludes the faulty sensors from each iteration and fuses the rest of the measurements, resulting in a more accurate positioning.
6. **Communication**: The receiving/transmission of "*inter*" measurements data requires the establishment of a connection between the inter-seeing robots in order to grant the access, and obtain/transfer the informational contribution of the Lidar. Any fault detected from an "*inter*" measurement correction must be reported back to the emitting robot as a feedback of the communication step between the robots.

2.2 Information Filtering

In order to estimate the pose of the robots, B-CIIF is used. It consists of two steps: *(a) Prediction* and *(b) Update* where the "*intra*" measurements and "*inter*" measurements between the robots are used to correct the robots position prediction.

(a) *Prediction:* The state vector of each robot of the team is composed of the position and the orientation of the robot "*a*" with respect to a fixed frame:

$$X^a = [x^a \quad y^a \quad \theta^a]^T \tag{1}$$

The propagation model is non linear, it is applied as shown below:

$$X_k^a = f(X_{k-1}^a, u_{k-1}^a) + W_k^a \tag{2}$$

where: u_{k-1}^a is the input vector,

and W_k^a is the process state noise modeled as uncorrelated white noise with covariance matrix Q^a.
The covariance matrix is built as the following:

$$P_{k|k-1}^{R_a} = \left(P_{k|k-1}^{a} \cdots P_{k|k-1}^{a\leftarrow b} \cdots P_{k|k-1}^{a\leftarrow n}\right) \tag{3}$$

where the notation $a \leftarrow b$ is the interaction between robot "a" and robot "b".
The information matrix is then calculated to be used in the correction step. It is obtained by the inverse of the covariance matrix of the global system:

$$\Upsilon_{k|k-1}^{R_a} = (P_{k|k-1}^{R_a})^{-1} \tag{4}$$

(b) *Correction*: The multi-sensor data fusion is accomplished by the summation of the multiple informational contributions of the sensors used in the configuration. We consider the case where robots are able to localize themselves relatively to others using the "*inter*" sensors (Lidar) providing the position of robot "b" relatively to the frame of robot "a" and "*intra*" sensors: gyroscope that provides a correction for the orientation angle $[\theta^{g^a}]$, and *marvelmind* that provides a correction for the position of the robot $[x^{m^a}, y^{m^a}]$.
The observation vector $Z_k^{obs^a}$ holds the same components as the state vector:

$$Z_k^{obs^a} = [x^{obs^a} \quad y^{obs^a} \quad \theta^{obs^a}]^T \tag{5}$$

Where $obs^a \in \{m^a, g^a, L^{a\leftarrow b}, \ldots\}$. In order to deal with the *data incest* problem, the (B-CIIF) separates the observations into dependent and independent values in the correction step.
The update of the information matrix and of the information vector are deduced obtained as follows [11]:

$$\Upsilon_{k|k}^{R_a^{CIF}} = (\Upsilon_{k|k-1}^{R_a} + \sum_{obs_{In}^a} I_k^{obs_{In}^a} + \sum_{obs_D^a} \omega_{obs_D^a} I_k^{obs_D^a}) \tag{6}$$

Where $I_k^{obs_{In}^a}$ denotes the information contribution (Eq. (7)) of the independent observations (in our case, *marvelmind* and gyroscope), and $I_k^{obs_D^a}$ denotes the dependent values "*inter*" observation that use the odometry value to calculate the absolute value of the pose from a relative one. $\omega_{obs_D^a}$ are the weights assigned to the dependent observation such as $\sum_{obs_D^a} \omega_{obs_D^a} \in [0, 1]$.

$$I_k^{obs^a} = (H_k^{obs^a})^T (R_k^{obs^a})^{-1} H_k^{obs^a} \tag{7}$$

2.3 Hybrid Fault Detection and Exclusion

After the information filtering step, the diagnosis step aiming to check the reliability of the obtained positioning solution is applied. It is applied by fusing the model-based and data-driven methods, and learning the behaviour of the system from the divergence obtained.

2.3.1 Information Theory and Divergence

This approach is statistical and non deterministic. Each sensor observation is considered as a distribution, and their comparison is done by calculating the divergence between them. In this paper, the Jensen Shannon divergence D_{JS} is used [12]. It is a symmetric form of the kullback-Leibler divergence. For a two d-dimensional Gaussian distributions, the kullback-Leibler divergence is the sum of three terms [13]: compactness, mutual information and Mahalanobis distance between two distributions.

Two kinds of indicators are derived from this divergence: detection and isolation residuals. The detection residual (gD^a_{JS}) obtained by calculating the divergence between the prediction $P^{R_a}_{k|k-1}$ and joined correction $P^{R_a}_{k|k}$, obtained as follows:

$$gD^a_{JS} = D_{JS}(\mathcal{N}(X^a_{k|k-1}, P^{R_a}_{k|k-1})||\mathcal{N}(X^a_{k|k}, P^{R_a}_{k|k})) \tag{8}$$

Once a fault is detected, the isolation residuals are calculated in order to localise the faulty sensor. These residuals JS^{obs^a}, $obs^a \in \{m^a, g^a, L^{a\leftarrow b}, \ldots\}$ are obtained by calculating the divergence between the prediction $P^{R_a}_{k|k-1}$ and the joined correction $P^{obs^{R_a}}_{k|k}$ deprived of one correction once then twice:

$$JS^{obs^a} = D_{JS}(\mathcal{N}(X^a_{k|k-1}, P^{R_a}_{k|k-1})||\mathcal{N}(X^{obs^a}_{k|k}, P^{obs^{R_a}}_{k|k})) \tag{9}$$

2.3.2 Machine Learning for Thresholding

In order to learn the behaviour of the system and delimit the behaviour of the system in a non faulty case, a MLP is used. It is a feed-forward artificial neural network that deals with cases where the mapping between inputs and output is non-linear [14]. In this work, the input of the MLP is the residual of the system, along with the prior probability of the no fault hypothesis (P_0) calculated by ratio of the non-faulty cases count over the overall iterations. $P_0 = C_0/(C_0 + C_1)$ where C_0 denotes the accumulated sum of the non-faulty cases and C_1 denotes that of the faulty case. The output is the sensor fault cases presented in the system at each iteration. These detected faults are then excluded from the fusion process to lead to a much precise solution.

The models are optimized and trained on a central processing unit, with data from the 3 identical robots, to generate one model that is used on each of the robot. In order to decide the architecture chosen for the detection and isolation (MLP), an optimization step is applied, using scikit-learn's *GridSearchCV* on the Learning rate, number of hidden layers, optimizer, activation function and Loss function.

For the detection MLP, it consists of (150, 100, 50, 50) hidden layers, using the 'relu' activation function, of constant learning rate of 1e-06, for a maximum iteration of 100.

As for the isolation MLP, input data PCA was applied reducing from 8 parameters to 2 with variance ratios of 80.903 and 18.699%. it consists of (200, 100, 100, 50, 25) hidden layers, using the 'tanh' activation function, of constant learning rate of 1e-06, for a maximum iteration of 300.

3 Experimentation and Results

3.1 Robotic Platform

To generate the database used to validate the proposed approach, an experimental platform consisting of three turtlebot 3 Burger (Fig. 2b) under Robotic Operating System (ROS) is established. These robots are equipped with wheel encoders, embedded OpenCR IMU, A1 RPLIDAR and *marvelmind* system (indoor localization system). The system is observed by an optitrack system, which is an accurate localization system used for ground truth and performance study (see Fig. 2a).

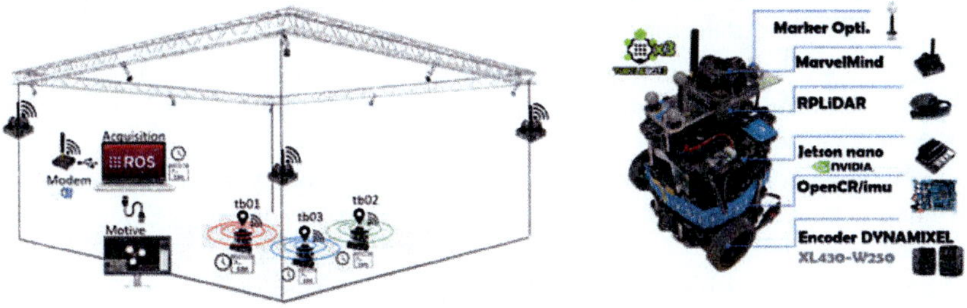

(a) The optitrack configuration (b) The robot used and its sensor placement

Fig. 2 The experimental platform

3.2 Results

The training was done over 5 different sensor fault scenarios for 3 robots evolving in each of the 6 different trajectories.

To compare the performance of the system, the accuracy is calculated for detection and isolation MLP, and for both training and testing. It is the set of labels predicted for a sample must exactly match the corresponding set of labels from the testing data set.

$$Accuracy(y, \hat{y}) = 1/n_{samples} \times \sum_{i=1}^{n_{samples}-1} 1(\hat{y}_i = y_i) \tag{10}$$

The faults occurring in this system have 2 types: injected and natural faults. Multiple sensor fault scenarios have been generated for 6 differently shaped trajectories, by generating the number of repetitions, duration and module for each fault.

Given the nature of the used sensors, three types of errors are implemented: bias errors on marvelmind data, drift error on the accumulating sensor gyroscope, drift error on the accumulating sensor encoder, and RPLIDAR faults that are either absence of information caused by the environment or the technology or bias errors on the intensity of the sensor beam.

The Learning is done in centralized way. The data from the 3 robots are transfered to a central processing unit that optimizes the architecture of the MLP for detection and isolation and trains it then diffuses it to the robots. The testing is done on a sensor fault scenario for each of the trajectories, and the results are shown in Table 1.

The accuracy changes given the shape of the trajectory. For the detection MLP, the training resulted an accuracy of 73.15%. As for the isolation MLP, the training resulted an accuracy of 66.524%. This means that not all the detected fault are isolated. For circular trajectories $\{1, 2\}$ it performs better than transversal trajectories $\{4, 5, 6\}$. The third trajectory has a transversal then a circular trajectory (carrefour scenario), and it has the best overall accuracy for detection and isolation. In transversal cases, there are some robots that are not able to detect nor isolate a big majority of the faults occurring.

Training with the PCA coordinates outperforms the one with all coordinates in most cases except in trajectory 5 where training with the 8 coordinates gives a better accuracy.

One trajectory ($traj_2$) is visualized in Fig. 3. This figure shows the evolution of trajectory for each robot and the corresponding generalized residual. The red color indicates the detection of a fault in the system. Those faults are also marked on the generalized residual by a vertical line given their occurrence for each robots, where some were detected, and others were missed. The accuracy of detection for this trajectory as shown in Table 1 is 83.52% for Robot 1, with 12.9% of true negative (miss detection (md)) and 3.52% false positive (false alarm (fa)), that could be coming from the real fault that occurred during the acquisition of the data. Those two values do not exceed 1/5 of the whole trajectory. In this table, the case "$traj_i detect$" denotes

Table 1 Accuracy of the MLP for detection and isolation

	Robot 1	Robot 2	Robot 3
$traj_1 detect$	0.6829	0.5975	0.6951
$traj_1 isolpca$	0.5853	0.5365	0.5731
$traj_1 isol p\bar{c}a$	0.5975	0.5121	0.6097
$traj_2 detect$	0.8352	0.6882	0.7529
$traj_2 isolpca$	0.776	0.6411	0.7
$traj_2 isol p\bar{c}a$	0.7823	0.6764	0.7529
$traj_3 detect$	0.8078	0.8374	0.7635
$traj_3 isolpca$	0.8029	0.8128	0.7241
$traj_3 isol p\bar{c}a$	0.7487	0.8128	0.7635
$traj_4 detect$	0.5212	0.6063	0.7553
$traj_4 isolpca$	0.5106	0.7127	0.6808
$traj_4 isol p\bar{c}a$	0.4893	0.7340	0.6914
$traj_5 detect$	0.6857	0.4857	0.6857
$traj_5 isolpca$	0.2	0.2857	0.2285
$traj_5 isol p\bar{c}a$	0.3142	0.4571	0.2571
$traj_6 detect$	0.7319	0.7010	0.6082
$traj_6 isolpca$	0.4948	0.6804	0.53608
$traj_6 isol p\bar{c}a$	0.4329	0.6288	0.5154

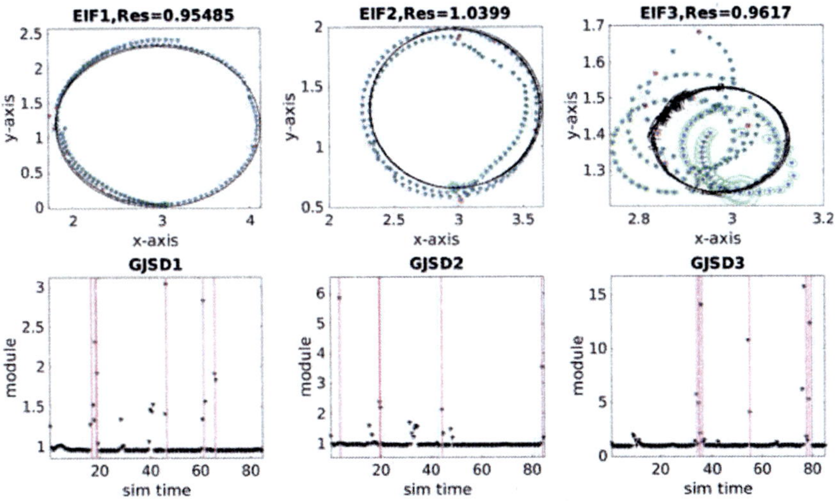

Fig. 3 Detected faults from the detection MLP

the detection MLP for the "$traj_i$". As for "$traj_i isolpca$", the isolation MLP with input's dimension reduced using pca, and "$traj_i isol\bar{pca}$" with full input, without applying pca.

The detection MLP was able to detect the majority of the faults that were injected. As for isolation, for the marvel sensor, there were 0.58% (md), 8.23% (fa) and an accuracy of 91.17%. For the Lidar 2.941% (md), no (fa) and an accuracy of 97.05%. For the gyroscope, 7.05% (md) 1.17% (fa) and an accuracy of 91.76%. Finally for the odometry, 3.5% (md) no (fa) and an accuracy of 96.47%.

4 Conclusion and Perspectives

This paper presented a study of the fusion of the model-based and data-driven methods for the thresholding phase of the diagnosis of a multi-vehicle cooperative localisation system. This approach is based on the observation of the system using multiple sensor and fusing the data that has validated the diagnosis step without implying the presence of an error.

The main benefit of this approach is the fact that it doesn't need prior knowledge on the system. A non-faulty case for the trajectory cannot always be available in order to base the study on, because the occurrence of real errors cannot be prevented during acquisition.

This method can be further developed to take into account the evolution factor of a residual in a trajectory. This can be done by establishing a memory in the learning of the model. Another way to look at it would be to study the evolution of the value of the residual instead of just its value, and apply regression to create a signature function that describes and predicts the occurrence of a fault given the current behavior of the system.

Acknowledgements This work benefits from the financial support of the ANR french Research Agency within the project LOCSP No 2019-CE22-0011.

References

1. Kurazume, R., Nagata, S., Hirose, S.: Cooperative positioning with multiple robots. In: Proceedings of the 1994 IEEE International Conference on Robotics and Automation, vol. 2, pp. 1250—1257 (1994)
2. Roumeliotis, S.I., Bekey, G.A.: Distributed multirobot localization. IEEE Trans. Robot. Autom. **18**(5), 781–795 (2002)
3. Al Hage, J., El Najjar, M.E., Pomorski, D.: Multi-sensor fusion approach with fault detection and exclusion based on the Kullback-Leibler divergence: application on collaborative multi-robot system. Inf. Fusion **37**, 61–76 (2017)
4. Abci, B., El Najjar, M.E.B., Cocquempot, V.: Sensor and actuator fault diagnosis for a multi-robot system based on the Kullback-Leibler divergence. In: 2019 4th Conference on Control and Fault Tolerant Systems (SysTol), pp. 68–73 (2019)

5. El Mawas, Z., Cappelle, C., El Najjar, M.E.B.: Fault tolerant cooperative localization using diagnosis based on Jensen Shannon divergence. In: 2022 25th International Conference on Information Fusion (FUSION), pp. 1–8 (2022)
6. Durrant-Whyte, H., Henderson, T.C.: Multisensor Data Fusion, pp. 585–610. Springer Berlin Heidelberg, Berlin, Heidelberg (2008)
7. Goel, P., Dedeoglu, G., Roumeliotis, S.I., Sukhatme, G.S.: Fault detection and identification in a mobile robot using multiple model estimation and neural network. In: Proceedings 2000 ICRA. Millennium Conference. IEEE International Conference on Robotics and Automation. Symposia Proceedings (Cat. No. 00CH37065), vol. 3, pp. 2302–2309 (2000). https://doi.org/10.1109/ROBOT.2000.846370
8. Wang, C., Shang, W., Sun, D.: Monitoring malfunction in multirobot formation with a neural network detector. Proc. Inst. Mech. Eng., Part I: J. Syst. Control. Eng. **225**(8), 1163–1172 (2011)
9. Christensen, A.L., O'Grady, R., Birattari, M., et al.: Fault detection in autonomous robots based on fault injection and learning. Auton. Robot. **24**, 49–67 (2008)
10. Frank, P.M.: Fault diagnosis in dynamic systems via state estimation—a survey. In: Tzafestas, S., Singh, M., Schmidt, G. (eds.) System Fault Diagnostics, Reliability and Related Knowledge-Based Approaches, pp. 35–98. Springer, Netherlands, Dordrecht (1987)
11. Julier, S., Uhlmann, J.K.: General decentralized data fusion with covariance intersection (CI). In: Hall, D., Llinas, J. (eds.) Multisensor Data Fusion, vol. 3, pp. 319–344. CRC Press (2009)
12. Basseville, M.: Information: entropies, divergences et moyennes (1996)
13. Do, M.N.: Fast approximation of Kullback-Leibler distance for dependence trees and hidden Markov models. IEEE Signal Process. Lett. **10**(4), 115–118 (2003)
14. Goodfellow, I., Bengio, Y., Courville, A.: Deep Learning. The MIT Press (2016)

Development of a Hybrid Safety System Based on a Machine Learning Approach Using Thermal and RGB Data

Nicolas Jathe, Hendrik Stern, and Michael Freitag

Abstract Using liquefied natural gas (LNG) as an alternative to conventional marine fuels can contribute to reaching carbon neutrality. Handling LNG requires tight safety measures due to the risk of explosion and frostbite. The present paper introduces a live monitoring approach to detect safety violations (persons in dangerous areas) and LNG leakage. The system utilizes a dual-camera system with regular RGB, and thermal vision and a machine learning approach suitable for the naval environment and movements like roll, pitch, and yaw. In lab tests, we showed a reliable detection of gas leakage and improved person detection.

1 Introduction and Motivation

Liquified natural gas (LNG) is refrigerated natural gas that can be used for transportation due to the significantly higher density in the cryogenic state compared to its natural state. LNG can be transported to or from regions efficiently that cannot be connected to pipeline systems [1]. This transport and supply of LNG requires adequate infrastructure, and vessels must be equipped for this purpose [2]. Here, safety requirements must be observed, mainly because of the risk of explosion and the extreme danger posed by cryogenic gases to living beings (freezing) [1].

Several requirements must be fulfilled at facilities and vessels to obtain adequate LNG safety. First, preventive safety measures can be executed, e.g., adequate

N. Jathe (✉) · M. Freitag
BIBA - Bremer Institut für Produktion und Logistik GmbH at the University of Bremen, Hochschulring 20, 28359 Bremen, Germany
e-mail: jat@biba.uni-bremen.de

M. Freitag
e-mail: fre@biba.uni-bremen.de

H. Stern
Faculty of Production Engineering, University of Bremen, Badgasteiner Str. 1, 28359 Bremen, Germany
e-mail: hstern@uni-bremen.de

D. Theilliol et al. (eds.), *Recent Developments in Model-Based and Data-Driven Methods for Advanced Control and Diagnosis*, Studies in Systems, Decision and Control 467,
https://doi.org/10.1007/978-3-031-27540-1_24

maintenance of pipe systems or vessels, to lower the risk of failures. Here, digital assistance systems can be used [3]. Second, further safety measures such as primary containment, secondary containment, safeguard systems, and separation distance are used. They apply throughout the whole LNG value chain, from production, liquefaction, and shipping to storage and re-gasification [4, 5]. This paper aims to contribute to LNG safety in the field of safeguard systems. It describes a novel monitoring approach for ensuring safety during LNG transfer operations (e.g., bunkering process). It is based on a combined RGB and thermal (IR) camera, which enables automatic hazard detection within the use cases (detection of persons in the hazardous area as well as detection of gas leaks during the bunkering process) in combination with Deep Learning algorithms. The monitoring approach was subsequently evaluated in laboratory tests.

2 State of the Art: Safety Systems for Leakage and Person Detection at LNG Vessels

The handling of LNG can potentially lead to hazardous events that adequate safety measures must prevent. Potential hazard events may be attributed to the specific properties of LNG. Such events can be explosions, gas venting, frostbite, or vaporization [4, 5]. As described above, various safety requirements for LNG systems are designed to prevent these hazardous events. In addition to structural design measures that serve to store and move the LNG in safe and stable tanks and pipe systems (primary and secondary containment), safeguard and safety systems are of great importance. In this case, the aim is to rapidly detect a possible gas leak utilizing detectors (e.g., gas, liquid, or fire sensors) and thus keep its effects as minor as possible. If a gas leak is detected, such a system would initiate an emergency shutdown and warn the crew [4, 5]. In addition, a fourth safety measure is to maintain safe distances [4, 5].

Our focus is on the consideration of safety systems. Various research efforts exist to provide additional protection against hazardous events through more advanced sensor technology (e.g., infrared sensors or cameras). In research work by Lin et al. [6], piezoelectric acoustic sensors were used to detect gas leaks using characteristic acoustic signals that differ from acoustic signals in the non-leaking state. In addition infrared sensors or cameras have been used in various works to detect gas leaks in conjunction with image processing by detecting concentration changes of the leaking gas [7, 8]. Use cases have been leakage from pipelines [7] or gas-powered household appliances in conjunction with an audible alarm [8]. In studies using machine learning approaches, gas detection was implemented using RGB images [9] or IR images [10] for smoke detection. Here, the focus was on leaks at pipelines [9] or inside gas production power plants [10].

3 Results

In this paper, the two main tasks are a monitoring approach for covering the bunkering process and fast and reliable detection of gas leakage (Sect. 3.1) and a dual camera approach to object detection used for detecting persons (Sect. 3.2), with an additional focus on safety violation prediction approach for moving workers via object tracking (Sect. 3.2.4).

3.1 Leakage Detection

Fast leakage detection is crucial in safely handling LNG. This section describes several algorithms for detecting fast and reliable potential leakages in maritime environments. Wong et al. [11] have introduced a threshold-based approach to detect temperature events. If the temperature reaches the threshold, the detector is triggered. The approach takes only temperature values converted to color pixels as input, which could be problematic if the conversion is done in a normalizing way. Further, since a leak of LNG would take time to cool down the surrounding area, relying only on a threshold approach for the temperature appears not to be fast enough to prevent hazardous events. Other approaches are based on leakage detection on IR images. Here, Bin et al. [10] used a ResNet [12] approach to detect leaking gases visible in the infrared band. A similar approach was performed by Xu et al. [13] with a Fast-RCNN [14] architecture. However, in maritime environments strong winds and fog could make detecting leakage only via detecting the gas itself unreliable. Consequently, the following section illustrates a novel approach to detecting the emersion point of the leakage.

3.1.1 Basic Approach: Leakage Detection on Thermal Images

To detect leakage via IR imaging, the following definition is used: Let $I_t \in \mathbb{R}^{w \times h}$ be an IR image of size $w \times h$ at timestamp $t \in \mathbb{N}$ and $I_{t,i,j}$ with $i \in [1, \ldots, w]$ and $j \in [1, \ldots, h]$ its corresponding values, then the Boolean leak-indicator $L_{t,i,j}$ at the location i, j and time t is defined as

$$L_{t,i,j} = \begin{cases} 1, & \text{if} \quad I_{t,i,j} < r_{\min} \\ & \& \quad (I_{t,i,j} - I_{t-1,i,j})/t_\Delta < c_{\min} \\ 0, & \text{else.} \end{cases} \tag{1}$$

with $r_{\min}$ as IR threshold, the time t_Δ between t and $t-1$ and the IR change rate threshold $c_{\min}$. Leaking gas is cooling the leaking spot rapidly and is detected if the temperature change is greater than the IR change rate threshold. The value $c_{\min}$ is always a trade-off between sensitivity and specificity and should be chosen for

Fig. 1 IR image of the demonstrator with detected leakage points illustrated in red. Different scenarios were tested by varying the air pressure inside the pipe system and by controlling a valve. The system had leaking points between the linked parts detected by the monitoring approach. The value for $r_{\min}$ was adjusted accordingly. (All the used IR data in this paper have a resolution of 0.04 mk and are stored on a 16-bit basis. With the conversion formula of $T°C = \text{rawvalue} * \text{resolution} - 273,15\,°\text{C}$ this results in a temperature range between −273, 15° and 2348, 29 °C. Therefore, every displayed temperature image here is color adjusted to its full color range)

each combination of gas and surrounding conditions. The value for $r_{\min}$ should be higher than the temperature of the monitored gas. For LNG the value for $r_{\min}$ has to be greater than −161 °C to apply. Also $r_{\min}$ should remain at a reasonable level to reduce the false positives rate.

A demonstrator was created to evaluate the monitoring approach's functionality for detecting leakage. It is designed as a system of pipe connections, which can be cooled down significantly compared to the environment by adding solid carbon dioxide. At an ambient temperature of 20 °C, up to −75 °C was reached in the pipe system. In prepared areas, the demonstrator has leaks where cold gas could leak out. Thus, the demonstrator intends to replicate a (faulty) LNG bunkering system without posing a risk to the operators involved in the tests. Figure 1 shows the demonstrator and the detected leakage as described in Eq. 1.

3.1.2 Extended Approach: Improved Leakage Detection on Thermal Images

The experiments performed with the demonstrator indicated that the monitoring approach described in Eq. 1 could detect leakage in a closed lab experiment. However, for the leakage detection approach to work reliably, the location covered by $I_{t,i,j}$ needs to be the same as in $I_{t+1,i,j}$, which is not given for moving parts or camera movement. Also, the values in $I_{t,i,j}$ and $I_{t+1,i,j}$ need to be related to the same object. Otherwise false positive detections could occur when objects are temporally occluded in one frame and visible in another.

The following section describes a neural network architecture to overcome both mentioned problems. The used architecture is an enhanced U-Net [15] with a Spatial Transformer Network (STN) [16], with τ_θ as a 2D affine transformation. The STN concatenated with the original first skip-connection (SC) used in [15] formed the used connection to link the first stages of the contracting and the expansive path. Further SCs are equal to the original. The input size is $2 \times 252 \times 336$, representing the frames I_t and I_{t+1}. The output size is $1 \times 252 \times 336$. Hence the convolution layers use padding to preserve the resolution. The U-Net is mainly used for detecting occlusions, and the STN is used to reverse potential movement between the frames.

The neural network was trained on captured data (69500 IR images) of the leakage detection at the demonstrator with a 5-fold cross-validation split. Here, we applied data augmentation consisting of random vertical and horizontal flipping and random affine transformations with a fill value of the minimum of the frame. Additional noise was not used since the detection via Eq. 1 is highly sensitive to temperature changes.

The training procedure described above to detect the leakage caused training instability due to the extreme imbalance of leakage and non-leakage. Thus, to get the training to be more stable, the network's main goal was not to detect the leakage but to transform the data so that the output frame was congruent to the first frame and the pixels occluded were equal to the first image. That way, no false detection via Eq. 1 is made. Figure 3 shows a chart of the approach. The loss function was a weighted mix between the MeanSquaredError (MSE) for the complete image and the MSE for the center part in a 1:10 ratio. In Fig. 2 an example of the augmented input, the target, and the prediction is shown. Reliable leakage detection using only the approach from Eq. 1 is not possible in actual use cases with moving parts. The data was split into five equally sized parts with added artificial movement to test the approach utilizing the neural network described in this section. Then a five-fold cross-validation was used to evaluate the network's performance: The data was split into five equally sized parts, with four parts used for training and the remaining part used for validation. This was repeated five times, i.e., until each part was used for validation. That approach achieved over all five folds a mean F1 score for leakage detection of 0.78 with a minimum of 0.73. Here, the detection via Eq. 1 was used as ground truth without the added movement.

3.2 *Object Detection*

For security reasons, some areas are prohibited to people during an ongoing bunkering process. The following section describes a surveillance solution to effective monitor these zones under challenging illumination and different environmental conditions based on RGB and IR cameras.

3.2.1 Approach: Object Detection on RGB Data

For person detection on RGB images, the yolox [17] algorithm is a good choice. The authors provide pretrained weights for 300 epochs on COCO [18] on their Github page, which achieves state-of-the-art results. The performance on the chosen test set is provided in Sect. 3.2.3 and covers the RGB part of the proposed dual camera

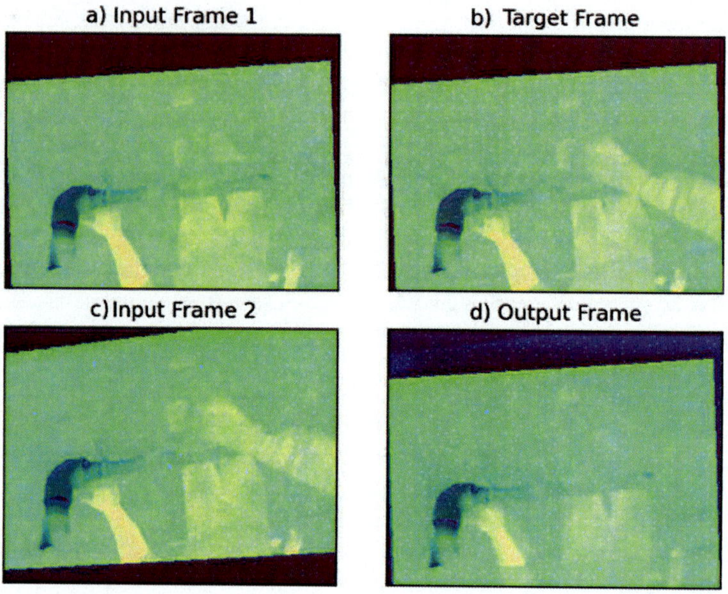

Fig. 2 Example images for the network: **a** First input frame I_t augmented with random flip and an affine transform. **b** The target frame I_{t+1} with the same flip and transform as for the first input frame. The target frame is unknown to the network. **c** Second input frame resulting from the target frame with an additional affine transform. **d** Output frame of the network. The network solved the first problem by correctly reverting the second affine transform. Further, the network extracted the occluding arm on the right side of the second input frame from the image, reducing the possibility of false positive detections

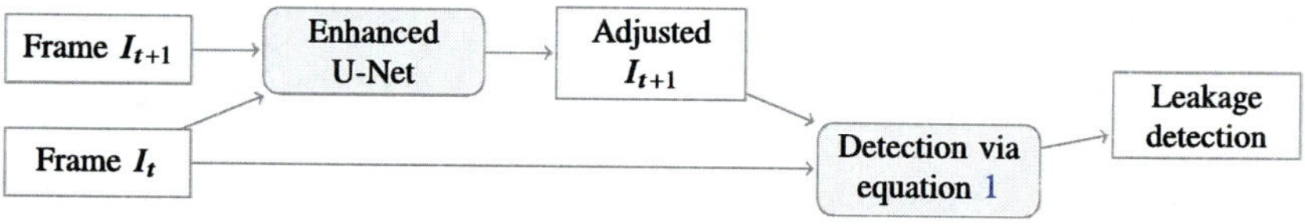

Fig. 3 The enhanced U-Net detects possible movement from frame I_t to frame I_{t+1} and adjusts frame I_{t+1} accordingly. The adjusted frame I_{t+1} and the normal frame I_t should now be congruent, and Eq. 1 is used to detect possible leakage

surveillance system. Akula and Sardana [19] indicate that the high variability of IR signatures is problematic for handcrafted feature detectors and propose to use CNN-based methods. Thus, this paper proposes an approach based on an adapted yolox architecture for the IR part.

3.2.2 Approach: Object Detection on Thermal Data

The algorithm uses the raw data of an IR camera and is based on the yolox [17] approach for object detection in computer vision. The following changes and additions were made to the initial algorithm:

Architecture adjustments The architecture uses the yolox backbone and has only one input channel since no color information is given. The head consists of the decoupled one proposed in yolox [17]. Since the scenario only requires the class *person*, the output size was lowered accordingly. Anchors were not adapted to the training data.

Training and preprocessing details The architecture was trained on the raw IR data of the training part of the Teledyne FLIR Free ADAS Thermal Dataset v2.[1] The dataset contains annotated RGB, thermal raw, and thermal to greyscale images captured in various locations and under different conditions. For this paper only annotated objects of the category *persons* of the 15 provided categories are relevant. The preprocessing of the raw data consists of a normalizing step to keep the gradients low and improve learning speed. The unnormalized input of raw data stored as 16-bit greyscale images resulted in unstable learning at later training stages and weaker performance. Mapping the image extrema to 0 and 255 and linear interpolating in between made the network uncertain in cases where variance was high. Artificially changing the extrema but not altering an object and its surroundings in the same image resulted in different output confidence values for the network for that object, which introduces uncertainty due to the normalizing method. The used normalization of the raw data was the subtraction by the training data's mean and division by its standard deviation. The loss function was the same composition of weighted loss functions provided by yolox [17]. The weights got randomly initialized. The training consisted of 300 epochs using cosine annealing for learning rate reduction.

Insights into data augmentation techniques Despite having 10742 training images, the following data augmentation methods were used in the training step for better results: Applied noise consisting of Gaussian noise with $\sigma = 1$ °C and salt and pepper noise with values from within image range. Further, randomly applied perspective transforms ranging between 0° and 10° were performed. Additionally, we used simple augmentation as random cropping and vertical flipping. Out of [20], Mosaic, and MixUp were considered. An adapted Mosaic, originally a method to combine four images into one new training image by stitching cropped versions of the original images together, was used for training. The adaption randomly combined two, three, or four images in one or two rows and up to three columns. MixUp is an advanced augmentation method mainly introduced for visual object detection classification. It combines two images of different objects by (weighted) cross-fading them into each other so that both objects are still visible and changes the target score from one hot encoded to one representing the weights of the object classes to enable more efficient learning. However, blending two IR images would result in losing the IR spectrum characteristic of the object, and only the shape characteristics would remain.

[1] https://www.flir.com/oem/adas/adas-dataset-agree/.

Table 1 Average precision at 50% intersection over union for the FLIR V2 dataset on the corresponding validation sets for detecting persons. The architectures used the weights provided by Ge et al. [17] on the colored and the thermal to greyscale images. The architectures handling the raw IR data used the weights after the training described in Sect. 3.2.2

Used architectures	AP_{50} on FLIR V2		
	RGB	Greyscale	Raw
Yolox_nano	11.6	8.8	17.8
Yolox_x	39.6	30.9	60.1

The training process showed that convergence was only possible without applying MixUp. However, it might be possible to use this technique on trained networks to boost robustness.

3.2.3 Application and Validation

For comparability, the performance of the proposed algorithms is measured on the validation part of the Teledyne FLIR Free ADAS Thermal Dataset v2. Table 1 shows the performance of the used detections. The time needed for inferencing and non-max suppression on a Nvidia GTX 1080ti for the yolox_nano is 5 ms (~200 fps) and for the yolox_x 42 ms (~24 fps).

3.2.4 Combined Approach: Movement Monitoring to Predict Safety Violations

To further improve person detection and tracking under various conditions, fusing the detection on the RGB and the IR data together by using all bounding boxes of the RGB and the bounding boxes of the IR detection, and applying non-max-suppression, taking the offset of the cameras into account, makes the overall detection more robust. This combined approach achieved a slightly better performance than the two individual approaches on a small intern dataset with known offset. The dataset is not big enough to quantify the results but indicates that the system can perform a domain shift, since the data source is different and the training data does not contain any part of this dataset. With the knowledge of where persons are, the system links on each timestamp the detected persons with the nearest bounding boxes of the previous frame. With this information of movement, the system predicts via Kalman filtering over the midpoints of the bounding boxes the most likely path of the person and can trigger a warning if the path violates predefined prohibited zones. Also, the system performs an alarm if the person is located in the prohibited zone. This approach is simple and was successfully tested in a lab environment with different persons as test subjects. In Fig. 4 the proposed system is illustrated.

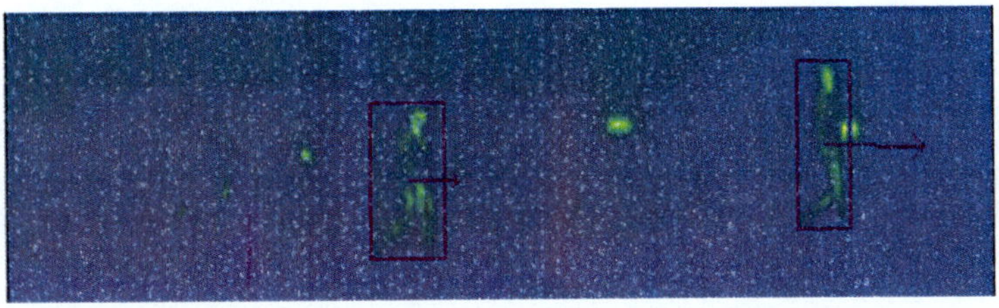

Fig. 4 Bounding boxes marking detected person and the arrows indicating the predicted path, calculated by a Kalman filter of the previous movement

4 Summary and Conclusion

This research work developed and evaluated a monitoring solution for ensuring safety during LNG transfer operations. The monitoring solution is based on a combined RGB and IR camera, which, combined with Deep Learning algorithms, enables automatic hazard detection within the use cases (detection of persons in the hazardous area and detection of gas leaks during the bunkering process).

Extensive laboratory tests showed that the selected approach enables reliable detection of the use cases based on IR images and thus represents a promising alternative to a monitoring approach based on RGB images due to its robustness to maritime environmental conditions.

Acknowledgements The authors would like to thank the German Federal Ministry of Economic Affairs and Energy (BMWi) for their support within the project "LNG Transfer—LNG Safety" (grant number 16KN062731).

References

1. Mokhatab, S., Economides, M.: Onshore LNG Production Process Selection (2006). https://doi.org/10.2118/102160-MS
2. Mokhatab, S., Mak, J.Y., Valappil, J., Wood, D.A.: Handbook of Liquefied Natural Gas. Gulf Professional Publishing (2013)
3. Stern, H., Leder, R., Lütjen, M., Freitag, M.: Human-centered development and evaluation of an AR-assistance system to support maintenance and service operations at LNG ship valves, vol. 09, pp. 272–294 (2021). ISBN 978-3-95545-396-1. https://doi.org/10.30844/wgab_2021_17
4. Alderman, J.A.: Introduction to LNG safety. Process. Saf. Prog. **24**(3), 144–151 (2005). https://doi.org/10.1002/prs.10085
5. Foss M.M., Delano, F., Gülen, G., Makaryan, R.: LNG safety and security, Center for Energy Economics (CEE) (2003)
6. Lin, W., Jiang, L., Haiyan, W.: Study of non-intrusive gas pipeline leak detection with acoustic sensor. IFAC Proc. Vol. **46**(20), 27–32 (2013). https://doi.org/10.3182/20130902-3-CN-3020.00039
7. Liu, B., Ma, H., Zheng, X., Peng, L., Xiao, A.: Monitoring and detection of combustible gas leakage by using infrared imaging. In: 2018 IEEE International Conference on Imaging Systems and Techniques (IST), pp. 1–6 (2018). https://doi.org/10.1109/IST.2018.8577102

8. Fraiwan, L., Lweesy, K., Bani-Salma, A., Mani, N.: A wireless home safety gas leakage detection system. In: 1st Middle East Conference on Biomedical Engineering (2011). https://doi.org/10.1109/MECBME.2011.5752053
9. Marshall, Park J.-S., Song, J.-K.: FCN based gas leakage segmentation and improvement using transfer learning. In: 2019 IEEE Student Conference on Electric Machines and Systems (SCEMS 2019), pp. 1–4 (2019). https://doi.org/10.1109/SCEMS201947376.2019.8972635
10. Bin, J., Rahman, C.A., Rogers, S., Liu, Z.: Tensor-based approach for liquefied natural gas leakage detection from surveillance thermal cameras: a feasibility study in rural areas. IEEE Trans. Ind. Inform. **17**(12), 8122–8130 (2021). https://doi.org/10.1109/TII.2021.3064845
11. Wong, W.K., Tan, P.N., Loo, C.K., Lim, W.S.: Machine condition monitoring using omnidirectional thermal imaging system. In: 2009 IEEE International Conference on Signal and Image Processing Applications, pp. 299–304 (2009). https://doi.org/10.1109/ICSIPA.2009.5478665
12. He, K., Zhang, X., Ren, S., Sun, J.: Deep residual learning for image recognition. In: 2016 IEEE Conference on Computer Vision and Pattern Recognition, pp. 770–778 (2016). https://doi.org/10.1109/CVPR.2016.90
13. Xu, K., Yuan, Z., Zhang, J., Ji, Y., He, X., Yang, H.: SF6 gas infrared thermal imaging leakage detection based on faster-RCNN. In: 2019 International Conference on Smart Grid and Electrical Automation (ICSGEA), pp. 36–40 (2019). https://doi.org/10.1109/ICSGEA.2019.00017
14. Girshick, R.B.: Fast R-CNN. CoRR, abs/1504.08083 (2015)
15. Ronneberger, O., Fischer, P., Brox, T.: U-net: convolutional networks for biomedical image segmentation. CoRR, abs/1505.04597 (2015)
16. Jaderberg, M., Simonyan, K., Zisserman, A., Kavukcuoglu, K.: Spatial transformer networks. CoRR, abs/1506.02025 (2015)
17. Ge, Z., Liu, S., Wang, F., Li, Z., Sun, J.: YOLOX: exceeding YOLO series in 2021. CoRR, abs/2107.08430 (2021)
18. Lin, T.-Y., Maire, M., Belongie, S.J., Bourdev, L.D., Girshick, R.B., Hays, J., Perona, P., Ramanan, D., Dollár, P., Zitnick, C.L.: Microsoft COCO: common objects in context. CoRR, abs/1405.0312 (2014)
19. Akula, A., Sardana, H.K.: Deep CNN-based feature extractor for target recognition in thermal images. In: TENCON 2019 - 2019 IEEE Region 10 Conference, pp. 2370–2375 (2019). https://doi.org/10.1109/TENCON.2019.8929697
20. Bochkovskiy, A., Wang, C.-Y., Liao, H.-Y.M.: Yolov4: optimal speed and accuracy of object detection. CoRR, abs/2004.10934 (2020)

A Comparative Study on Damage Detection in the Delta Mooring System of Spar Floating Offshore Wind Turbines

Christos S. Sakaris, Anja Schnepf, Rune Schlanbusch, and Muk Chen Ong

Abstract The most common type of Floating Offshore Wind Turbine (FOWT) installed in Norway, is the spar FOWT in which a delta mooring system (DMS) is used. The mooring system of a FOWT is an essential part for its station-keeping, whose loss can lead to the collapse of the FOWT and the endangerment of the human safety. Thus, early detection of damages in the mooring system is vital. In this study, damage detection in the DMS of a spar FOWT under varying environmental conditions (ECs) is investigated through a comparison of the Multiple Model-AutoRegressive (MM-AR) method, the Multiple Model-Power Spectral Density (MM-PSD) method and the Functional Model Based Method (FMBM). The MM-AR and the MM-PSD methods are based on multiple PSD based or AR models and the FMBM on a single Functional Model (FM) for the description of the healthy FOWT's dynamics under varying ECs. The results show that successful and precise damage detection in a spar FOWT's DMS can be achieved through the employed statistical methods as the MM-AR method and the FMBM detect all the examined 7 healthy and 16 damage cases whereas the MM-PSD method misses only one damage case. The results also show that the parametric AR models and FM describe more precisely the FOWT dynamics under varying ECs in comparison to the non-parametric PSDs.

C. S. Sakaris (✉) · R. Schlanbusch
Technology Department, Norwegian Research Centre, Jon Lilletuns vei 9 H, 3. et, 4879 Grimstad, Norway
e-mail: csak@norceresearch.no

R. Schlanbusch
e-mail: rusc@norceresearch.no

A. Schnepf · M. C. Ong
Department of Mechanical and Structural Engineering and Materials Science, University of Stavanger/CoreMarine, Stavanger, Norway
e-mail: anja.schnepf@uis.no

M. C. Ong
e-mail: muk.c.ong@uis.no

D. Theilliol et al. (eds.), *Recent Developments in Model-Based and Data-Driven Methods for Advanced Control and Diagnosis*, Studies in Systems, Decision and Control 467,
https://doi.org/10.1007/978-3-031-27540-1_25

1 Introduction

Floating Offshore Wind Turbines (FOWTs) supported by a spar type platform, are most commonly used in Norway. Currently, the only operational spar FOWT in Norway is the Zephyros FOWT, formerly Hywind Demo, which has been developed by Equinor and it is connected to the onshore grid [1]. However, a floating wind farm consisted of spar FOWTs, named Hywind Tampen, is under construction and it will be connected to power offshore oil and gas facilities [2]. The station-keeping system of the aforementioned spar FOWTs is a delta mooring system (DMS), also called crowfoot mooring system, consisting of three chain mooring lines.

The mooring lines constitute a critical part in maintaining the station-keeping of a FOWT and thus early detection of damages in this part is vital. Otherwise, failure to detect a damage may lead to an increase in the mooring lines' tension, loss of stability, high maintenance cost, and possibly to the FOWT's collapse. However, damage detection in the mooring lines of FOWTs has been investigated in very few studies under constant [3] and varying [4–6] environmental conditions (ECs) and through methods based on data-based models (developed with acquired signals from the structure) such as Power Spectral Density (PSD) [3, 5] and Neural Network [4, 6]. In the baseline (training) phase of some of these methods, the use of a large number of data records under different ECs is required [6]. Additionally, there are methods in which it is not clear how the different ECs are handled in the baseline phase [5]. Also only damage cases are examined in the inspection (real time) phase of some methods, thus leaving the methods' effectiveness in damage detection unclear [4, 5]. These methods treat damage detection as a classification problem in which an unknown state is classified as a healthy or a damage state.

Recently, damage detection in the tendons of the multibody TELWIND platform of a 10 MW FOWT under varying mean wind velocity (MWV) and significant wave height (SWH), has been investigated through the Functional Model Based Method (FMBM) [7].

The aim of the present study is the investigation of damage detection in a spar FOWT's DMS through a comparison of data-based statistical methods. These are the Multiple Model-AutoRegressive (MM-AR) method, Multiple Model-PSD (MM-PSD) method [8] and the FMBM [7] which have been applied for damage detection on different structures successfully and they are part of the Statistical Time Series (STS) methods. The STS methods constitute an effective tool for robust damage detection and they present a number of advantages such as needing no physics-based models, achieving damage detection in a statistical decision context and with partial data-based models developed with response signals from a limited number of sensors, taking into account of various types of uncertainties through statistical tools and getting applied without interrupting the inspected structure's normal operation [9].

The spar FOWT used in the present study constitutes a modification of the artificial 5 MW OC3-Hywind FOWT [10]. Various cases with the healthy and damaged

FOWT under varying MWV and SWH are examined and the simulated damages are considered at different locations in a single delta line and of different magnitudes.

2 The Spar FOWT and the Simulations

The spar FOWT. In the current study, the examined spar FOWT is a modified 5 MW OC3-Hywind FOWT which is based on the Hywind concept [10]. The artificial OC3-Hywind FOWT has been studied extensively in the literature [11, 12] with its original mooring system being a plain catenary mooring system and consisting of three quasi-static catenary lines and an artificial yaw stiffness.

The modification consists of the replacement of the original mooring system with the DMS which is part of the design of the spar FOWTs Zephyros and Hywind Tampen. In the DMS, each of the three mooring lines consists of two delta lines and a catenary line that are connected through a delta plate. The delta lines are also connected to two fairleads on the spar platform whereas the catenary line is anchored at a water depth of 320 m, as shown in Fig. 1a. The properties of the catenary lines of the modified spar FOWT are the same as for the original OC3-Hywind FOWT's catenary lines [10], whereas the properties of the delta lines are the half of the catenary lines' properties.

Simulations and damages. MWV and SWH are considered as varying ECs in this study and various healthy and damage cases are examined for seven pairs of MWVs/SWHs shown in Table 1. Each considered damage case is simulated by a stiffness reduction of a specific magnitude (%) at a specific location in a delta line of a mooring line. The considered damage magnitudes correspond to 30 and 70% stiffness reduction, whereas the considered damage locations correspond to 3 and 16% of the delta line's length (see Fig. 1b). Due to the dependence of SWH on MWV, each healthy case is represented as F^w with w denoting the WV and each examined damage case is represented as $F^w_{m,q}$ with m being the damage magnitude and q the damage location.

The numerical model of the modified OC3-Hywind FOWT is set up in the numerical software OrcaFlex according to [13]. Based on this model, heave acceleration signals are generated under the considered healthy and damage cases. In relation to the varying MWV and SWH considered during the simulations, the MWV is computed based on the power law velocity profile in [14]. The Pierson-Moskowitz spectrum is used for generating fully developed sea states in deep water and the

Table 1 Values of the varying mean wind velocity (MWV) and significant wave height (SWH)

MWV (m/s)	8	10.3	11.4	13.8	15	17.3	18
SWH (m)	0.85	1.4	1.72	2.52	2.98	3.96	4.29

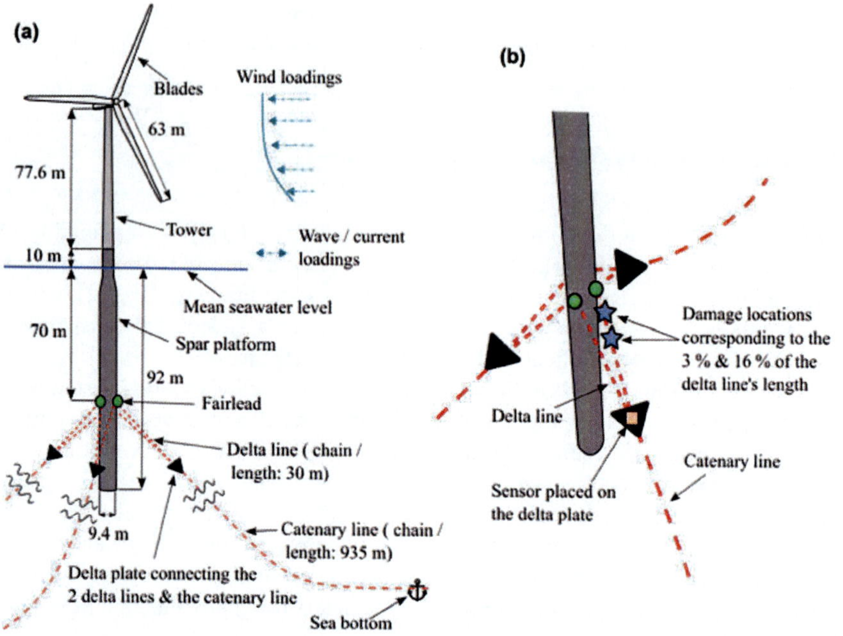

Fig. 1 **a** The Spar FOWT and **b** the position of the sensor and the considered damages

values of the spectrum's SWH and the peak period are obtained with the empirical formulations in [15, pp. 150–155]. The surface current velocity is 0.5 m/s and the mean current velocity throughout the entire water depth is calculated based on the power law profile in [16].

During each simulation, an acceleration signal is generated based on the location of one sensor placed on the delta plate, as shown in Fig. 1b. Table 2 presents the details and the numbers of the conducted simulations. It must be noted that the signals corresponding to the 7 baseline phase simulations under the healthy FOWT are used for the training of each method and they are different from their counterparts in the inspection phase. Each signal is mean corrected and normalized based on the sample's standard deviation.

Effects of varying MWV, SWH and damages on FOWT's dynamics. The MWV variability leads to a change in the FOWT's dynamics. Moreover, a comparison of PSDs in Fig. 2 shows that damages of different magnitude $F_{30,3}^{11.4}$, $F_{70,3}^{11.4}$ and the healthy state under the same MWV have almost the same effects on the dynamics. This shows that some damages have small effects and their detection is very difficult. The PSDs are estimated based on the Welch estimator [7] (Window length 400 samples, 0% overlap, Hamming window, frequency resolution of 0.025 Hz).

Table 2 Number of simulations per healthy/damage state and details of vibration signals

Structural state	No. of simulations—baseline phase	No. of simulations—inspection phase
Healthy	7 (one per F^8, $F^{10.3}$, $F^{11.4}$, $F^{13.8}$, F^{15}, $F^{17.3}$, F^{18})	7 (one per F^8, $F^{10.3}$, $F^{11.4}$, $F^{13.8}$, F^{15}, $F^{17.3}$, F^{18})
Damaged	–	9 (one per $F^{10.3}_{30,16}$, $F^{10.3}_{30,3}$, $F^{10.3}_{70,16}$, $F^{10.3}_{70,3}$, $F^{8}_{30,3}$, $F^{8}_{70,3}$, $F^{11.4}_{30,3}$, $F^{11.4}_{70,3}$, $F^{17.3}_{30,3}$)
Sampling frequency: $f_s = 10$ Hz, Signal bandwidth: [0–1.4] Hz		
Signal length: $N = 20000$ samples ($\approx$1000 s)		

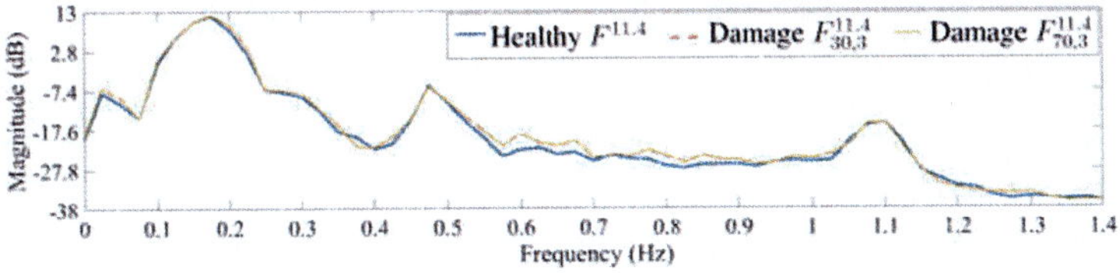

Fig. 2 Comparison of Welch-based PSD estimates using heave acceleration signals for healthy & damage states under MWV 11.4 m/s

3 Damage Detection Methods

In the present study, unsupervised damage detection in the DMS of a spar FOWT under varying MWV and SWH is investigated through a comparison of three data-based methods. In these methods, the healthy FOWT's dynamics under varying ECs is represented by features in a proper subspace and damage detection is achieved by checking if the current dynamics resides in the subspace (see Fig. 3). In the MM-AR and MM-PSD methods, the subspace is described by multiple models such as Auto-Regressive (AR) models and PSDs correspondingly and the features are the AR parameters or PSD magnitude values [8]. In the FMBM, the subspace is described by a single Functional Model (FM) and the features are the FM's parameters [7]. The methods consist of two phases, the baseline phase which is performed based on data from known structural states and the inspection phase which is based on current data while the structure is under an unknown health state.

3.1 *MM-AR and MM-PSD Methods [8]*

Baseline phase. Initially in this phase, *M* response signals corresponding to a sample of the considered MWVs, are acquired from the healthy FOWT. Due to the dependence between MWV and SWH, only MWV is used. The sampled MWVs cover the

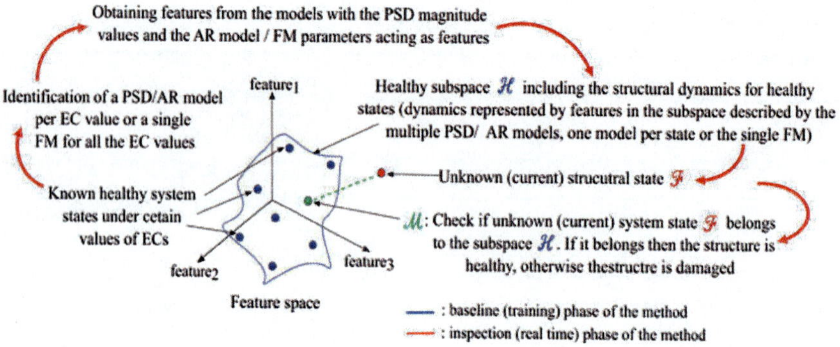

Fig. 3 General concept of damage detection through the methods MM-PSD, MM-AR and FMBM

range $[w_{min}, w_{max}]$ via the discretization $w_r \in w_1, w_2, \ldots, w_M$. Then each signal characterized by a specific MWV w_r, is used for estimating a PSD for the MM-PSD method [7] or an AR model for the MM-AR method via standard identification procedures [7]. A feature vector $\mathbf{m}_{o,w_r}$ is obtained from each model and it contains the PSD magnitude values of the frequencies in the considered bandwidth or the AR model parameters.

The effectiveness of the MM-AR method may be improved if the AR parameters with variance over a user-defined threshold are excluded from the method's feature vector, as the increased variances indicate that these parameters are more sensitive to the varying MWV. The variance of each parameter is based on the AR models corresponding to the considered MWVs. The same procedure can be followed for the improvement of the MM-PSD method where the frequencies with PSD variance over a user-defined threshold are excluded.

Inspection phase. A new response signal is acquired under the FOWT's current (unknown) health state and it is used for estimating a new model (PSD or AR model) and obtaining a new feature vector $\mathbf{m}_u$. Then the distance f_r between $\mathbf{m}_u$ and each feature vector $\mathbf{m}_{o,w_r}$ from the baseline phase, is calculated and it is used for obtaining the distance metric Q with Q= $\min_r f_r$ [8]. Feature vectors $\mathbf{m}_u$ and $\mathbf{m}_{o,w_r}$ must have the same length. Finally, a damage in the FOWT is detected when Q is greater than a user-defined limit that is selected based on the values of Q in the baseline phase. In the current study, two distances f_r are used, the Euclidean distance in the MM-PSD method and the Mahalanobis distance in the MM-AR method.

3.2 *FMBM [7]*

Baseline phase. In this initial phase, a Functionally Pooled-AutoRegressive (FP-AR) model is estimated based on M response signals from the healthy FOWT corresponding to a sample of the considered MWVs [7] (the same M signals used in the baseline phase in Sect. 3.1). The structure of a FP-AR$(na)_p$ model is the following [7]:

$$y_{w_r}[t] + \sum_{i=1}^{na} a_i(w_r) \cdot y_{w_r}[t-i] = e_{w_r}[t], e_{w_r}[t] \sim \text{iid}\mathcal{N}\big(0, \sigma_e^2(w_r)\big) \quad (1)$$

with na designating the AR order, respectively, $y_{w_r}[t]$ ($t = 1, \ldots, N$) the response signal with length of N samples under WV w_r and $e_{w_r}[t]$ the disturbance signal that is a white (serially uncorrelated), normal distributed ($\mathcal{N}$), zero-mean signal with variance $\sigma_e^2(w_r)$. The AR parameters $a_i(w_r) = \sum_{j=1}^{p} a_{i,j} \cdot G_j(w_r)$, are explicit functions of w_r by belonging to a p-dimensional functional subspace spanned by the (mutually independent) functions $G_1(w_r), G_2(w_r), \ldots, G_p(w_r)$. These functions are polynomials of one variable (like Chebyshev, Legendre, Jacobi etc.). The $a_{i,j}$ designates the AR coefficients of projection. The model is estimated through identification procedures for the selection of the model orders, the determination of the model's functional subspace dimensionality p and the model validation [7]. In the context of the p determination, the $a_{i,j}$ are estimated based on linear regression techniques such as Ordinary Least Squares (OLS) [7] or Weighted Least Squares (WLS) [17].

Inspection phase. A response $y_u[t]$ signal corresponding to a known/measured MWV w, is acquired under the FOWT's current (unknown) state. Then $y_u[t]$ and w are used in the re-parametrized FP-AR$(na)_p$ model (from the baseline phase) in terms of w and the residual signal $e[t, w]$ is obtained [7]:

$$y_u[t] + \sum_{i=1}^{na} a_i(w) \cdot y_u[t-i] = e[t, w] \quad (2)$$

Damage detection is achieved by checking the uncorrelatedness (whiteness) of $e[t, w]$ through a statistical hypothesis test. If $e[t, w]$ is white then the FOWT is in a healthy state, otherwise a damage is detected. In the current study, the Pena-Rodriguez hypothesis test is employed for the whiteness check and a damage is detected when the test's normal distributed quantity D exceeds the distribution's critical limits, $-Z_{1-\alpha}$ >D and/or $Z_{1-\alpha}$ <D ($-Z_{1-\alpha}$, $Z_{1-\alpha}$ the critical limits, α the risk level) [7].

4 Damage Detection Results

MM-AR and MM-PSD methods. In the MM-PSD method's baseline phase, a Welch-PSD estimate (*Matlab function: pwelch.m*) is obtained based on each of the $M = 7$ acceleration signals from the healthy FOWT and under a different MWV (see Table 2). The considered MWVs cover the range $w_r \in [8, 18]$ m/s through the

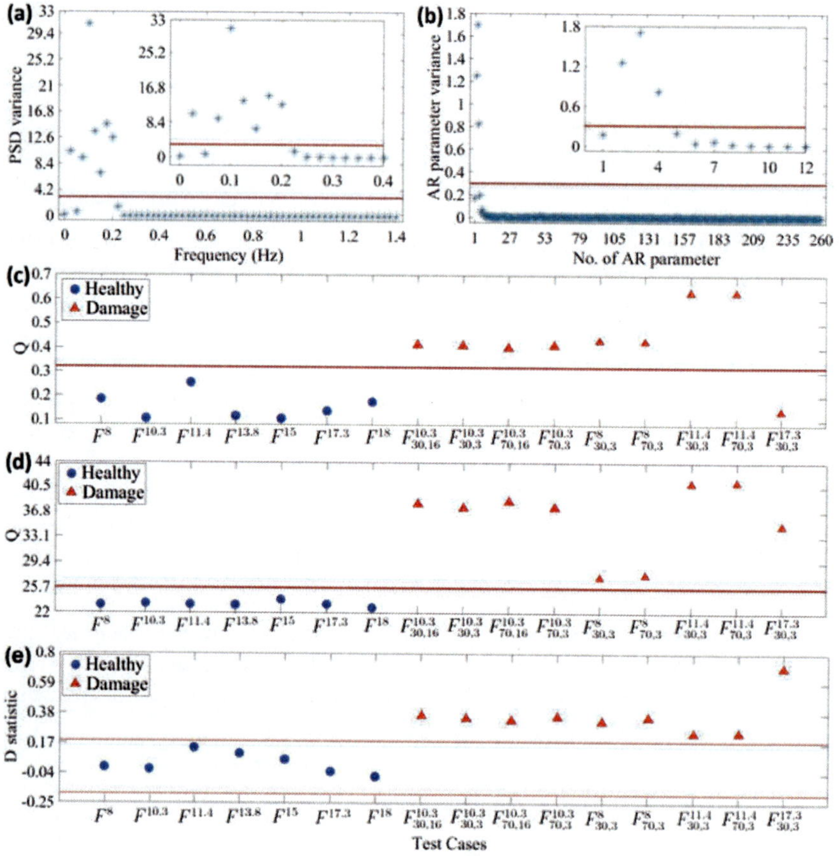

Fig. 4 **a** PSD variance per frequency and **b** AR parameter variance per parameter. Damage detection results based on the **c** MM-PSD & **d** MM-AR methods and **e** the FMBM with limits (—) based on risk level $\alpha = 4.25 \cdot 10^{-1}$

discretization $w_r \in [8, 10.3, 11.4, 13.8, 15, 17.3, 18]$ m/s. Based on these PSDs, the PSD variance per frequency is determined and the feature vectors of reduced dimension are obtained. In the MM-AR method's baseline phase, an AR(260) model is estimated (*Matlab function: arx.m*) based on each of the $M = 7$ signals, the model parameters and their covariance matrices are extracted and the reduced feature vectors are obtained to be used in the Mahalanobis distance. Figure 4a, b depict the PSD variance per frequency and the variance per AR parameter with the excluded frequencies and parameters corresponding to variances above the designated limits (horizontal lines). The details of the employed models and reduced feature vectors are shown in Table 3.

In the inspection phase of both of the methods, damage detection is examined based on 7 healthy cases and 9 damage cases (see Table 2). The damage detection

Table 3 Baseline phase details of the damage detection methods

Detection method	Estimated model	Estimation method	No.of signals	No.of models	Feature vector dimension		Distance
					Initial	Reduced	
MM-PSD	PSD	Welch	$M = 7$	7	$[57 \times 1]$	$[50 \times 1]$	Euclidean
MM-AR	AR(260)	OLS	$M = 7$	7	$[260 \times 1]$	$[255 \times 1]$	Mahalanobis
FMBM	FP-AR$(140)_6$	WLS	$M = 7$	1	–	–	–

PSD Welch-based estimation: segment length 400 samples, overlap 0%, window type Hamming, frequency resolution 0.025 Hz

Condition number: $6.54 \cdot 10^6$–$2.19 \cdot 10^9$ (AR models), $2.97 \cdot 10^{13}$ (FP-AR model)

FP-AR$(140)_6$ model's functional subspace: $p = 6$ univariate Shifted Legendre polynomials

results are presented in terms of Q in Fig. 4c, d with no false alarms occurring in both of the methods. As for missed damages, the MM-PSD method fails to detect only one damage case whereas the MM-AR method is able to detect all of the damage cases.

FMBM. In the *baseline phase*, an FP-AR(140) model is estimated based on the $M = 7$ acceleration signals (see **MM-AR and MM-PSD methods** of Sect. 4). The selected functional subspace consists of $p = 6$ univariate Shifted Legendre polynomials. Thus, the identified model may represent the healthy FOWT under any MWV in the considered range [8–18] m/s. Details of the FP-AR$(140)_6$ model are presented in Table 3.

In the inspection phase of the FMBM, damage detection is examined based on the same 16 cases used in the **MM-AR and MM-PSD methods** of Sect. 4. Figure 4e depicts the damage detection results in terms of D and no false alarms or missed damages occur.

5 Conclusions

Damage detection in the DMS of a spar FOWT has been investigated in the present study. This investigation has been conducted through a comparison of three data-based statistical methods: the MM-AR method, the MM-PSD method and the FMBM. Based on the examined healthy and damage cases under varying MWV and SWH, the MM-AR method and FMBM have been able to achieve a perfect damage detection whereas the MM-PSD method has missed only one damage. This has happened because the AR models (MM-AR) and FMs (FMBM) have been able to describe the FOWT dynamics more precisely in comparison to the PSDs (MM-PSD) and that some damages have had small effects on the structural dynamics. Future steps include the examination of damage detection and quantification in the DMS through MWVs considered only in the inspection phase (not in the baseline phase) and the use of a larger variety of damage magnitudes and data-based methods.

Acknowledgements This research work has been funded by the Research Council of Norway through the projects: (i) Analytics for asset Integrity Management of Windfarms (AIMWind), a collaboration between University of Agder, Norwegian Research Center AS and Technical University Delft, grant no. 312486. Origo Solutions is included as advisory partner. (ii) Design Optimisation of Power Cable, Shared Electrical Line and Mooring configurations for Floating Offshore Wind Turbines, grant no. 320902.

References

1. Driscoll, F., Jonkman, J., Robertson, A., Sirnivas, S., Skaare, B., Nielsen, F.: Validation of a FAST model of the Statoil-hywind demo floating wind turbine. Energy Procedia **94**, 3–19 (2016)
2. Equinor: Hywind Tampen, PL050 - PL057 - PL089, PUD del II - Konsekvensutredning. Tech. rep., Statoil (2019)
3. Jamalkia, A., Ettefagh, M., Mojtahedi, A.: Damage detection of TLP and spar floating wind turbine using dynamic response of the structure. Ocean Eng. **125**, 191–202 (2016)
4. Dehkharghani, P., Ettefagh, M., Hassannejad, R.: Mooring damage identification of floating wind turbine using a non-probabilistic approach under different environmental conditions. J. Mar. Sci. Appl. **20**, 156–169 (2021)
5. Liu, Y., Ferrari, R., Wu, P., Jiang, X., Li, S., Wingerden, J.: Fault diagnosis of the 10MW floating offshore wind turbine benchmark: a mixed model and signal-based approach. Renew. Energy **164**, 391–406 (2021)
6. Gorostidi, N., Nava, V., Aristondos, A., Pardo, D.: Predictive maintenance of floating offshore wind turbine mooring lines using deep neural networks. J. Phys.: Conf. Ser. **2257** (2022)
7. Sakaris, C., Yang, Y., Bashir, M., Michailides, C., J., W., Sakellariou, J., Li, C.: Structural health monitoring of tendons in a multibody floating offshore wind turbine under varying environmental and operating conditions. Renew. Energy **179**, 1897–1914 (2021)
8. Kaliorakis, N., Iliopoulos, I., G., V., Sakellariou, J., Fassois, S., Deloukas, A., Leoutsakos, G., Chronopoulos, E., Mamaloukakis, C., Katsiana, K.: On the on-board random vibration-based detection of hollow worn wheels in operating railway vehicles. Lect. Notes Civ. Eng. **128**, 480–489 (2020)
9. Fassois, S.D., Sakellariou, J.: Statistical time series methods for structural health monitoring. In: Boller, C., Chang, F.K., Fujino, Y. (eds.) Encyclopedia of Structural Health Monitoring, pp. 443–472. Wiley, Chichester (2009)
10. Jonkman, J.: Definition of the floating system for phase IV of OC3. Tech. Rep. NREL/TP-500-47535, National Renewable Energy Laboratory, U.S. Department of Energy Office of Energy Efficiency & Renewable Energy (2010)
11. El Beshbichi, O., Xing, Y., Ong, M.: An object-oriented method for fully coupled analysis of floating offshore wind turbines through mapping of aerodynamic coefficients. Mar. Struct. **78**, 102979 (2021)
12. Xu, X., Day, S.: Experimental investigation on dynamic responses of a spar-type offshore floating wind turbine and its mooring system behaviour. Ocean Eng. **236** (2021)
13. Schnepf, A., Lopez-Pavon, C., Devulder, A., Johnsen, Ø., Ong, M.: Suspended power cable configurations for floating offshore wind turbines in deep water powering an FPSO. In: Proceedings of the 41st International Conference on Ocean, Offshore & Arctic Engineering (OMAE). Hamburg, Germany (2022)

14. Li, H., Ong, M., Leira, B., Myrhaug, D.: Effects of soil profile variation and scour on structural response of an offshore monopile wind turbine. J. Offshore Mech. Arct. Eng. **140**(4), 042001 (2018)
15. Holthuijsen, L.: Waves in Oceanic and Coastal Waters. Cambridge University Press (2007)
16. Orcina: OrcaFlex. https://www.orcina.com/webhelp/OrcaFlex/ (2022)
17. Sakellariou, J., Fassois, S.: Functionally pooled models for the global identification of stochastic systems under different pseudo-static operating conditions. Mech. Syst. Signal Process. **72–73**, 785–807 (2016)

Robust and Fault Tolerant Control

A Cooperative Centralized MPC for Collision Avoidance of a Fleet of AVs

Manel Ammour, Rodolfo Orjuela, and Michel Basset

Abstract An original centralized model predictive controller (CMPC) with time varying Sigmoid-based safety constraints is developed for the control of a fleet of autonomous vehicles (AVs) and presented in this paper. The defined CMPC framework generates the appropriate control inputs for each vehicle in the network. This formulation allows each vehicle to attain its objective (reference lane tracking and driving at the desired speed) while maintaining a safety inter vehicular distance. The results of an illustrative scenario are presented to show the consistency of the proposed algorithm.

1 Introduction

Multi-agent systems have gained great interest in recent years due to their potential in enhancing traffic safety and efficiency in executing their missions such as area exploration [12], surveillance [1], rescue [6], formation control on unmanned ground vehicles (UGVs) [2], etc.

Cooperative control requires the consideration of the behavior of each agent in the network. To this aim, many approaches have been investigated and classified into three main schemes: centralized, distributed, and decentralized. For the centralized scheme, a single controller receives the information from all the agents and computes the necessary control inputs for each of them [7]. In the distributed scheme, each agent has its own controller which uses its state information and that of its neighbors to compute its appropriate control inputs [7]. Finally, in the decentralized scheme,

M. Ammour (✉) · R. Orjuela · M. Basset
Institut de Recherche en Informatique, Mathématique, Automatique et Signal (IRIMAS UR 7499), 68093 Mulhouse, France
e-mail: manel.ammour@uha.fr

R. Orjuela
e-mail: rodolfo.orjuela@uha.fr

M. Basset
e-mail: michel.basset@uha.fr

D. Theilliol et al. (eds.), *Recent Developments in Model-Based and Data-Driven Methods for Advanced Control and Diagnosis*, Studies in Systems, Decision and Control 467,
https://doi.org/10.1007/978-3-031-27540-1_26

each agent has its own controller and measures only its own states [11]. In this paper, a centralized architecture is adopted to take into account the states of all the agents and to avoid conflicts in the generated control actions.

The presence of more than one agent in a small area to accomplish a specific mission invokes the consideration of collision avoidance to ensure the successful completion of the agents' missions. Consequently, collision avoidance in crowded and dynamic environments is one of the important topics addressed by researchers. A quadratic programming is formulated in [15] to handle collision avoidance of a multi-robot system, in which centralized safety barrier certificates are established to dictate how the nominal controller can be changed to avoid pairwise robot collisions. Then, a decentralization of the safety barriers is performed by making individual robots consider only nearby robots while guaranteeing conservativeness of solution. The problem of formation control is investigated in [9] based on a distributed MPC for multiple linear second order agent systems with collision avoidance and safety constraints while ensuring the closed loop stability. Decentralized algorithms are also employed to develop collision avoidance strategies where each agent computes its avoid set with respect to the others and takes action when one or more agent approaches its boundary [10]. A group of autonomous flying vehicles is controlled through a distributed cooperative MPC control framework in [8] where the cost function of the MPC is designed to integrate the mission objectives (collision avoidance, reaching the targets, traveling as a flock). A different strategy is adopted in [13], where a message passing algorithm for cooperative collision avoidance is developed. Each vehicle in the network is controlled by a MPC controller so that it can achieve its objective and avoid colliding with a nearby vehicle. Convoy control of autonomous vehicle is presented in [14] using a two level MPC architecture where the high level MPC is devoted to the convoy formation and the generation of the reference trajectory for each vehicle and the low level MPC is for tracking purposes.

This study is an extension of our prior work [5], where a single vehicle is controlled in the presence of neighboring vehicles. In this work, a networked model predictive control (Net-MPC) strategy is proposed to control a fleet of autonomous vehicles. Typically, a cooperative centralized architecture is developed. The aim here is to control the vehicles so that each vehicle follows its reference without colliding with other vehicles. The use of Net-MPC is motivated by its predictive nature and its ability to forecast the positions and velocities of all vehicles. Hence, networked-based trajectory planning is reactive since the provided control inputs are based on the current and the predicted behaviors of the agents. The main contribution of this paper is the proposal of an online cooperative collision avoidance trajectory planning based on a centralized networked model predictive control architecture, using a set of Sigmoid-based safety constraints to drive each vehicle to a collision free zone when the safety inter vehicular distance is not kept. The suggested formulation guarantees the safe navigation and the online generation of a collision free trajectory for each vehicle.

The paper is organized as follows: the description of the problem treated in this paper is presented in Sect. 2. The plant modelling is described in Sect. 3 while the centralized networked MPC formulation is introduced in Sect. 4. The simulation results are discussed in Sect. 5 and the conclusion is stated in Sect. 6.

2 Problem Formulation

his paper deals with the autonomous driving of a fleet of autonomous vehicles. The aim is to ensure the cooperative motion while guaranteeing a safety inter-vehicular distance for all vehicles. This study considers a network of N autonomous vehicles driving on a two-lane one-way road. An illustration of a network composed of four autonomous vehicles is presented in Fig. 1.

Each vehicle is desired to drive at its own target speed on its reference lane while respecting the safety distance. For this purpose, the vehicles must communicate their states to the center. The latter then computes the appropriate control actions to control each vehicle in the network and drive them to a collision free zone, when necessary. The consistency of the solutions provided by the center depends highly on the cooperation of the vehicles. This feature is implicitly taken into account by the controller adopted in this work. It is handled by the fact that the predicted positions and reference trajectories, as well as the current states of all the agents, are transmitted to the Net-MPC controller. Moreover, these predictions are taken into account by the Net-MPC controller when computing the control actions to guarantee the safe navigation of the agents. In this work, the problem is addressed through a centralized architecture. The modeling of this architecture can be made using graph theory.

Definition 1 A graph $\mathcal{G} = (\mathcal{V}, \mathcal{E})$ is a pair that consists of a set of vertices $\mathcal{V}$ and a set of edges $\mathcal{E} \in \mathcal{V} \times \mathcal{V}$. The number of vertices of a graph is denoted by $\mathcal{N} = |\mathcal{V}| \in \mathbb{N}$ and the number of edges by $\mathcal{E} = |\mathcal{E}| \in \mathbb{N}$. A vertex is denoted by $v_i \in \mathcal{V} = v_1, v_2, \ldots, v_N$ and an edge by $e_{ij} = (v_i, v_j) \in \mathcal{E}$, where $v_i, v_j \in \mathcal{V}$ [3].

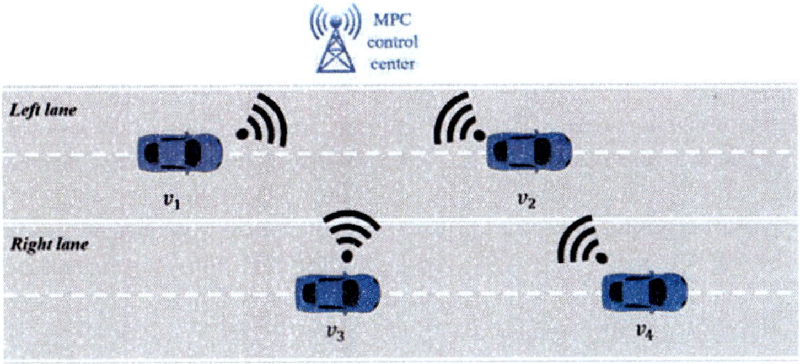

Fig. 1 A network of autonomous vehicles on a two-lane one-way road

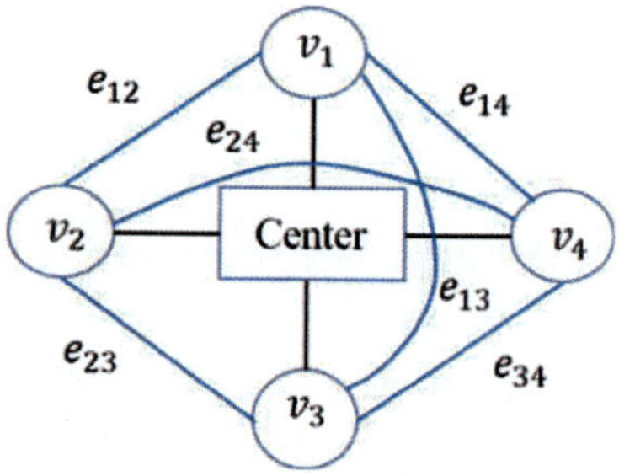

Fig. 2 Example of the CMPC graph of the considered vehicle network

Each vehicle in the network represents a vertex v_i in the graph. The communication between the vehicles and the center is represented by edges. The interaction between agents is also represented by edges (coupling), as depicted in Fig. 2. In this paper, Sigmoid based safety constraints [5] form the coupling between the agents.

These safety constraints are integrated into the Centralized MPC (CMPC) framework to generate the adequate solution to the optimization problem. These safety constraints allow the convexification of the driving areas of each agent by prohibiting it from driving in the vicinity of another agent occupying the same lane. The detailed description of this formulation will be presented in the next section.

It is worth to mention that the following assumptions are considered: (1) all the vehicle states are measurable, (2) the communication between the agents is well established, (3) the study focuses on the case where a lane change maneuver is possible, i.e., no lane is blocked, (4) each vehicle must take action with respect to its leading vehicle. Hence, the leader of each lane is supposed to be greedy and does not change its behavior.

3 AV Fleet Modelling

The purpose of the proposed algorithm is to control the motion of a fleet of autonomous vehicles and to ensure their navigation in a collision free area. To this end, the kinematic of each vehicle in the network is described by a point mass model. Consider a set of N agents. Let x_{veh_i}=$[x_i, y_i, v_{x_i} v_{y_i}]^T$ denotes the state vector of each agent $i \in N$, where x_i, y_i are the longitudinal and lateral positions respectively, and v_{x_i}, v_{y_i} are the longitudinal and the lateral velocities, respectively. Let $u_{veh_i} = [a_{x_i}, a_{y_i}]$ is the control input vector of the ith vehicle in the network, where a_{x_i}, a_{y_i} are the longitudinal and lateral accelerations, respectively. Initially, each agent has its desired center-line $y_{ref_i} \in \{y_{l_1}, y_{l_2}\}$. This reference may be changed depending on the lateral gap between the y_i position and the closest lane of the ith vehicle during a collision avoidance maneuver.

In a state space form, the longitudinal and lateral motions of an agent $v_i \in \mathcal{V}$ at each time step ΔT are described as follows:

$$x_{veh_i}(k+1) = A_i x_{veh_i}(k) + B_i u_{veh_i}(k), \qquad i = 1, \ldots, N \tag{1a}$$

$$Y_i(k) = C_i x_{veh_i}(k) \tag{1b}$$

where:

$$A_i = \begin{bmatrix} 1 & 0 & \Delta T & 0 \\ 0 & 1 & 0 & \Delta T \\ 0 & 0 & 1 & 0 \\ 0 & 0 & 0 & 1 \end{bmatrix}, \quad B_i = \begin{bmatrix} 0 & 0 \\ 0 & 0 \\ \Delta T & 0 \\ 0 & \Delta T \end{bmatrix}, \quad C_i = \begin{bmatrix} 0 & 1 & 0 & 0 \\ 0 & 0 & 1 & 0 \end{bmatrix}$$

The centralized augmented model used to predict the future states of the vehicles is:

$$X_{veh_{tot}}(k+1) = \begin{bmatrix} A_c & 0 \\ 0 & \mathrm{I} \end{bmatrix} \begin{bmatrix} X_{veh}(k) \\ Y_{ref}(k) \end{bmatrix} + \begin{bmatrix} B_c \\ 0 \end{bmatrix} U(k) \tag{2}$$

with:

$$A_c = \begin{bmatrix} A_1 & & \\ & \ddots & \\ & & A_N \end{bmatrix}, B_c = \begin{bmatrix} B_1 & & \\ & \ddots & \\ & & B_N \end{bmatrix} \tag{3}$$

$$X_{veh} = \begin{bmatrix} x_{veh_1} \\ \vdots \\ x_{veh_N} \end{bmatrix}, Y_{ref} = \begin{bmatrix} y_{ref_1} \\ \vdots \\ y_{ref_N} \end{bmatrix}, U = \begin{bmatrix} u_{veh_1} \\ \vdots \\ u_{veh_N} \end{bmatrix} \tag{4}$$

4 Networked Model Predictive Control Path Planning

The objective of this work is to ensure the safe driving of each vehicle in the network. In fact, each vehicle in the network aims to: (1) follow its reference center-line y_{ref_i}, (2) drive at the desired speed v_{des_i}, (3) navigate in a collision free area and avoid colliding with the neighboring vehicles, (4) satisfy vehicle dynamic constraints and road limits.

The objectives 1 and 2 are formulated as a minimization of a cost function within the MPC framework, described as follows:

$$\min_{u(.)} J = \sum_{i=1}^{N} \Big(\sum_{k=0}^{N_p-1} \alpha_i (y_i^k - y_{ref_i}^k)^2 + \beta_i (v_{x_i}^k - v_{x_{des_i}}^k)^2 + \sum_{k=0}^{N_c-1} \gamma_i (\Delta u_{veh_i})^2 \Big)$$

where $\alpha_i, \beta_i, \gamma_i$ are positive weighting coefficients, N_p and N_c are the prediction and control horizon, respectively. The control input $\Delta u_i = [\Delta a_{x_i}, \Delta a_{y_i}]^T$ denotes the increment in the longitudinal and lateral acceleration of the ith agent in the network. The first term $\alpha_i(y_i^k - y_{ref_i}^k)^2$ is used to penalize the deviation from the lateral coordinate of the reference center-line and achieve objective 1, while the term $\beta_i(v_{x_i}^k - v_{x_{des_i}}^k)^2$ satisfies objective 2. The objectives 3 and 4 are satisfied by the constraints explained in the following section.

4.1 Dynamic Constraints

Many limitations should be handled to have realistic and feasible control actions provided by the CMPC. The considered constraints are presented hereafter. $\forall v_i \in \mathcal{V}$:

Actuator constraints. The following conditions are taken into account to ensure that the control inputs are within the admissible range and to attain objective 4:

$$a_{x_{i_{min}}} < a_{x_i} < a_{x_{i_{max}}} \tag{5a}$$
$$\Delta a_{x_{i_{min}}} < \Delta a_{x_i} < \Delta a_{x_{i_{max}}} \tag{5b}$$

$$a_{y_{i_{min}}} < a_{y_i} < a_{y_{i_{max}}} \tag{5c}$$
$$\Delta a_{y_{i_{min}}} < \Delta a_{y_i} < \Delta a_{y_{i_{max}}} \tag{5d}$$

Velocity limits. Each vehicle should drive with a speed within the allowed range specified by the traffic rules and vehicle capabilities. To this end, the following constraint is established for each agent:

$$v_{x_{i_{min}}} < v_{x_i} < v_{x_{i_{max}}} \tag{6}$$

4.2 Safety Constraints

The safety constraints aim to ensure that each vehicle can navigate the road without colliding with its boundaries and nearby agents.Therefore, to achieve objective 3, the following constraints are imposed. $\forall v_i \in \mathcal{V}$:

Road boundaries. Each vehicle should stay within the road limits and avoid hitting the road edges by respecting the following constraint:

$$y_{i_{min}} + \frac{w_i}{2} < y_i < y_{i_{max}} - \frac{w_i}{2} \tag{7}$$

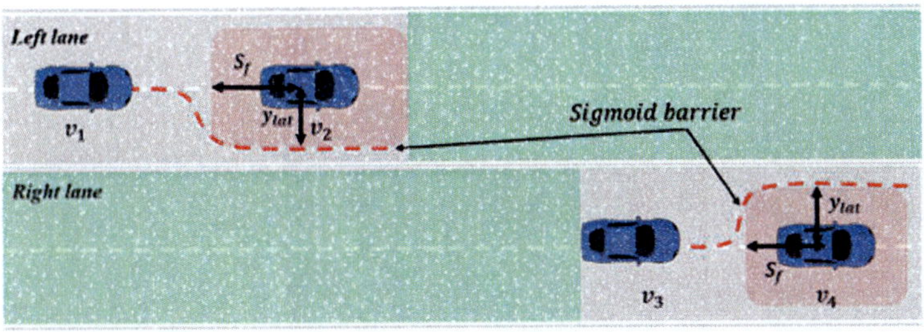

Fig. 3 The delimited navigable as well as non navigable zones for each vehicle in the network using Sigmoid barriers

where the term $\frac{w_i}{2}$ refers to the half width of vehicle v_i to take into account the vehicle size.

Sigmoid-based safety constraints. The proposed collision avoidance strategy is based on Sigmoid safety barriers. These constraints encounter a non navigable zone around an agent as depicted in Fig. 3.

More precisely, a lane change maneuver and a navigation to a collision-free zone are required when a vehicle $v_i \in \mathcal{V}$ approaches its lead vehicle $v_j \in \mathcal{V}$ and tends not to maintain the required safe distance. Each pair of vehicles within a network lane must meet the Sigmoid-based safety constraint. Therefore, $\forall v_i, v_j \in \mathcal{V}$:

$$\Delta y_i \geq \pm \frac{y_{lat}}{1 + e^{-\zeta_i(-\Delta x_i + S_{f_i})}} \tag{8}$$

with : $\Delta y_i = y_i - y_{ref_i}$ and $\Delta_{x_i} = x_j - x_i$ the longitudinal distance between vehicle v_i and its leading vehicle v_j. The sign of the Sigmoid-based barrier depends on which lane vehicle v_i is located. It is positive (+) when the vehicle is in the right lane and it is desirable to execute a left lane change maneuver and negative (−) when it is in the left lane and wants to perform a right lane change maneuver.

The desired lateral gap y_{lat} represents the distance that should be kept between two agents in the case where the longitudinal safety distance is not maintained and a collision avoidance maneuver is initiated. The shape of the lane change maneuver executed by an agent v_i to avoid colliding with its leading vehicle v_j is adjusted by the geometric parameter ζ_i in (8). This parameter depends on vehicle dynamics and road state (icy, wet and dry) and the longitudinal velocity to guarantee the feasibility of the lane change maneuver. It is modified in real-time according to the driving situation. The tuning of this parameter is detailed in the previous work [4]. The reader is encouraged to consult the mentioned work to have more details about the derivation of this parameter. The element S_{f_i} in (8) is time varying longitudinal safety distance that should be kept between each pair of vehicles. This safety distance depends on the speed of the considered agent and the inter vehicular time (TIV) that must be maintained. It is stated here, as follows:

$$S_{f_i} = v_{x_i}\text{TIV} \tag{9}$$

When a vehicle executes a lane change maneuver, its reference lane is changed to the nearest lane.

5 Simulation and Results

Numerous scenarios with different initial positions and speeds, as well as various reference lanes and target speeds, were tested to assess the proposed algorithm. An illustrative scenario of the proposed CMPC collision free strategy for a fleet of four autonomous vehicles is selected and presented in this section. Each vehicle is driving in its reference center-line and is assigned a desired speed. The initial positions and velocities of each vehicle are summarized in Table 1 and the design parameters of the centralized MPC are reported in Table 2. In this scenario, vehicles 1 and 2 are driving in the left lane and are following $y_{l_1} = 5.25$ m, whereas vehicles 3 and 4 are in the right lane and have $y_{l_2} = 1.75$ m as starting center-line reference. Initially, vehicle 1 is moving at higher speed than vehicle 2. They are separated by 30 m, hence the required safety distance $S_{f_1} = TIV v_{x_1} = 60$ m is not maintained. In order to avoid collision, vehicle 1 performs a right lane change and modifies its reference lane to reach the collision free zone delimited by its associated Sigmoid barrier. Additionally, it decelerates to reach its desired speed of 25 m/s as depicted in Figs. 4 and 5.

Vehicle 3 is 40 m behind vehicle 4, the safety distance $S_{f_3} = \text{TIV}v_{x_3} = 50$ m is thus not kept. Therefore, vehicle 3 changes the lane while accelerating (see Fig. 5) to reach its desired velocity $v_{x_{des_3}} = 27$ m/s and becomes the leader of vehicle 2 in the left lane. But since vehicle 2 drives faster than vehicle 3 ($v_{x_2} = 30 > v_{x_3} = 27$) m/s and gets closer to its leader, it changes the lane to avoid collision. The leading

Fig. 4 The corresponding trajectory and longitudinal velocity of each vehicle

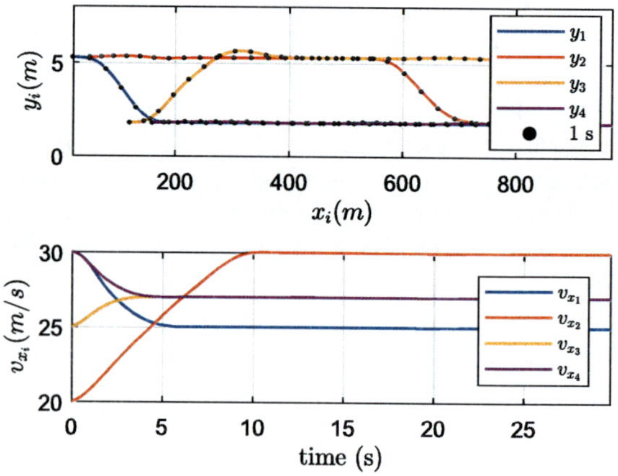

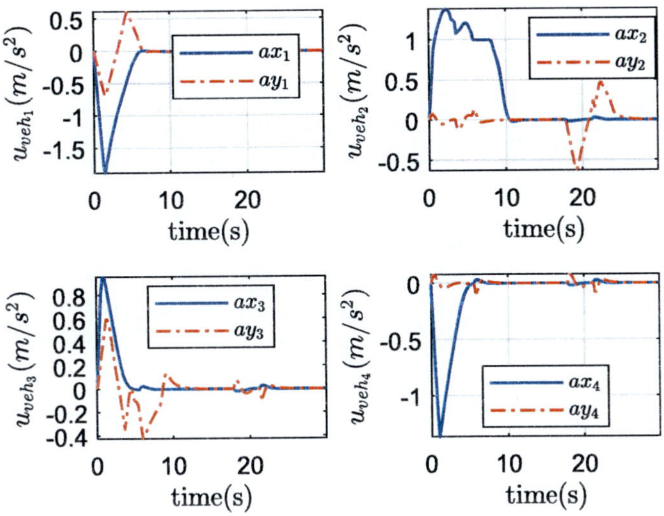

Fig. 5 The longitudinal and lateral accelerations of each vehicle

vehicle 4 of the left lane maintains its reference lane and decelerate to reach its target speed. As shown in the presented results, the generated accelerations are within the limits imposed during the design of the centralized MPC (see Table 2). Furthermore, the executed lane change maneuvers are comfortable since the lateral acceleration is $a_{y_i} < 0.2$ g for all the vehicles. The proposed CMPC-based collision avoidance approach manages successfully in defining the drivable zone of each agent to prevent collisions.

6 Conclusion

In this paper, a new collision avoidance strategy based on the Sigmoid function barrier is presented. Each agent in the network receives the proper control action from the centralized MPC control center in order to avoid collision with the vehicle in front of it. Each vehicle manages to drive in a collision free area and keep the appropriate safety distance with its neighbors. The proposed algorithm considers the case when it is possible to perform a lane change but this is not always the case. In fact, it is possible to have near agents in the target lane that prevent the lane change. In this case, it is necessary to perform some braking actions. This matter will be addressed in future works.

Appendix

Table 1 Initial conditions and desired velocities

Vehicle	x_i [m]	y_i [m]	v_{x_i} [m/s]	v_{y_i} [m/s]	$v_{x_{des_i}}$ [m/s]
1	20	5.25	30	0	25
2	50	5.25	20	0	30
3	120	1.75	25	0	27
4	160	1.75	30	0	27

Table 2 Design parameters

$y_{min} = 0$ m $v_{x_{min}} = 0$ m/s $w_i = 1$ m	$y_{max} = 7$ m $v_{x_{max}} = 40$ m/s $y_{lat} = 3.5$ m	TIV= 2 s $N_c = 3$ $\alpha = 1$	$\Delta T = 0.2$ s $N_p = 20$ $\beta = 10$
$a_{cx_{min}} = -4$ m/s^2 $a_{cy_{min}} = -2$ m/s^2 $\Delta a_{cx_{min}} = -1.5$ m/s^2 $\Delta a_{cy_{min}} = -0.5$ m/s^2	$a_{cx_{max}} = 2$ m/s^2 $a_{cy_{max}} = 2$ m/s^2 $\Delta a_{cx_{max}} = 1.5$ m/s^2 $\Delta a_{cy_{max}} = 0.5$ m/s^2	$\gamma_1 = 1$	$\gamma_2 = 5$

References

1. Acevedo, J.J., Arrue, B.C., Maza, I., Ollero, A.: A decentralized algorithm for area surveillance missions using a team of aerial robots with different sensing capabilities. In: IEEE International Conference on Robotics and Automation (ICRA), pp. 4735–4740. China, Hong Kong (2014)
2. Ailon, A., Zohar, I.: Control strategies for driving a group of nonholonomic kinematic mobile robots in formation along a time-parameterized path. IEEE/ASME Trans. Mechatron. **17**(2), 326–336 (2012)
3. Alrifaee, B.: Networked model predictive control for vehicle collision avoidance. Ph.D. thesis, RWTH Aachen University, Germany (2017)
4. Ammour, M., Orjuela, R., Basset, M.: Trajectory reference generation and guidance control for autonomous vehicle lane change maneuver. In: 28th Mediterranean Conference on Control and Automation (MED), pp. 13–18. Saint-Raphaël, France (2020)
5. Ammour, M., Orjuela, R., Basset, M.: Collision avoidance for autonomous vehicle using MPC and time varying sigmoid safety constraints. IFAC-PapersOnLine **54**(10), 39–44 (2021)
6. Beck, Z., Teacy, W., Jennings, N., Rogers, A.: Online planning for collaborative search and rescue by heterogeneous robot teams. In: 15th International Conference on Autonomous Agents and Multi-Agent Systems (2016)
7. Belkadi, A.: Conception de commande tolérante aux défauts pour les systèmes multi-agents. Theses, Université de Lorraine, France (2017)
8. Bertrand, S., Marzat, J., Piet-Lahanier, H., Kahn, A., Rochefort, Y.: MPC strategies for cooperative guidance of autonomous vehicles. AerospaceLab J. 8, 18 (2014)
9. Dai, L., Cao, Q., Xia, Y., Gao, Y.: Distributed MPC for formation of multi-agent systems with collision avoidance and obstacle avoidance. J. Frankl. Inst. **354**(4), 2068–2085 (2017)

10. Hoffmann, G.M., Tomlin, C.J.: Decentralized cooperative collision avoidance for acceleration constrained vehicles. In: 2008 47th IEEE Conference on Decision and Control, pp. 4357–4363. Cancun, Mexico (2008)
11. Lindqvist, B., Sopasakis, P., Nikolakopoulos, G.: A scalable distributed collision avoidance scheme for multi-agent UAV systems. In: IEEE/RSJ International Conference on Intelligent Robots and Systems (IROS). Prague, Czech Republic (2021)
12. Nieto-Granda, C., Rogers, J.G., Christensen, H.I.: Coordination strategies for multi-robot exploration and mapping. Int. J. Robot. Res. **33**(4), 519–533 (2014)
13. Ong, H.Y., Gerdes, J.C.: Cooperative collision avoidance via proximal message passing. In: 2015 American Control Conference (ACC), pp. 4124–4130. Chicago, IL, USA (2015)
14. Qian, X., de La Fortelle, A., Moutarde, F.: A hierarchical model predictive control framework for on-road formation control of autonomous vehicles. In: 2016 IEEE Intelligent Vehicles Symposium (IV), pp. 376–381 (2016)
15. Wang, L., Ames, A.D., Egerstedt, M.: Safety barrier certificates for collisions-free multirobot systems. IEEE Trans. Robot. **33**(3), 661–674 (2017)

Fault Tolerant Control of Markov Jump Systems Using an Asynchronous Virtual Actuator

Damiano Rotondo

Abstract This paper investigates a fault tolerant control (FTC) strategy using virtual actuators for a class of discrete-time Markov jump systems (MJSs). The main idea behind this FTC method is to add a block that masks the actuator fault from the controller/observer point of view, so that the nominal gains can be kept in the loop without need for retuning them. In line with recent developments in the field of MJSs, it is assumed that state-feedback controller, state observer, and virtual actuator operate asynchronously, which means that their active mode might differ from the controlled plant's active mode. Based on a stochastic Lyapunov function approach, design conditions to ensure the fault tolerant stabilization are proposed in the form of linear matrix inequalities (LMIs). The efficacy of the proposed method is demonstrated using a numerical example that shows the achieved tolerance against the complete loss of an actuator due to a fault.

1 Introduction

Markov jump systems (MJS) is the name given to a particular class of stochastic switching systems for which the underlying switching mechanism corresponds to a Markov process [1]. These systems are suitable for describing plants that experience sudden changes in their model, and have found applications in many fields, such as biology [2] and power grids [3], among many others.

Recent years have seen a surge in the development of theoretical results for asynchronous MJSs that go beyond the more *classical* synchronous design, as performed by, e.g., [4]. In a synchronous design, the control system elements (typically the controller and, possibly, the state observer) switch along with the plant. However, in the real world, due to uncertainties about the current mode of operation, the presence of communication delays, and the possibility of loss of data packets, the switching of

D. Rotondo (✉)
Department of Electrical Engineering and Computer Science (IDE), University of Stavanger, 4009 Stavanger, Norway
e-mail: damiano.rotondo@uis.no

D. Theilliol et al. (eds.), *Recent Developments in Model-Based and Data-Driven Methods for Advanced Control and Diagnosis*, Studies in Systems, Decision and Control 467,
https://doi.org/10.1007/978-3-031-27540-1_27

the control system elements might be only partially related to the current operational mode of the plant. Hence, there is the need to account for this asynchronicity during the design phase, to avoid the overall stability and performance to be affected in unexpected and undesired ways [5, 6].

In many control applications, faults can occur and deteriorate the system performance up to the loss of stability. For this reason, research on fault detection and isolation (FDI), fault estimation (FE) and fault tolerant control (FTC) have been a hot topic during the past few decades. The ongoing research considers many classes of systems, among which MJSs [7, 8]. The FTC methods developed so far can be classified into different families, based on the underlying design philosophy. The fault-hiding paradigm attempts at hiding the fault from the control system perspective by activating a dedicated block. When the activated block deals with actuator faults, then it is called *virtual actuator*, a concept that was first formalized by Jan Lunze in [10]. Initially developed for LTI systems, virtual actuators have been applied successfully to many other systems, such as Lure [11], interval [12] and descriptor [13] systems. However, to our best knowledge, FTC for MJSs via virtual actuator has not been studied yet.

Inspired by the research mentioned above, the main contribution of this paper is to describe how asynchronous virtual actuators can be used to achieve fault tolerance in MJSs. It is shown that the overall system comprising the plant to be controlled, the state-feedback controller, the state observer and the virtual actuator can be brought into an equivalent representation using a change of coordinate frame. This representation is suitable for obtaining analysis conditions that can be used as a first step towards design conditions in the form of LMIs, which can be solved to calculate all the required gains. The overall approach relies on a candidate Lyapunov function that allows deriving sufficient conditions for mean square exponential stability of the reconfigured system.

2 Problem Formulation

Let us consider the following discrete-time Markov Jump System (MJS):

$$\begin{cases} x(k+1) = A_{r(k)}x(k) + B_{r(k)}u(k) \\ y(k) = C_{r(k)}x(k) \end{cases} \tag{1}$$

where $x(k) \in \mathbb{R}^n$ represents the state vector, $u(k) \in \mathbb{R}^m$ denotes the control input and $y(k) \in \mathbb{R}^p$ are the sensor outputs. The state-space matrices appearing in (1) are governed by a Markov jump process through the stochastic variable $r(k) \in \mathcal{L} = \{1, 2, \ldots, L\}$. The jump probability for different values of $r(k)$ is given by:

$$\Pr\{r(k+1) = j | r(k) = i\} = \pi_{ij} \tag{2}$$

with $\pi_{ij} \in [0, 1]$ and $\sum_{j=1}^{L} \pi_{ij} = 1$ for all combinations of $i, j \in \mathcal{L}$.

In this paper, we will consider actuator faults that affect the system (1), so that under fault occurrence the input matrices become $B^f_{r(k)}$, thus transforming (1) into:

$$\begin{cases} x(k+1) = A_{r(k)}x(k) + B^f_{r(k)}u(k) \\ y(k) = C_{r(k)}x(k) \end{cases} \tag{3}$$

We will consider severe faults, in the sense that given the nominal and faulty matrices corresponding to a specific mode $r(k)$, the following holds:

$$\text{rank}\left(B^f_{r(k)}\right) < \text{rank}\left(B_{r(k)}\right) \tag{4}$$

At the same time, we assume that the system (3) does not lose its stabilizability properties, so that it can still be controlled satisfactorily, albeit the necessary changes in the overall control loop are implemented. It is worth stating that the physical interpretation of the rank condition (4) is that the faults cannot be tolerated by means of a simple redistribution of the control inputs. The main objective of this paper is to design a fault tolerant control scheme which employs virtual actuators. The main idea of this FTC method is to reconfigure the faulty plant so that the nominal controller and state observer can be kept in the loop without retuning them. This is achieved by adding a dynamic system whose purpose is to mask the fault. When an actuator fault appears, the virtual actuator computes an appropriate action $u(k)$ using the nominal controller's output $u_c(k)$, taking into account the available information about the fault.

The expression for the virtual actuator is given as follows:

$$\begin{cases} x_{va}(k+1) = \left(A_{\tau(k)} + B^\star_{\tau(k)}M_{\tau(k)}\right)x_{va}(k) + \left(B_{\tau(k)} - B^\star_{\tau(k)}\right)u_c(k) \\ u(k) = \left(B^f_{\tau(k)}\right)^\dagger B_{\tau(k)}\left(u_c(k) - M_{\tau(k)}x_{va}(k)\right) \\ y_c(k) = y(k) + C_{\tau(k)}x_{va}(k) \\ B^\star_{\tau(k)} = B^f_{\tau(k)}\left(B^f_{\tau(k)}\right)^\dagger B_{\tau(k)} \end{cases} \tag{5}$$

where $x_{va} \in \mathbb{R}^n$ is the virtual actuator state and $M_{\tau(k)}$ is the virtual actuator gain to be designed. The stochastic variable $\tau(k)$ is a probabilistic mapping of $r(k)$ from space $\mathcal{L}$ to space $\mathcal{S} = \{1, \ldots, S\}$ with transition probabilities:

$$\Pr\{\tau(k) = q | r(k) = i\} = \theta_{iq} \tag{6}$$

with $\theta_{iq} \in [0, 1]$ and $\sum_{q=1}^{S}\theta_{iq} = 1$ for all combinations of $i \in \mathcal{L}$ and $q \in \mathcal{S}$.

The overall fault tolerant control system is completed by the following nominal state observer and estimate-feedback control law:

$$\begin{cases} \hat{x}(k+1) = A_{\tau(k)}\hat{x}(k) + B_{\tau(k)}u_c(k) + L_{\tau(k)}\left(C_{\tau(k)}\hat{x}(k) - y_c(k)\right) \\ u_c(k) = K_{\tau(k)}\hat{x}(k) \end{cases} \tag{7}$$

where $\hat{x} \in \mathbb{R}^n$ is the estimated state, $L_{\tau(k)}$ is the observer gain and $K_{\tau(k)}$ is the feedback gain.

Let us define the following mapping between the state variables in a new coordinate frame $x_1(k), x_2(k), x_3(k)$ and the state variables in the original coordinate frame $x(k), \hat{x}(k), x_{va}(k)$:

$$\begin{cases} x_1(k) = \hat{x}(k) - x(k) - x_{va}(k) \\ x_2(k) = x(k) + x_{va}(k) \\ x_3(k) = x_{va}(k) \end{cases} \Leftrightarrow \begin{cases} x(k) = x_2(k) - x_3(k) \\ \hat{x}(k) = x_1(k) + x_2(k) \\ x_{va}(k) = x_3(k) \end{cases} \tag{8}$$

Then, the overall system obtained as the interconnection of (3), (5) and (7) is described by (the dependence of $r(k)$ and $\tau(k)$ on the current sample k is omitted to ease the notation):

$$\begin{cases} x_1(k+1) = \hat{A}_{r\tau}^{11} x_1(k) + \hat{A}_{r\tau}^{12} x_2(k) + \hat{A}_{r\tau}^{13} x_3(k) \\ x_2(k+1) = \hat{A}_{r\tau}^{21} x_1(k) + \hat{A}_{r\tau}^{22} x_2(k) + \hat{A}_{r\tau}^{23} x_3(k) \\ x_3(k+1) = \hat{A}_{r\tau}^{31} x_1(k) + \hat{A}_{r\tau}^{32} x_2(k) + \hat{A}_{r\tau}^{33} x_3(k) \end{cases} \tag{9}$$

where:

$$\begin{aligned}
\hat{A}_{r\tau}^{11} &= A_\tau + L_\tau C_\tau + \left(B_\tau^\star - B_r^f \left(B_\tau^f\right)^\dagger B_\tau\right) K_\tau \\
\hat{A}_{r\tau}^{12} &= A_\tau - A_r + L_\tau \left(C_\tau - C_r\right) + \left(B_\tau^\star - B_r^f \left(B_\tau^f\right)^\dagger B_\tau\right) K_\tau \\
\hat{A}_{r\tau}^{13} &= A_r - A_\tau + L_\tau \left(C_r - C_\tau\right) + \left(B_r^f \left(B_\tau^f\right)^\dagger B_\tau - B_\tau^\star\right) M_\tau \\
\hat{A}_{r\tau}^{21} &= \left(B_\tau + B_r^f \left(B_\tau^f\right)^\dagger B_\tau - B_\tau^\star\right) K_\tau \\
\hat{A}_{r\tau}^{22} &= A_r + \left(B_\tau + B_r^f \left(B_\tau^f\right)^\dagger B_\tau - B_\tau^\star\right) K_\tau \\
\hat{A}_{r\tau}^{23} &= A_\tau - A_r + \left(B_\tau^\star - B_r^f \left(B_\tau^f\right)^\dagger B_\tau\right) M_\tau \\
\hat{A}_{r\tau}^{31} &= \hat{A}_{r\tau}^{32} = \left(B_\tau - B_\tau^\star\right) K_\tau \\
\hat{A}_{r\tau}^{33} &= A_\tau + B_\tau^\star M_\tau
\end{aligned}$$

It is worth highlighting that, due to the assumption that the control system blocks are asynchronous with respect to the plant ($r(k) \neq \tau(k)$), the augmented system (9) does not appear in a block-triangular structure.

3 Analysis Conditions

In this section, we obtain analysis conditions that allow checking if an MJS reconfigured via virtual actuators is mean square exponentially stable or not. These conditions will serve later the purpose of deriving design conditions that can be used to compute the gains M_q, L_q and K_q, $q \in \mathcal{S}$. In the following, we will denote $r(k) = i$ and $\tau(k) = j$, respectively.

Theorem 1 *The system (9) is mean square exponentially stable if there exist positive definite matrices $Q_{l,iq} \in \mathbb{S}^n$ and $P_{l,i} \in \mathbb{S}^n$, $l = 1, 2, 3$, such that:*

$$\sum_{q=1}^{S} \theta_{iq} \text{diag}\left\{Q_{1,iq}, Q_{2,iq}, Q_{3,iq}\right\} \prec \text{diag}\left\{P_{1,i}, P_{2,i}, P_{3,i}\right\} \tag{10}$$

$$\begin{bmatrix} -Q_{1,iq} & \star & \star & \star & \star & \star \\ 0 & -Q_{2,iq} & \star & \star & \star & \star \\ 0 & 0 & -Q_{3,iq} & \star & \star & \star \\ \hat{A}^{11}_{iq} & \hat{A}^{12}_{iq} & \hat{A}^{13}_{iq} & -\check{P}^{-1}_{1,i} & \star & \star \\ \hat{A}^{21}_{iq} & \hat{A}^{22}_{iq} & \hat{A}^{23}_{iq} & 0 & -\check{P}^{-1}_{2,i} & \star \\ \hat{A}^{31}_{iq} & \hat{A}^{32}_{iq} & \hat{A}^{33}_{iq} & 0 & 0 & -\check{P}^{-1}_{3,i} \end{bmatrix} \prec 0 \tag{11}$$

for any $i \in \mathcal{L}$, $q \in \mathcal{S}$.

Proof Let us consider the following candidate Lyapunov function:

$$V(k) = \sum_{l=1}^{3} V_l(k) = \sum_{l=1}^{3} x_l(k)^T P_{l,i} x_l(k) \tag{12}$$

and let us define the matrices $\check{P}_{l,i}$, $l = 1, 2, 3$, $i = 1, \ldots, L$ as follows:

$$\check{P}_{l,i} = \sum_{j=1}^{L} \pi_{ij} P_{l,j} \tag{13}$$

Then, by taking into account (9), the following holds:

$$E\left\{\Delta V(k)\right\} = \sum_{l=1}^{3} \bar{x}(k)^T \left(\sum_{q=1}^{S} \theta_{iq} \hat{\mathcal{A}}^T_{l,iq} \check{P}_{l,i} \hat{\mathcal{A}}_{l,iq} \right) \bar{x}(k) - \sum_{l=1}^{3} x_l(k)^T P_{l,i} x_l(k) \tag{14}$$

where $\bar{x}(k) = \left[x_1(k)^T, x_2(k)^T, x_3(k)^T\right]^T$ and $\hat{\mathcal{A}}_{l,iq} = \left[\hat{A}^{l1}_{iq}\ \hat{A}^{l2}_{iq}\ \hat{A}^{l3}_{iq}\right]$. Then, using (10), one gets that:

$$E\{\Delta V(k)\} < \sum_{q=1}^{S} \theta_{iq} \bar{x}(k)^T \left(\sum_{l=1}^{3} \hat{A}_{l,iq}^T \check{P}_{l,i} \hat{A}_{l,iq} \right) \bar{x}(k)$$
$$- \sum_{q=1}^{S} \theta_{iq} \bar{x}(k)^T \text{diag}\left\{ Q_{1,iq}, Q_{2,iq}, Q_{3,iq} \right\} \bar{x}(k) \quad (15)$$

By ensuring that:

$$\sum_{l=1}^{3} \hat{A}_{l,iq}^T \check{P}_{l,i} \hat{A}_{l,iq} - \text{diag}\left\{ Q_{1,iq}, Q_{2,iq}, Q_{3,iq} \right\} \prec 0 \quad (16)$$

one gets that $E\{\Delta V(k)\} < 0$, which ensures mean square exponential stability. Finally, (11) is obtained from (16) using Schur complements, which completes the proof. □

4 Design Conditions

Let us make the assumption that the output-related matrices are all constant, that is, $C_i = C$. A consequence of this fact is that the matrices $\hat{A}_{r\tau}^{12}$ and $\hat{A}_{r\tau}^{13}$ that appear in (9) get simplified, as follows:

$$\hat{A}_{r\tau}^{12} = A_\tau - A_r + \left(B_\tau^\star - B_r^f \left(B_\tau^f \right)^\dagger B_\tau \right) K_\tau$$
$$\hat{A}_{r\tau}^{13} = A_r - A_\tau + \left(B_r^f \left(B_\tau^f \right)^\dagger B_\tau - B_\tau^\star \right) M_\tau$$

Under these conditions, the following theorem provides conditions for designing the unknown gains M_q, L_q and K_q, $q \in \mathcal{S}$.

Theorem 2 *Let $C = V_1 \left[V_2 \; 0 \right] V_3^T$ be the singular value decomposition (SVD) of the output matrix C, with $V_1 V_1^T = I$ and $V_3 V_3^T = I$. The system (9) is mean square exponentially stabilizable if there exist positive definite matrices $\bar{Q}_{l,iq} \in \mathbb{S}^n$ and $\bar{P}_{l,i} \in \mathbb{S}^n$, $l = 1, 2, 3$, and matrices $\Lambda_q \in \mathbb{R}^{n \times p}$, $\Gamma_q \in \mathbb{R}^{m \times n}$, $W_q \in \mathbb{R}^{m \times n}$, $Z_1 \in \mathbb{R}^{p \times p}$, $Z_2 \in \mathbb{R}^{(n-p) \times p}$, $Z_3 \in \mathbb{R}^{(n-p) \times (n-p)}$ such that:*

$$\sum_{q=1}^{S} \theta_{iq} \text{diag}\left\{ \bar{Q}_{1,iq}, \bar{Q}_{2,iq}, \bar{Q}_{3,iq} \right\} \prec \text{diag}\left\{ \bar{P}_{1,i}, \bar{P}_{2,i}, \bar{P}_{3,i} \right\} \quad (17)$$

$$\begin{bmatrix} -\bar{Q}_{1,iq} & \star & \star & \star & \star & \star \\ 0 & -\bar{Q}_{2,iq} & \star & \star & \star & \star \\ 0 & 0 & -\bar{Q}_{3,iq} & \star & \star & \star \\ \bar{A}_{iq}^{11} & \bar{A}_{iq}^{12} & \bar{A}_{iq}^{13} & \Upsilon_{1,i} & \star & \star \\ \bar{A}_{iq}^{21} & \bar{A}_{iq}^{22} & \bar{A}_{iq}^{23} & 0 & \Upsilon_{2,i} & \star \\ \bar{A}_{iq}^{31} & \bar{A}_{iq}^{32} & \bar{A}_{iq}^{33} & 0 & 0 & \Upsilon_{3,i} \end{bmatrix} \prec 0 \tag{18}$$

for any $i \in \mathcal{L}$, $q \in \mathcal{S}$, where:

$$\Upsilon_{l,i} = \sum_{j=1}^{L} \pi_{ij} \bar{P}_{l,j} - Z - Z^T, \qquad l = 1, 2, 3 \tag{19}$$

$$Z = V_3 \begin{bmatrix} V_2^{-1} V_1^T Z_1 & 0 \\ Z_2 & Z_3 \end{bmatrix} V_3^T \tag{20}$$

$$\bar{A}_{iq}^{11} = A_q Z + \left[\Lambda_q \ 0\right] V_3^T + \left(B_q^\star - B_i^f \left(B_q^f\right)^\dagger B_q\right) \Gamma_q \tag{21}$$

$$\bar{A}_{iq}^{12} = A_q Z - A_i Z + \left(B_q^\star - B_i^f \left(B_q^f\right)^\dagger B_q\right) \Gamma_q \tag{22}$$

$$\bar{A}_{iq}^{13} = A_i Z - A_q Z + \left(B_i^f \left(B_q^f\right)^\dagger B_q - B_q^\star\right) W_q \tag{23}$$

$$\bar{A}_{iq}^{21} = \left(B_q + B_i^f \left(B_q^f\right)^\dagger B_q - B_q^\star\right) \Gamma_q \tag{24}$$

$$\bar{A}_{iq}^{22} = A_i Z + \left(B_q + B_i^f \left(B_q^f\right)^\dagger B_q - B_q^\star\right) \Gamma_q \tag{25}$$

$$\bar{A}_{iq}^{23} = A_q Z - A_i Z + \left(B_q^\star - B_i^f \left(B_q^f\right)^\dagger B_q\right) W_q \tag{26}$$

$$\bar{A}_{iq}^{31} = \bar{A}_{iq}^{32} = \left(B_q - B_q^\star\right) \Gamma_q \tag{27}$$

$$\bar{A}_{iq}^{33} = A_q Z + B_q^\star W_q \tag{28}$$

Then, the stabilizing controller, observer, and virtual actuator gains can be obtained as:

$$K_q = \Gamma_q Z^{-1}, \ M_q = W_q Z^{-1}, \ L_q = \Lambda_q Z_1^{-1} \tag{29}$$

Proof Let us define the matrices $\bar{Q}_{l,iq}$ and $\bar{P}_{l,i}, l = 1, 2, 3$, as follows:

$$\bar{Q}_{l,iq} = Z^T Q_{l,iq} Z \tag{30}$$

$$\bar{P}_{l,i} = Z^T P_{l,i} Z \tag{31}$$

Then, by pre-and post-multiplying (10) by Z^T and Z, respectively, it follows that (10) and (17) are equivalent. By applying appropriately the Young's relation, the following holds:

$$\sum_{j=1}^{L} \pi_{ij} \bar{P}_{l,j} - Z - Z^T = Z^T \left(\sum_{j=1}^{L} \pi_{ij} P_{l,j} \right) Z - Z - Z^T \succ - \left(\sum_{j=1}^{L} \pi_{ij} P_{l,j} \right)^{-1} = -\check{P}_{l,i}^{-1} \quad (32)$$

Let us pre-and post-multiply (11) by diag$\{Z^T, Z^T, Z^T, I, I, I\}$ and its transpose which, after taking into account (30)–(31) and the inequality (32), leads to:

$$\begin{bmatrix} -\bar{Q}_{1,iq} & \star & \star & \star & \star & \star \\ 0 & -\bar{Q}_{2,iq} & \star & \star & \star & \star \\ 0 & 0 & -\bar{Q}_{3,iq} & \star & \star & \star \\ \hat{A}_{iq}^{11} Z & \hat{A}_{iq}^{12} Z & \hat{A}_{iq}^{13} Z & \Upsilon_{1,i} & \star & \star \\ \hat{A}_{iq}^{21} Z & \hat{A}_{iq}^{22} Z & \hat{A}_{iq}^{23} Z & 0 & \Upsilon_{2,i} & \star \\ \hat{A}_{iq}^{31} Z & \hat{A}_{iq}^{32} Z & \hat{A}_{iq}^{33} Z & 0 & 0 & \Upsilon_{3,i} \end{bmatrix} \prec 0 \quad (33)$$

where the matrices $\Upsilon_{1,i}$, $\Upsilon_{2,i}$ and $\Upsilon_{3,i}$ are defined as in (19). Let us structure the matrix Z as in (20). Then, by introducing the auxiliary matrices $\Gamma_q \triangleq K_q Z$, $W_q \triangleq M_q Z$ and $\Lambda_q \triangleq L_q Z_1$, the following is true:

$$L_q C Z = L_q V_1 \begin{bmatrix} V_2 & 0 \end{bmatrix} V_3^T V_3 \begin{bmatrix} V_2^{-1} V_1^T Z_1 & 0 \\ Z_2 & Z_3 \end{bmatrix} V_3^T = \begin{bmatrix} L_q Z_1 & 0 \end{bmatrix} V_3^T = \begin{bmatrix} \Lambda_q & 0 \end{bmatrix} V_3^T \quad (34)$$

Then, one has $A_{iq}^{jk} Z = \bar{A}_{iq}^{jk}$ for $j, k = 1, 2, 3$, which completes the proof. □

5 Numerical Example

Let us consider an MJS as in (1) with two different modes ($L = 2$) described by matrices A_1, B_1, C_1 and A_2, B_2, C_2, respectively, with $C_1 = C_2$ to satisfy the assumption introduced in Sect. 4:

$$A_1 = \begin{bmatrix} 1 & 0.3 \\ 0.1 & 1.1 \end{bmatrix} \quad B_1 = \begin{bmatrix} 3.7 & 2.8 \\ 1 & 0.5 \end{bmatrix} \quad C_1 = \begin{bmatrix} 1 & 1 \end{bmatrix}$$

$$A_2 = \begin{bmatrix} 1 & 0.1 \\ 0.1 & 0.5 \end{bmatrix} \quad B_2 = \begin{bmatrix} 3.5 & 2.1 \\ 0.8 & 1.8 \end{bmatrix} \quad C_2 = \begin{bmatrix} 1 & 1 \end{bmatrix}$$

Note that both modes are open-loop unstable due to eigenvalues of the state matrix A in 1.23 and 1.02, respectively. The probabilistic mappings (2) and (6) are described by the following matrices:

$$\pi = \begin{bmatrix} 0.8 & 0.2 \\ 0.2 & 0.8 \end{bmatrix} \qquad \theta = \begin{bmatrix} 0.7 & 0.3 \\ 0.1 & 0.9 \end{bmatrix}$$

The considered fault scenario is a complete loss of the second actuator, which results in faulty input matrices:

$$B_1^f = \begin{bmatrix} 3.7 & 0 \\ 1 & 0 \end{bmatrix} \qquad B_2^f = \begin{bmatrix} 3.5 & 0 \\ 0.8 & 0 \end{bmatrix}$$

At first, a virtual actuator-less design has been considered, which means using a reduced version of Theorem 2 without decision variables $\bar{Q}_{3,iq}$, $\bar{P}_{3,i}$ and W_q, where the third row/column of the matrices in (17), the third and sixth rows/columns of the matrix in (18), and the terms $\left(B_q^\star - B_i^f\left(B_q^f\right)B_q\right)\Gamma_q$ appearing in (21)–(22) and (24)–(25) have been removed. Note that such design is akin to the one considered in [6], albeit the difference in that a H_∞ performance metric for disturbance rejection is not considered. This design will be later referred to as *without FTC*. On the other hand, the straight application of Theorem 2 has returned controller/observer/virtual actuator gains that will be referred to in the following as *with FTC*. The simulation results have been obtained with initial conditions $x(0) = [1, 1]^T$, $\hat{x}(0) = [0, 0]^T$ and $x_{va}(0) = [0, 0]^T$ and with asynchronicity between the jump modes of the plant and those of the control system. The simulated fault scenario is a sudden switch from matrices B_1, B_2 to faulty matrices B_1^f, B_2^f occurring at sample $k = 500$. The comparison between simulation results without and with the proposed FTC strategy is shown in Fig. 1. For visualization purposes, a logarithmic scale is used on the y-axis as it enables comparing convergent and divergent trends of the state variables. It can be clearly seen that the proposed FTC strategy is able to maintain closed-loop mean square exponential stability in spite of the actuator fault, whereas the more traditional design without FTC fails in achieving so.

6 Conclusions

This paper has investigated the FTC of discrete-time MJSs by designing an asynchronous virtual actuator. The proposed FTC system is able to preserve mean square exponential stability in spite of severe actuator faults, such as the complete loss of an actuator. This has been demonstrated using simulations of a numerical example. Future research will focus on developing an FDI algorithm that can be integrated with this FTC strategy.

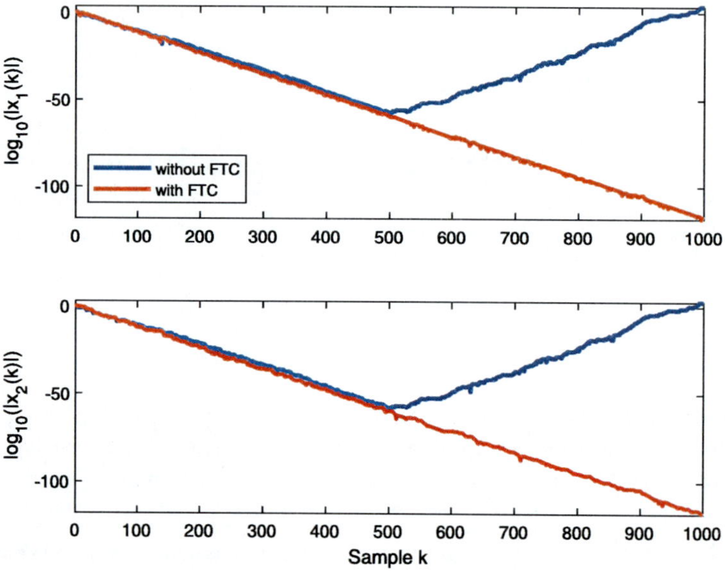

Fig. 1 System states x_1 and x_2

References

1. Zhao, P., Kang, Y., Zhao, Y.-B.: A brief tutorial and survey on Markovian jump systems: stability and control. IEEE Systems, Man, and Cybernetics Magazine. **5**(2), 37-C3 (2019)
2. Wilkinson, D.J.: Stochastic modelling for quantitative description of heterogeneous biological systems. Nat. Rev. Genetics. **10**(2), 122–133 (2009)
3. Lai, J., Lu, X., Yu, X., Monti, A.: Stochastic distributed secondary control for AC microgrids via event-triggered communication. IEEE Trans. Smart Grid. **11**(4), 2746–2759 (2020)
4. Hou, T., Ma, H.: Exponential stability for discrete-time infinite Markov jump systems. IEEE Trans. Autom. Control. **61**(12), 4241–4246 (2015)
5. Wang, L., Wu, Z.-G., Shen, Y.: Asynchronous mean stabilization of positive jump systems with piecewise-homogeneous Markov chain. IEEE Trans. Circuits Syst. II: Express Briefs. **68**(10), 3266–3270 (2021)
6. Zhang, M., Shen, C., Wu, Z.-G.: Asynchronous observer-based control for exponential stabilization of Markov jump systems. IEEE Trans. Circuits Syst. II: Express Briefs. **67**(10), 2039–2043 (2019)
7. Li, X., Ahn, C.K., Lu, D., Guo, S.: Robust simultaneous fault estimation and nonfragile output feedback fault-tolerant control for Markovian jump systems. IEEE Trans. Syst. Man Cybern. Systems. **49**(9), 1769–1776 (2018)
8. Yang, H., Yin, S., Kaynak, O.: Neural network-based adaptive fault-tolerant control for Markovian jump systems with nonlinearity and actuator faults. IEEE Trans. Syst. Man Cybern. Systems. **51**(6), 3687–3698 (2020)
9. Xue, M., Yan, H., Wang, M., Shen, H., Shi, K.: Lstm-based intelligent fault detection for fuzzy Markov jump systems and its application to tunnel diode circuits. IEEE Trans. Circuits Syst. II: Express Briefs. **69**(3), 1099–1103 (2021)
10. Lunze, J., Steffen, T.: Control reconfiguration by means of a virtual actuator. IFAC Proc. Volumes. **36**(5), 131–136 (2003)

11. Pedersen, A.S., Richter, J.H., Tabatabaeipour, M., Jóhannsson, H., Blanke, M.: Fault tolerant emergency control to preserve power system stability. Control. Eng. Practice. **53**, 151–159 (2016)
12. Rotondo, D., Cristofaro, A., Johansen, T.A.: Fault tolerant control of uncertain dynamical systems using interval virtual actuators. Int. J. Robust Nonlinear Control. **28**(2), 611–624 (2018)
13. Wang, Y., Rotondo, D., Puig, V., Cembrano, G.: Fault-tolerant control based on virtual actuator and sensor for discrete-time descriptor systems. IEEE Trans. Circuits Syst. I: Regul. Papers. **67**(12), 5316–5325 (2020)
14. Zhou, K., Khargonekar, P.P.: Robust stabilization of linear systems with norm-bounded time-varying uncertainty. Syst. & Control. Letters. **10**(1), 17–20 (1988)

Robust Safety Control for Automated Driving Systems with Perception Uncertainties

Yan Feng Yu, Kaveh Nazem Tahmasebi, Peng Su, and Dejiu Chen

Abstract Safety assurance and trustworthiness guarantees represent some of the most challsenging problems in the development of next generation automated driving and driving assistance systems. A systematic approach, with measures ranging from development-time modeling and simulation support to operation-time mechanisms for situation-awareness and adaptation, becomes necessary for tackling the problems. This paper presents a novel approach to safety control for automated driving under the condition of uncertain perception due to emergent properties in the environment or sensor faults. Based on the theory of optimal control, the safety controller is built upon a control barrier function and a model predictive control function. The effectiveness of the proposed strategy is evaluated on a simulation scenario created in the open-source autonomous driving simulator CARLA.

Keywords Autonomous driving system · Control barrier function · Model predictive control · Measurement uncertainties · Kalman filter

1 Introduction

Currently, increasingly novel technologies are being integrated into automotive vehicles to enable Advanced Driver Assistance Systems (ADAS) or fully Automated Driving Systems (ADS). In the development of such systems, safety assurance and

Y. F. Yu · K. Nazem Tahmasebi (✉) · P. Su · D. Chen
Unit of Mechatronics and Embedded Control System, KTH Royal Institute of Technology, Brinellävgen 83, 10044 Stockholm, Sweden
e-mail: kavent@kth.se

Y. F. Yu
e-mail: yfyu@kth.se

P. Su
e-mail: pensu@kth.se

D. Chen
e-mail: chendj@kth.se

D. Theilliol et al. (eds.), *Recent Developments in Model-Based and Data-Driven Methods for Advanced Control and Diagnosis*, Studies in Systems, Decision and Control 467,
https://doi.org/10.1007/978-3-031-27540-1_28

trustworthiness guarantees represent some of the most challenging tasks. The goal is to at least assure that, as long as other road users act according to certain assumed behaviors, an automated driving (AD) vehicle will not crash or be able to mitigate collisions by unpredictable behaviors of other road users. Therefore, one fundamental requirement in the development of ADAS/ADS is to provide fail-safe/fail-operational capabilities so that the system will attain minimal risk conditions in all operational situations. To this end, one key issue is related to the support for qualified perception of complex operational conditions, where the presence of measurement uncertainties and unexpected system anomalies could result in hazardous situations, resulting potentially in serious injuries or fatality to people involved. Nevertheless, challenges have been indicated by recent traffic accident statistics. In the latest report concerning the period July 2021 to May 2022 by the US National Highway Traffic Safety Administration, it appears that 392 collisions with ADAS have been reported. These have resulted in six people being killed and five seriously injured [1].

This paper presents an approach to safety control by integrating Model Predictive Control (ä) and Control Barrier Function (CBF) for effectively attaining the minimal risk conditions. The result of the proposed safety control method is evaluated by a case study of Adaptive Cruise Control (ACC) in a highway condition simulated with CARLA, an open-source AD simulator [2]. The reminder of this paper is organized as follows: Sect. 2 introduces the related concepts and technologies. Section 3 elaborates the design concepts and theoretical foundations for the proposed control approach. Section 4 presents the implementation of the proposed safety control system in different cases regarding state estimation method based on CARLA. Finally, the conclusion is given in Sect. 5.

2 Related Work

MPC [3, 4] as one advanced control framework is widely used in robotics and autonomous systems. One special strength of MPC is its capability of predicting future behaviors of dynamic process based on linear and nonlinear time-varying models given a set of inputs and constraints. For AD vehicles, MPC can be used for a range of control functions, such as ACC [5] and Lane-Keeping Control [6]. The common strategy is to treat these control functions as the tasks of solving model-based optimization problems over certain predefined prediction horizon based on parameters of vehicle dynamics (i.e., positions of throttle, brake, steering) and environmental constraints (i.e., minimal safe distances to other vehicles).

The motivation behind the choice of CBF algorithm has been its support for stipulating the properties required for system safety assurance [7]. By safety, we refer to the system's capability of being protected from hazardous operational conditions, i.e. posing unacceptable risks to the AD vehicle and its environment. The control objective is related to risk mitigation by actively avoiding unsafe regions of the state space [8]. To apply CBF to real-time control problems, it is necessary to transform continuous-time CBF to discrete-time CBF [9]. This also makes it possible to integrate CBF into the MPC framework [10, 11] as a useful method for specifying additional constraints. One benefit by such an integration is that the control policy can become less greedy as future state information can be more effectively taken into account. For ADS, through the unification of MPC and CBF, the vehicle can plan necessary actions for risk mitigation at earlier stages without requiring a large horizon as typically expected. This helps to reduce the load of computation, which is considered to be one of the main challenges for successful MPC implementation

For AD vehicles, the perception of environmental conditions by sensors suffers from uncertainties from external disturbances and measurement perturbations [12]. Such uncertainties, if improperly treated, could cause biased control decisions with negative impact on system performance and safety. To avoid the system robustness loss under measurement disruptions, Kalman Filter (KF) [13], as one conventional method for coping with noisy measurements, has been adopted for providing operation condition (e.g. vehicle state) estimation for the control actions. KF produces predictions of system states based on a joint probability distribution over the state variables for each timeframe with a series of measurements observed over time. One further method to promote control robustness against external disturbances on system dynamics is the Zeroing Control Barrier Function (ZCBF) method [14]. By this paper, we investigate how the ZCBF method can be integrated in MPC by considering the uncertainty caused by common faults in the perception system.

3 Key Design Concepts

In this section, we elaborate on necessary preliminaries to the safety optimal control method integrated with an estimation algorithm for disrupted systems. The notations employed in this paper are explained below.

Notation Given an arbitrary set $\mathcal{S}$, the boundary and interior of this set are denoted by $\partial\mathcal{S}$ and $\text{Int}(\mathcal{S})$, respectively. $\mathcal{A}$ transpose matrix is denoted by $\mathcal{A}^{\top}$, and the expectation of a random variable X is denoted by $\mathbb{E}(\mathbb{X})$. An identity matrix is shown by $\mathcal{I}$. A set $\mathcal{S}$ is called *(forward) invariant* for a dynamic system $\dot{x} = f(x)$ if for every $x(0) \in \mathcal{S} \Rightarrow x(t) \in \mathcal{S}$ for all $t \in \mathbb{R}^{+}$. The *class $\mathcal{K}$ function* is defined as a strictly increasing continuous function $\alpha : [0, a) \rightarrow \mathbb{R}^{+}$, with $a > 0$ and $\alpha(0) = 0$. This class $\mathcal{K}$ function α belongs to *class $\mathcal{K}_{\infty}$* if $a = \infty$ and $\lim_{r\rightarrow\infty} \alpha(r) = \infty$.

The discrete-time linear uncertain control system is defined as follows:

$$x_{t+1} = Ax_t + Bu_t + w_t \tag{1}$$

$$y_t = Cx_t + v_t \tag{2}$$

where $x_t \in \mathbb{R}^n$, $u_t \in \mathbb{R}^m$, and $y_t \in \mathbb{R}^k$ denote the state, control input, and output of the system at time step t, respectively. $w_t \in \mathbb{R}^n$ is the exogenous process noise with zero mean that generates the nominal covariance matrix $Q' = \mathbb{E}(w_t w_t^{\top})$ as a process noise uncertainty. $v_t \in \mathbb{R}^k$ is the exogenous measurement/output noise with zero mean that generates the nominal covariance matrix $R' = \mathbb{E}(v_t v_t^{\mathsf{T}})$, as a measurement uncertainty. The matrices $A \in \mathbb{R}^{n\times n}$, $B \in \mathbb{R}^{n\times m}$, and $C \in \mathbb{R}^{k\times n}$ are the state transition, control input, and system output matrices, respectively.

Control Barrier Function: Considering [9] and [10], we formulate a ZCBF as a robust CBF approach. A continuously differentiable function $h : \mathbb{R}^n \to \mathbb{R}$ is defined such that the corresponding superlevel set $\mathcal{C}$ of this function would be our safe set:

$$\begin{aligned} \mathcal{C} &= \left\{x \in \mathcal{X} \subset \mathbb{R}^n : h(x) \geq 0\right\} \\ \partial\mathcal{C} &= \left\{x \in \mathcal{X} \subset \mathbb{R}^n : h(x) = 0\right\} \\ \mathrm{Int}(\mathcal{C}) &= \left\{x \in \mathcal{X} \subset \mathbb{R}^n : h(x) > 0\right\}. \end{aligned} \tag{3}$$

Function h is a CBF if it satisfies (4), where there exists a control input $u \in \mathcal{U} \subset \mathbb{R}^m$ and an extended class $\mathcal{K}_\infty$ function α [15, 16].

$$\dot{h}(x, u) \geq -\alpha(h(x)) \Leftrightarrow \mathcal{C} \text{ is invariant.} \tag{4}$$

In discrete time domain, by using $\Delta h(x_k, u_k) := h(x_{k+1}) - h(x_k)$ and the simplified form of extended class $\mathcal{K}$ function $\alpha(h(x)) = \gamma h(x)$, we reformulate (4) as:

$$\Delta h(x_k, u_k) \geq -\gamma h(x_k) \Leftrightarrow \mathcal{C} \text{ is invariant.} \tag{5}$$

where scalar $\gamma > 0$ as a class $\mathcal{K}$ function, defined as a relaxation constant in this paper. A large γ increases the optimization problem feasibility but may result in the CBF constraint not being activated. A small γ imposes stricter conditions for safety based on CBF.

Kalman Filter: According to the uncertain control system model (1) and (2), which has the process and measurement noises, a simple linear KF algorithm is formulated to deal with these disturbances by estimating the measured and predicted variables. Besides, the process noise uncertainty, Q', and the measurement uncertainty, R', we also assume the state estimation uncertainties of the current state with a covariance matrix $P'_{t|t} = \mathbb{E}((x_t - \hat{x}_{t|t})(x_t - \hat{x}_{t|t})^{\mathsf{T}})$. Additionally, $P'_{t|t-1}$ denote the estimation uncertainties calculated by the previous filter estimation. Once KF [13, 17] is initialized with $\hat{x}_{0|0}$ and $P'_{0|0}$, it operates in a loop containing a time update (6) and a measurement update (7), where K_t is known as Kalman Gain.

$$\begin{aligned} \hat{x}_{t+1|t} &= A\hat{x}_{t|t} + Bu_t \\ P'_{t|t-1} &= AP'_{t-1|t-1}A^T + Q'_{t-1}. \end{aligned} \tag{6}$$

$$\begin{aligned} K_t &= P'_{t|t-1}C^T\left[CP'_{t|t-1}C^T + R'_t\right]^{-1} \\ \hat{x}_{t|t} &= \hat{x}_{t|t-1} + K_t\left[z_t - C\hat{x}_{t|t-1}\right] \\ P'_{t|t} &= P'_{t|t-1} - K_tCP'_{t|t-1}. \end{aligned} \tag{7}$$

Safety Optimal Control Algorithm (MPC-CBF) Formulation: The mathematical model of the proposed safety optimal control method, integration of MPC and CBF, subjected to the discrete-time linear control system (1) should be formulated in (8). Its solution is similar to MPC [18] except for an additional constraint of CBF, Δh, added to guarantee the associated safe set C to be forward invariant.

$$J^*(x_t) = \underset{u_{t:t+N-1}}{\text{minimize}} \sum_{k=0}^{N-1} [x_{t+k}^{\mathsf{T}} Q x_{t+k} + u_{t+k}^{\mathsf{T}} R u_{t+k}] + x_{t+N}^{\mathsf{T}} P x_{t+N} \tag{8a}$$

$$x_{t+k} \in \mathcal{X}, \quad u_{t+k} \in \mathcal{U} \qquad k = 0, \ldots, N-1 \tag{8b}$$

$$\Delta h\,(x_{t+k}, u_{t+k}) \geq -\gamma h\left(x_{t+k|t}\right) \qquad k = 0, \ldots, N-1$$

$$h(x_{t+k+1|t}) - h(x_{t+k|t}) \geq -\gamma h\left(x_{t+k|t}\right) \tag{8c}$$

$$x_{t+N} \in \mathcal{X}_f \tag{8d}$$

$$x_t \in \mathcal{X}_0 \tag{8e}$$

where x_t, x_{t+k}, x_{t+N}, and u_{t+k} are the initial state, the state at step k, the end of state at the terminal horizon step, and the predicted control action at step k obtained from time-step t, respectively. Q, R, and P are the positive semi-definite weight matrices. The value function $J^*(x_t)$, formulated as a quadratic optimization problem, is minimized subject to the control system model (1), and the constraints on; (i) state and input control (8a), (ii) terminal and initial constraints (8c) and (8d), and (iii) CBF constraint, (8b), as a safe constraint added to the conventional MPC constraints.

4 Implementation and Case Study

To evaluate the proposed safety control method with KF algorithm, a case study of adaptive cruise control (ACC) is designed based on the concepts described in Sect. 3. The overall control structure shown in Fig. 1.

Fault Modelling of Perception Components: The system uses multi-sensors for perceiving the environmental conditions. Learning-Enable Components (LEC) are used for the operational conditions estimation as describied in [19]. In particular, a RGB-camera is used to measure the relative distance between the ego and leading vehicles. To investigate the LECs' ability to compute the relative distance in

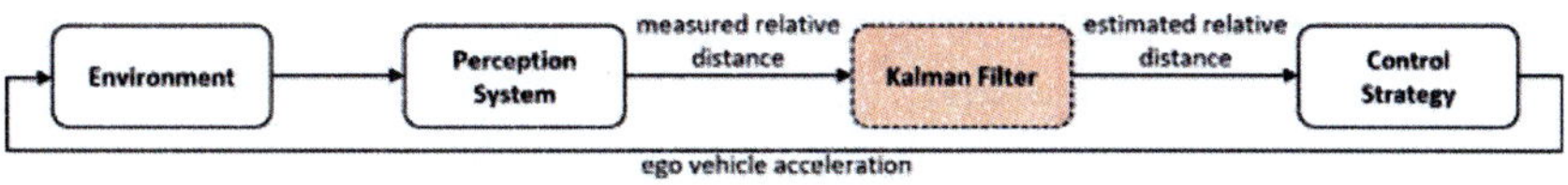

Fig. 1 Overall control structure for ACC. The Environment is implemented by CARLA, an open-source AD simulator. Noises/faults are injected into the Perception System block. The Control Strategy block provides an integration of MPC and CBF

Fig. 2 Different intensity salt-and-pepper noise injected into the RGB-camera

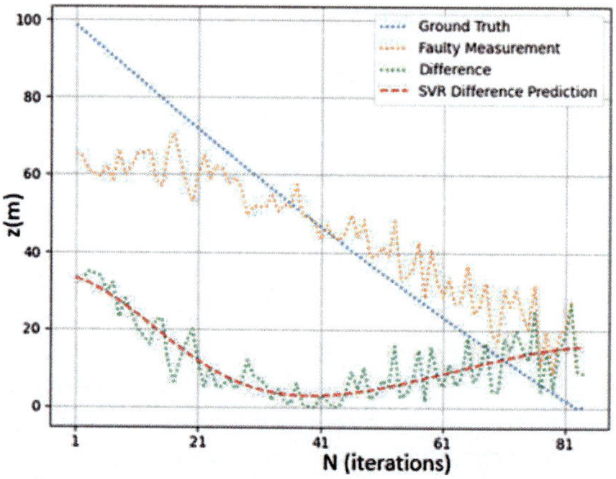

Fig. 3 The predicted difference between faulty and true relative distance by SVR-the sampling rate is set to 0.04 s

faulty situations, salt-and-pepper noise is injected with 80% failure probability with different intensities into the RGB-camera, as shown in Fig. 2. The effect of the corresponding faults are modeled using Support Vector Regression (SVR), as a predictor algorithm, based on the absolute value of the discrepancy between the ground truth and the measured distances. This fault model is used as an exogenous measurement noise in the control system model. The prediction result is shown in Fig. 3.

Safety Optimal Control Design: The vehicle dynamics is given by the following state variables for ACC, $x_k = \begin{bmatrix} v_f \ v_l \ z \end{bmatrix}^\intercal$,where v_f and v_l are the velocity of the AD and leading vehicles, respectively; and z is the relative distance between AD and leading vehicles. The corresponding control input variable $u_k = a_F$ is the AD vehicle acceleration. According to the vehicle kinematics behavior, matrices A and B in (1) are derived as follows:

$$A = \begin{bmatrix} 1 & 0 & 0 \\ 0 & 1 & 0 \\ -T_s & T_s & 1 \end{bmatrix}, \quad B = \begin{bmatrix} T_s \\ 0 \\ \frac{1}{2}T_s^2 \end{bmatrix}. \tag{9}$$

where $T_s = 0.2$s refers to the sampling time. The required parameters for the safety optimal control algorithm include: $N = 8$, referring to the control horizon; $\gamma = 5$, referring to the class $\mathcal{K}$ function; $Q = P = \mathcal{I}_3$, and $R = 2 \cdot \mathcal{I}_2$, referring to the weights matrices. For the basic vehicle dynamics model in Eq. (1), the following constraints on the states and inputs are defined :

$$\begin{aligned}
\mathcal{X} &= \left\{x_{t+k} \in \mathbb{R}^n : x_{t+k} \leq [20, 14, \infty]^\mathsf{T}\right\}, \\
\mathcal{X}_f &= \left\{x_{t+k} \in \mathbb{R}^n : [0.2, 0, 7]^\mathsf{T} \leq x_{t+k} \leq [20, 14, \infty]^\mathsf{T}\right\}, \\
\mathcal{U} &= \left\{u_{t+k} \in \mathbb{R}^m : -g \leq u_{t+k} \leq a_f g, \quad g = 9.8\text{m/s}^2\right\}.
\end{aligned} \tag{10}$$

The maximum vehicle acceleration is delimited to a fraction of g by $a_f = 0.3$. The lower bound on the terminal constraint refers to a slack value to ensure the AD vehicle stops before the leading vehicle. The AD vehicle desired velocity is set to 20 m/s, and the leading vehicle velocity varies between 0 and 14 m/s as indicated in the two first elements of the stage and terminal state upper-level constraints. The third element, maximum relative distance, is considered as infinity. In addition, the lower-level of state constraint is managed based on a CBF constraint, specifying a safe distance for ACC:

$$z \geq T_h v_f. \tag{11}$$

According to a general rule for keeping a minimum relative distance [20], the T_h is set to 1.8 s. It should be noticed this constraint is not activated during braking when both z and v_f become very small. Consequently, the intuitive choice of CBF would be $h(x) = z - T_h v_f$ which yields the admissible set $\mathcal{C}$ (3). Furthermore, to ensure that the CBF constraint does not conflict with other constraints, the CBF constraint is reformulated as in Equation (12) by means of a definiton of $h_F(x)$ [21], which yields a set $\mathcal{C}_F = \{x | h_F(x) \geq 0\}$ for all $x \in \mathcal{C}_F$.

$$h_F(x_k) = z - T_h v_f - \frac{1}{2}\frac{v_f^2 a_l - v_l^2 a_f}{a_f a_l g}. \tag{12}$$

Besides the CBF constraint mentioned above, $h_F(x)$ is exclusively employed in the situation where the leading vehicle start braking. The adopted constant $a_l = 0.3$ has the same role as the a_f to restrict the leading vehicle acceleration. During sudden braking situation, the terminal constraints $\mathcal{X}_f$ must ensure the collision avoidance.

Kalman Filter Design: By only considering the measurement noise added directly to z as the system output y_t, the KF parameters are set as follows: $R' = 25$, which decides a choice of nominal covariance matrix for measurement noise; $P'_{0|0} = \mathcal{I}_3$, which refers to the covariance matrix for estimation uncertainties; and $Q' = 0$, which refers to nominal covariance matrix for process noise.

Simulation Results: The evaluation of the proposed safety optimal control approach is supported by a software-in-the-loop simulation using Python and CARLA[1] The driving scenario is given by a situation where an AD vehicle follows another preceding vehicle on highway according to a pre-existing scenario model given by the

[1] The simulation model and the codes are available in https://github.com/PiggyMouth/MPC-CBF_for_ADS.

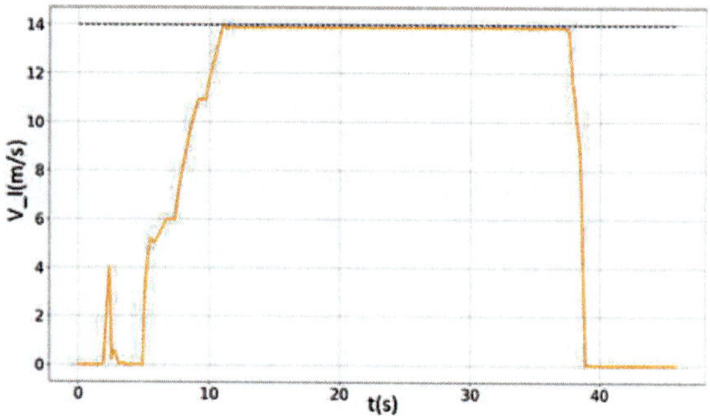

Fig. 4 Velocity of the preceding vehicle

ScenarioRunner of CARLA. The initial state is chosen to be $x_0 = [0, 0, 100]^\mathsf{T}$. The preceding vehicle velocity behaves with a pattern shown in Fig. 4. The computed control request, a_F, is fed to the longitudinal controller in CARLA as throttle/brake request after a normalization with the range between 0 and 1. According to the data derived from simulations of behaviors with different throttle/braking levels, the following throttle/braking function is defined:

$$
\begin{aligned}
\text{throttle} &= \frac{a_F}{10} + 0.5, && 0 \leq a_F, \text{brake} = 0.\\
\text{brake} &= \frac{a_F}{-10}, && -a_f g < a_F \leq 0, \text{throttle} = 0.\\
\text{full brake} &= 1, && -g < a_F \leq -a_f g, \text{throttle} = 0.
\end{aligned}
\tag{13}
$$

Case 1: Figure 5 shows the simulation results of the situation without any perception noise. The sudden drop of control input and velocity shows the ACC actions taken to ensure a safe distance to the leading vehicle by applying the full braking. The fluctuations of control input and vehicle velocity at the end of the simulation phase is due to the fact that the AD vehicle tries keeping 7 m safe distance as defined in the terminal constraint. As shown in Fig. 5a, initially AD vehicle reaches to its desired velocity, 20 m/s, and then starting reducing its velocity to follow the leading vehicle velocity at 14 m/s as their relative distance decreases.

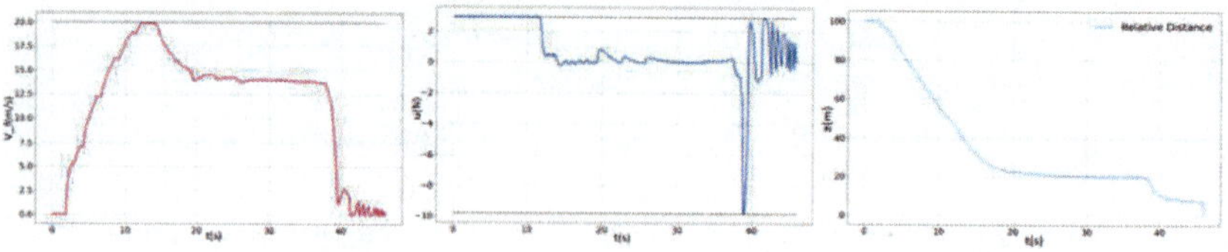

Velocity of following vehicle | Control input | Relative distance

Fig. 5 Simulation results without any faults

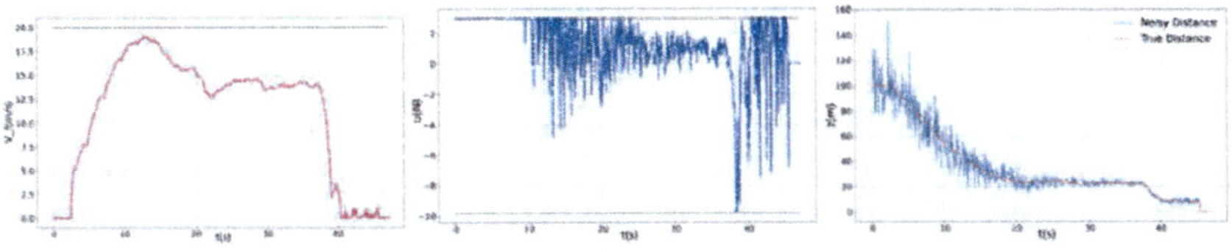

Velocity of following vehicle | Control input | Relative distance

Fig. 6 Simulation results with perception noise w/o KF-**c** Salt-and-pepper noise is injected into measured relative distance

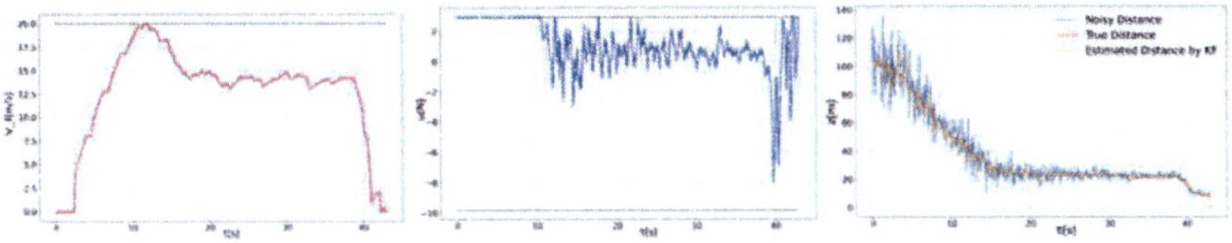

Velocity of following vehicle | Control input | Relative distance

Fig. 7 Simulation results with perception noise with KF-**c** Salt-and-pepper noise is injected into measured relative distance

Case 2: Here, the perception noise is added to the AD vehicle and its simulation results without KF are shown in Fig. 6. Since the perception noise directly influences the control performance, the control input varies more largely in order to follow the required acceleration for keeping the corresponding relative distance. Based on the outcomes, integration of MPC and ZCBF is relatively resilient to noises without KF.

Case 3: Case 3 covers the same situation in Case 2 by adding KF. The results are shown in Fig. 7. As shown in Fig. 7c, the effect of perception noise on the measured relative distance is diminished by applying KF. Consequently, the control input and the velocity become smoother. In addition, the AD vehicle velocity initially can reach its desired value and then remains about the leading vehicle velocity with slight variation compared to the previous case.

5 Conclusion and Future Work

In this paper, we have presented an approach to safety control of ADS under the uncertainties of perception sensors by exploiting an integration of ZCBF and MPC. A comparison of different experimental results showed the overall performance of the proposed control strategy, on the basis of fault injection and software-in-the-loop simulation. We have also described how the perception noise were injected into the system. Further work will enrich the operational situations by adding model-based specification of the environment and system uncertainties, and the system model by using nonlinear state-space to allow further investigations for controller design optimisation and refinement.

Acknowledgements This work is supported by 1. KTH Royal Institute of Technology with the industrial research project ADinSOS (2019065006); and 2. the Swedish government agency for innovation systems (VINNOVA) with the cooperative research project Trust-E (Ref: 2020-05117) within the programme EUREKA EURIPIDES.

References

1. NHTSA, Summary report: Standing general order on crash reporting for automated driving systems (2022). https://www.nhtsa.gov/sites/nhtsa.gov/files/2022-06/ADS-SGO-Report-June-2022.pdf
2. Dosovitskiy, A., Ros, G., Codevilla, F., Lopez, A., Koltun, V.: CARLA: an open urban driving simulator. In: Proceeding of the 1st Annual Conference on Robot Learning, pp. 1–16 (2017)
3. Rawlings, J., Mayne, D., Diehl, M.: Model Predictive Control: Theory, Computation, and Design. Nob Hill Publishing (2017)
4. Son, T.D., Nguyen, Q.: Safety-critical control for non-affine nonlinear systems with application on autonomous vehicle. In: 2019 IEEE 58th Conference on Decision and Control (CDC), pp. 7623–7628 (2019)
5. Moser, D., Schmied, R., Waschl, H., del Re, L.: Flexible spacing adaptive cruise control using stochastic model predictive control. IEEE Trans. Control. Syst. Technol. **26**(1), 114–127 (2018)
6. Xu,Y., Chen, B., Shan, X., Jia, W., Lu, Z., Xu, G.: Model predictive control for lane keeping system in autonomous vehicle. In: 2017 7th International Conference on Power Electronics Systems and Applications - Smart Mobility, Power Transfer Security (PESA), pp. 1–5 (2017)
7. Wieland, P., Allgöwer, F.: Constructive safety using control barrier functions. IFAC Proc. Vol. **40**(12), 462–467 (2007)
8. Ames, A.D., Coogan, S., Egerstedt, M., Notomista, G., Sreenath, K., Tabuada, P.: Control barrier functions: theory and applications. In: 2019 18th European Control Conference (ECC). pp. 3420–3431 (2019)
9. Agrawal, A., Sreenath, K.: Discrete control barrier functions for safety-critical control of discrete systems with application to bipedal robot navigation. In: Robotics: Science and System (2017)
10. Zeng, J., Zhang, B., Sreenath, K.: "Safety-critical model predictive control with discrete-time control barrier function. In: American Control Conference (ACC) 2021, pp. 3882–3889 (2021)
11. Zeng, v., Li, Z., Sreenath, K.: Enhancing feasibility and safety of nonlinear model predictive control with discrete-time control barrier functions. In: 2021 60th IEEE Conference on Decision and Control (CDC), pp. 6137–6144 (2021)
12. Jha, S., Banerjee, S.S., Cyriac, J., Kalbarczyk, Z.T., Iyer, R.K.: Avfi: Fault injection for autonomous vehicles. In: 48th Annual IEEE/IFIP International Conference on Dependable Systems and Networks Workshops (DSN-W), pp. 55–56 (2018)
13. Kordic, V.: Kalman Filter. InTech (2010)
14. Xu, X., Tabuada, P., Grizzle, J.W., Ames, A.D., Robustness of control barrier functions for safety critical control**this work is partially supported by the national science foundation grants 1239055, 1239037 and 1239085. IFAC-PapersOnLine. **48**(27), 54–61, (2015) Analysis and Design of Hybrid Systems ADHS
15. Grandia, R., Taylor, A., Singletary, A., Hutter, M., Ames, A.: Nonlinear model predictive control of robotic systems with control lyapunov functions. In: Robotics: Science and System XVI

16. Khalil, H.K.: Nonlinear Systems, 3rd edn. Prentice Hall (2002)
17. Huang, Y., Zhang, Y., Li, N., Wu, Z., Chambers, J.A.: A novel robust student's t-based kalman filter. IEEE Trans. Aerosp. Electron. Syst. **53**(3), 1545–1554 (2017)
18. Borrelli, F., Bemporad, A., Morari, M.: Predictive Control for Linear and Hybrid Systems. Cambridge University Press (2017)
19. Su, P., Chen, D.: Using fault injection for the training of functions to detect soft errors of dnns in automotive vehicles. In: Dependability and Complex Systems, pp. 308–318 (2022)
20. A comparison of headway and time to collision as safety indicators. Accid. Anal & Prev, **35**(3), 427–433 (2003)
21. Ames, A.D., Xu, X., Grizzle, J.W., Tabuada, P.: Control barrier function based quadratic programs for safety critical systems. IEEE Trans. Autom. Control. **62**(8), 3861–3876 (2017)

Algorithms and Methods for Fault-tolerant Control and Design of a Self-balanced Scooter

Ralf Stetter, Markus Till, and Marcin Witczak

Abstract Due to complex operation and surrounding conditions, faults are inevitable during the use of complex technical systems such as two-wheel scooters. This paper explains algorithms and methods which allow the accommodation of faults through the design and control of such systems on the functional level of product concretization. A main component of the approach is a fuzzy decision engine. This engine combines the evaluation of residuals and the current state and is based on specialist's expertise.

1 Introduction

The central topics of this paper are fault-tolerant design (FTD) and the concepts and algorithms which allow to realize FTD [11]. In the last decades, an enormous amount of research covered the topic fault-tolerant control (FTC), i.e. the capability of a control system to accommodate faults which inevitably occur in complex technical systems [1, 2, 9, 14]. Powerful algorithms, methods and systems were created that realize FTC—current research aspects are tracking control with ellipsoidal bounding [5], reliability aware model predictive control [4] and parameter identifiability [8]. However, this kind of control can be supported by certain design aspects—the design aspects referred to as fault-tolerant design [9]. In general, it is sensible to distinguish certain layers of fault-tolerant design based on the level of

R. Stetter (✉) · M. Till
Department of Mechanical Engineering, Ravensburg-Weingarten University (RWU), 88250 Weingarten, Germany
e-mail: ralf.stetter@rwu.de

M. Till
e-mail: markus.till@rwu.de

M. Witczak
Institute of Control and Computation Engineering, University of Zielona Góra, 65-246 Zielona, Góra, Poland
e-mail: m.witczak@issi.uz.zgora.pl

D. Theilliol et al. (eds.), *Recent Developments in Model-Based and Data-Driven Methods for Advanced Control and Diagnosis*, Studies in Systems, Decision and Control 467,
https://doi.org/10.1007/978-3-031-27540-1_29

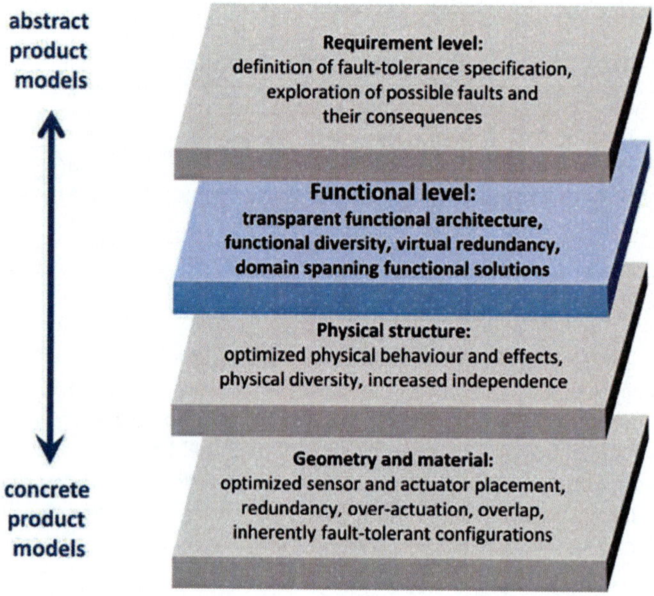

Fig. 1 Levels for characteristics of fault-tolerant design

abstraction of the involved models of the involved technical system (Fig. 1—adapted from [9]).

This paper discusses FTD on the functional level using the example of a two-wheel self-balanced scooter with a steering lever, which is one of the use cases of a current research project. In this large scale research project the digital product lifecycle (DiP) is explored. Several variants of the self-balanced scooter were developed and are used for the exploration of all elements of a product-lifecycle ranging from design over production to operation and recycling. Figure 2 shows two realized variants of the scooter (compare [7])—the variant shown on the left side was focused on cost efficiency while the variant shown on the right side disposes of a weight-optimised frame.

2 Fuzzy Decision Engine

The functional level describes, amongst others, the processes within a technical system which allow to fulfil the central requirements. Functions can be connected with the direct purpose of the system, e.g. for a self-balanced scooter, the central function is to transport a person. However, much more functions are realised in such a system such as a steering function, control functions and diagnosis functions. Fault-tolerant design on the function level can be achieved, amongst others, by means of

Fig. 2 Realised variants of a self-balanced scooter

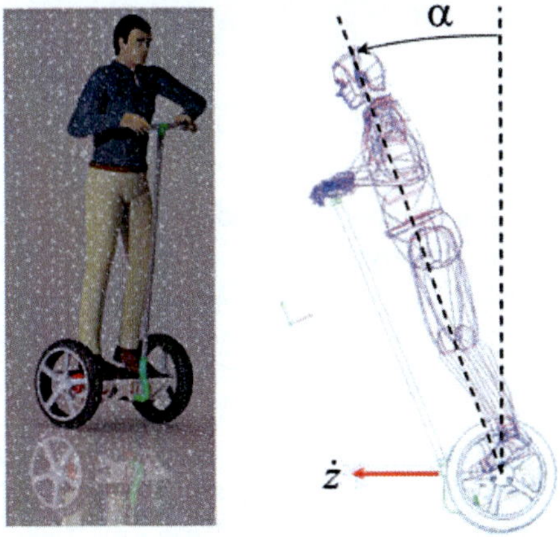

Fig. 3 Central parameters of a self-balanced scooter

virtual redundancy. Based on the given example of the scooter, a fuzzy decision engine, which is based on an analytical redundancy—a virtual sensor, is elucidated in this section [12].

A self-balanced scooter basically consists of a platform which disposes of two wheels that are independently driven by electric motors. For the further elaborations, it is concluded that the scooter disposes of two sensor arrays. A first sensor array is able to measure the inclination angle of the main platform of the scooter and, consequently, the inclination angle of the person (α). The individual velocity of both wheels of the scooter is measured by a second sensor array. From this information, the velocity of the complete scooter can be calculated ($\dot{z}$). For the elaborations in this section it is assumed that both wheels drive with identical speed and that the scooter is driving on a straight line. It is further assumed that both wheels are continuously in contact with the floor, that both wheels are not objected to slipping as well as that the driver is not behaving unusually or unexpectedly and can consequently be modelled as a rigid body. These two central parameters are represented in Fig. 3.

If one seeks to analyze the dynamics of self-balanced scooters, the basis can be a dynamic analysis of an inverted pendulum, because the main characteristics are similar. This kind of dynamic analysis can lead to a state space model building upon the proposal of [3] and employing the simplifications proposed by [15]; compare also [13]. From an initial continuous form, the discrete form can be created employing the Euler methods. It is possible to formulate a discrete state space model that includes disturbances and faults in the following form:

$$\boldsymbol{x}_{k+1} = \boldsymbol{A}_k \boldsymbol{x}_k + \boldsymbol{B}_k \boldsymbol{u}_k + \boldsymbol{B}_k f_{a,k} + \boldsymbol{W}_1 \boldsymbol{w}_{1,k}, \tag{1}$$

$$\boldsymbol{y}_k = \boldsymbol{C} \boldsymbol{x}_k + \boldsymbol{C}_f f_{s,k} + \boldsymbol{W}_2 \boldsymbol{w}_{2,k}, \tag{2}$$

with

$$\boldsymbol{A}_k = \boldsymbol{I} + T_s \cdot \boldsymbol{A}, \quad \boldsymbol{B}_k = T_s \cdot \boldsymbol{B}. \tag{3}$$

In these equations $\boldsymbol{x}$ stands for the state vector, $\boldsymbol{y}$ denotes the measurement vector and $\boldsymbol{u}$ stands for the input vector. The vector $f_{a,k}$ describes an actuator fault whereas the vector $f_{s,k}$ stands for a sensor fault. $\boldsymbol{w}_{1,k}$ denotes an exogenous disturbance vector (including the discretization error) with a distribution matrix $\boldsymbol{W}_1$ which is assumed as known. $\boldsymbol{w}_{2,k}$ stands for the measurement uncertainties with a distribution matrix $\boldsymbol{W}_2$ which is assumed as known. The state consists of the distance covered by the self-balanced scooter (z), the velocity of this scooter ($\dot{z}$), the inclination angle of this scooter (α) and the angular velocity of this angle ($\dot{\alpha}$). The input $\boldsymbol{u}$ denotes the voltage applied to the electrical drive motors on both sides. The system matrices of this state and the parameters in these state matrices are described in [12].

From the measurement of the state an input estimation can be achieved using unknown input estimation; compare for instance [10]. In the case of the self-balanced scooter, one has to accept that this system has an unknown input d_k that must be estimated for enabling the detection and identification of faults. Earlier research activities concerning a product platform lead to the proposal of an adaptive estimator for a comparable task [10]. The proposed estimator disposes of the capability to estimate the unknown input d_k and it does not need to rely on an estimation of the state of the system x_k. In the case of the self-balanced scooter, the measurement may be described using the subsequent equation:

$$\boldsymbol{y}_k = \boldsymbol{C}_k \boldsymbol{x}_k + \boldsymbol{V} \boldsymbol{v}_k. \tag{4}$$

In this equation, $\boldsymbol{v}_k$ denotes the measurement noise and $\boldsymbol{V}$ denotes its distribution matrix which is assumed as known. It is possible to employ an unknown input estimator [10]. The basis can be formulated for the given system in the subsequent form:

$$\hat{d}_{k-1} = \boldsymbol{M}_k(\boldsymbol{y}_k - \boldsymbol{C}_k\hat{x}_{k-1|k-1}). \tag{5}$$

In this equation, $\hat{x}_{k-1|k-1}$ stands for an estimate of $\boldsymbol{x}_{k-1}$. It is possible, to define the innovation $\tilde{y}_k$ by

$$\tilde{y}_k \triangleq \boldsymbol{y}_k - \boldsymbol{C}_k\hat{x}_{k-1|k-1}. \tag{6}$$

In the proposed estimation process, it is possible to estimate the unknown input $\hat{d}_{k-1}$ from the measurement $\boldsymbol{y}_k$ employing the matrix $\boldsymbol{M}_k$. Furthermore, it is possible to compare this estimation with the voltage actually applied and to synthesize a residual that may be applied for fault identification purposes:

$$z_{u,k} = \boldsymbol{u}_k - \hat{\boldsymbol{u}}_k. \tag{7}$$

The next step in this process is the application of a fuzzy decision engine. In this kind of intelligent system, the residual information resulting from the process steps described above may be combined with certain elements of the state, in the given case the inclination angle (α). The fuzzy decision engine is able to combine information that results from residuals with information concerning the state and specialist's expertise in order to support a well-founded decision whether a certain fault is present in the system. It is possible to expand a fuzzy decision engine into a fuzzy virtual actuator [10]. As explained above, the fuzzy decision engine can decide whether a certain fault is present.

It is possible to derive membership functions μ_z and μ_α for both the residual z and elements of the state (in the given case the inclination angle α). These membership functions μ_z and μ_α allow an primary evaluation of the available input information. Research with similar intentions identified trapezoidal membership functions as appropriate [6]. Their main strength is to represent an available input information in an efficient manner. Two possibilities to find the width of the membership functions exist. The first possibility is to analyse the maximum and minimum values of the residuals. The second possibility is to analyse the maximum and minimum values of the state variables. The advantages behind this is that it is possible to determine the values of the parameters which define the width and inclination experimentally.

In order to be able to accommodate process noises, disturbances and mismatches between the plant and the analytical models, it is sensible that the core is realised as an interval around zero; it is possible in this kind of problem to find a sensible size of this interval by means of the analysis of experimental data [6]. During the development process of the proposed fuzzy decision engine, bot design experts and control experts generally agreed that residuals in general can be a very powerful fault indication, but they also agreed that a combination with other measured state variables in a fuzzy logic system can be extremely fruitful. For the scooter, the specialist's expertise concerning two probable faults (first a sensor fault; second an actuator fault) could be represented employing three membership functions for the residual z_i and the state variable α—which both serve as inputs of the fuzzy system. In this example, the specialist's expertise concerning a certain fault (e.g. a sensor fault) could be formulated employing three membership functions for the residual

z_i, which is an input of the fuzzy inference system. The first input membership function μ_{z_1} was formulated as follows:

$$\mu_{z_1} = \begin{cases} 1, & z < c_1 \\ \frac{d_1 - z}{d_1 - c_1}, & c_1 \leq z \leq d_1 \\ 0, & z > d_1. \end{cases} \quad (8)$$

In these equations c_1 and d_1 stand for parameters which can be found based on the specialist's expertise or based on experimental data. The main intention of this membership function is to characterize the situation that the residual indicates a fault and that the value of the residual is negative. The next membership function μ_{z_2} indicates the case that the residual is close to zero, thus not indicating a fault. This membership function was formulated as follows:

$$\mu_{z_2} = \begin{cases} 0, & (z < a_2) \ \ or \ \ (z > d_2) \\ \frac{z - a_2}{b_2 - a_2}, & a_2 \leq z \leq b_2 \\ 1, & b_2 < z < c_2 \\ \frac{d_2 - z}{d_2 - c_2}, & c_2 \leq z \leq d_2 \end{cases} \quad (9)$$

In these equations, a_2, b_2, c_2 and d_2 stand for parameters which can be determined based on specialist's expertise or based on experimental data. The main intention of this membership function is to characterize the situation that the residual indicates no fault. The third membership function μ_{z_3} may be formulated in the subsequent manner:

$$\mu_{z_3} = \begin{cases} 0, & z < a_3 \\ \frac{z - a_3}{b_3 - a_3}, & a_3 \leq z \leq b_3 \\ 1, & z > b_3 \end{cases} \quad (10)$$

In these equations, a_3 and b_3 stand for parameters which can be determined based on specialist's expertise or based experimental data. The main intention of this membership function is to characterize the situation that the residual indicates a fault and that the value of the residual is positive. Similarly, also three membership functions may be defined for each output variable. The first one μ_{o11} may be formulated in the subsequent form:

$$\mu_{o11} = \begin{cases} 1, & afo_1 < c_{o11} \\ \frac{d_{o11} - afo_1}{d_{o11} - c_{o11}}, & c_{o11} \leq afo_1 \leq d_{o11} \\ 0, & afo_1 > d_{o11} \end{cases} \quad (11)$$

In these equations, afo_1 denotes the respective aggregated fuzzy output and c_{o11} and d_{o11} denote parameters which can be determined based on specialist's expertise or based on experimental data. A value in the given range indicates that another fault (e.g., an actuator fault) has occurred, but not the fault associated with this fuzzy output. The second output membership function μ_{o12} could be formulated in the subsequent form:

$$\mu_{o12} = \begin{cases} 0, & afo_1 < a_{o12} \;\; or \;\; afo_1 > d_{o12} \\ \frac{afo_1 - a_{o12}}{b_{o12} - a_{o12}}, & a_{o12} \leq afo_1 \leq b_{o12} \\ 1, & b_{o12} < afo_1 < c_{o12} \\ \frac{d_{o12} - afo_1}{d_{o12} - c_{o12}}, & c_{o12} \leq afo_1 \leq d_{o12} \end{cases} \quad (12)$$

In these equations, afo_1 denotes the respective aggregated fuzzy output and a_{o12}, b_{o12}, c_{o12} and d_{o12} stand for parameters which can be determined based on specialist's expertise or experimental data. A value in this range indicates that no fault has occurred. The third output membership function μ_{o13} may be formulated as follows:

$$\mu_{o3} = \begin{cases} 0, & afo_1 < a_{o13} \\ \frac{afo_1 - a_{o13}}{b_{o13} - a_{o13}}, & a_{o13} \leq afo_1 \leq b_{o13} \\ 1, & afo_{>} b_{o13} \end{cases} \quad (13)$$

In these equations, afo_1 denotes the respective aggregated fuzzy output and a_{o13} and b_{o13} are parameters which can be determined based on specialist's expertise or experimental data. A value in the given range indicates that the associated fault (the sensor fault) has occurred.

3 Evaluation

Two prominent fault scenarios can be distinguished in the given case of the self-balanced scooter:

- The first one consists of a sensor fault which results in an incorrect voltage reading.
- The second one consists of an actuator fault infesting one or even two drive motors—this fault will directly lead to an effect on the inclination angle.

Both faulty cases have been simulated. The residual z and the state variable α for a sensor fault are represented in Fig. 4 ([12] Reprint with permission from Springer).

The residual z and the state variable α for an actuator fault are represented in Fig. 5.

In Figs. 4 and 5 the blue line represents the residual z and the red line represents the inclination angle α for a fault happening at $k = 4000$. It is obvious that the residual might be similar in both cases, but that the state variable will present additional helpful information. Both are input information for the fuzzy decision engine. The results of this fuzzy decision engine are shown in Fig. 6 ([12] Reprint with permission from Springer) respectively Fig. 7.

In Figs. 6 and 7 the blue line represents the result of one output membership function—the one which aims to detect the specific sensor fault. For this output

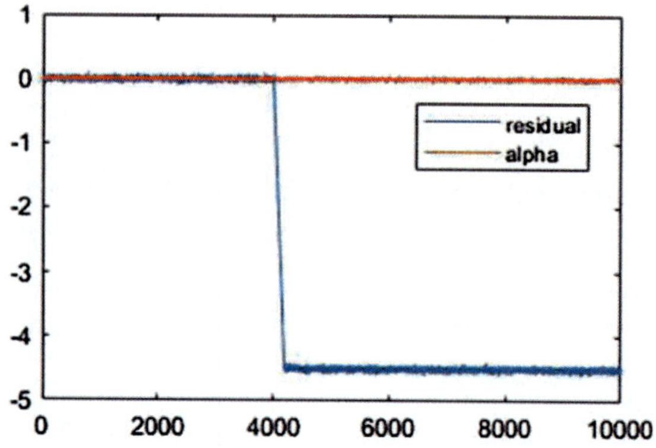

Fig. 4 Residual z and state variable α in the case of a sensor fault

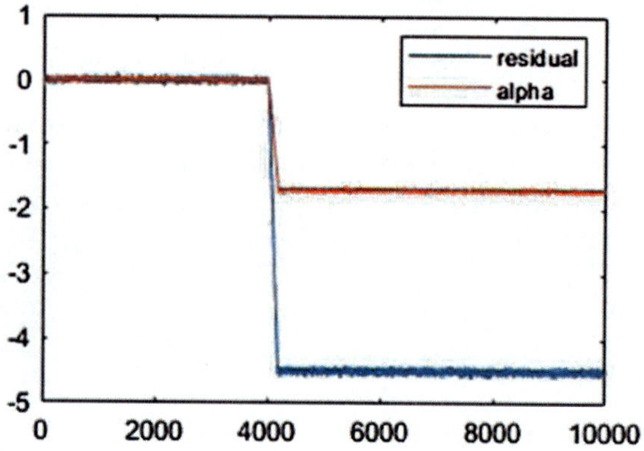

Fig. 5 Residual z and state variable α in the case of an actuator fault

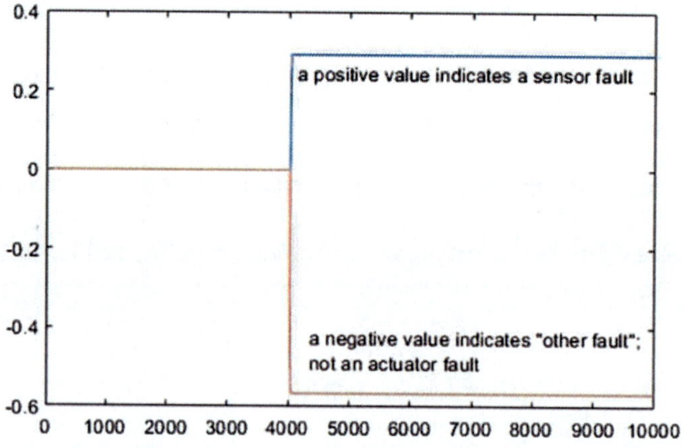

Fig. 6 Output of the fuzzy interference system for the sensor fault scenario

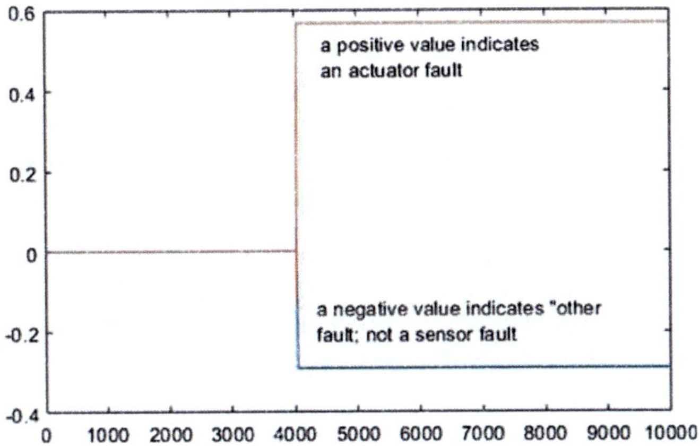

Fig. 7 Output of the fuzzy interference system for the actuator fault scenario

membership function, a positive value represents the information that—taking into consideration both the residual AND the state variable—a conclusion is achieved that the specific sensor fault is present (see Fig. 6). For this output membership function, a negative value represents the information that—taking into consideration both the residual AND the state variable—a conclusion is achieved that the specific sensor fault is not present and that another fault is present (see Fig. 7). In Figs. 6 and 7 the red line represents the result of the other output membership function—the one which aims to detect the specific actuator fault. For this output membership function, a positive value represents the information that—taking into consideration both the residual AND the state variable—a conclusion is achieved that the specific actuator fault is present (see Fig. 7). For this output membership function, a negative value represents the information that—taking into consideration both the residual AND the state variable—a conclusion is achieved that the specific actuator fault is not present, but that another fault is present (see Fig. 6). It is clearly visible that in Fig. 7 the opposite information is indicated than in Fig. 6; this result corresponds clearly to the simulated input scenario. It is also obvious that for the sensor fault and the actuator fault the presence of the fault is immediately detected; thus accentuating the effectiveness of the proposed fuzzy decision engine. The implementation of the specific realization of an analytical redundancy presents consequently an appropriate high level measure intended to increase the fault-tolerance of the self-balanced scooter.

4 Conclusions

Today's products in the industrial and consumer market enable an enormous functionality and performance. This is possible by means of more and more complex systems. The drawback of this complexity is the growing possibility for faults; one might say that faults are unpreventable in such systems. In the field of fault-tolerant control, powerful algorithms were developed in recent years; since some years they are supported by fault-tolerant design. Current investigations underline the immense importance of the connection of the algorithm development with the system development [11]. This paper presents a fuzzy decision engine which enables the identification and accommodation on the functional level of product concretization.

References

1. Blanke, M., Kinnaert, M., Lunze, J., Staroswiecki, M.: Diagnosis and Fault-Tolerant Control. Springer-Verlag, New York (2016)
2. Ding, S.X.: Model-based Fault Diagnosis Techniques: Design Schemes, Algorithms, and Tools. Springer-Verlag, Berlin/Heidelberg (2008)
3. Grasser, F., D'Arrigo, A., Colombi, S., Rufer, A.: Joe: a mobile, inverted pendulum. IEEE Trans. Ind. Electron. **40**(1), 107–114 (2002)
4. Khoury, B., Nejjari, F., Puig, V.: Reliability-aware zonotopic tube-based model predictive control of a drinking water network. Int. J. Appl. Math. Comput. Sci. **32**(2), 197–211 (2022)
5. Kukurowski, N., Mrugalski, M., Pazera, M., Witczak, M.: Fault-tolerant tracking control for a non-linear twin-rotor system under ellipsoidal bounding. Int. J. Appl. Math. Comput. Sci. **32**(2), 171–183 (2022)
6. Mendonca, L.F., Sousa, J.M.C., da Costa, J.M.G.: Fault isolation using fuzzy model-based observers. IFAC Proc. Volumes. **39**(13), 735–740 (2006)
7. Schuster, J., Pahn, F.: Entwicklung und bau zweier konzeptionell unterschiedlicher segways. Bachelor-Theses Ravensburg-Weingarten University (RWU) (2018)
8. Srinivasarengan, K., Ragot, J., Aubrun, C., Maquin, D.: Parameter identifiability for nonlinear lpv models. Int. J. Appl. Math. Comput. Sci. **32**(2), 255–269 (2022)
9. Stetter, R.: Fault-Tolerant Design and Control of Automated Vehicles and Processes. Insights for the Synthesis of Intelligent Systems. Springer-Verlag, Cham (2020)
10. Stetter, R.: A fuzzy virtual actuator for automated guided vehicles. Sensors. **20**(15), (2020)
11. Stetter, R.: Algorithms and methods for the fault-tolerant design of an automated guided vehicle. Sensors. **22**(12), (2022)
12. Stetter, R., Witczak, M., Till, M.: Fault-tolerant design of a balanced two-wheel scooter. In: Bartoszewicz, A., Kabziński, J., Kacprzyk, J. (eds.) Advanced Contemporary Control, pp. 1399–1410. Springer, Berlin (2020)
13. van der Veen, J.: Stabilization and trajectory tracking of a segway. University of Groningen, Faculty of Science and Engineering (2018)
14. Witczak, M.: Fault diagnosis and fault-tolerant control strategies for non-linear systems. Lecture Notes in Electrical Engineering, vol. 266. Springer International Publishing, Heidelberg, Germany (2014)
15. Younis, W., Abdelati, M.: Design and implementation of an experimental segway model. In: Proceedings of the 2nd Mediterranean Conference on Intelligent Systems and Automation, (2009)

Robust Control of a Customized Lane Change Maneuver

Benoit Vigne, Rodolfo Orjuela, Jean-Philippe Lauffenburger, and Michel Basset

Abstract A full acceptation of Autonomous Vehicles (AV) could be achieved by the customization of different driving services. One of these could consist of choosing the driving style and consequently its impact on the generated trajectory shapes, particularly in overtaking maneuvers. A previous work introduced a driving style adaptive path planning method able to reproduce different behaviors during overtaking maneuvers. This paper focuses on the integration of the aforementioned path planning algorithm in the control scheme of an autonomous vehicle. The generated adapted path is the reference of a robust feedback feed-forward robust controller. The overall path generation and control are validated in simulation as well as in real-world conditions. The obtained results show a good correlation between simulation and experimental results.

1 Introduction

Although road deaths have decreased significantly over the last 50 years, 3219 people still died in crashes in France in 2021 [8]. The main causes are alcohol and speed, but overtaking and lane-changing maneuvers also represent potential dangerous situations (833 crashes with 14 deaths and 60 hospitalized people in 2020). Advanced Driver Assistance Systems (ADAS) and automated driving vehicles offer promising solutions to break this scourge by providing more security. However, this interesting perspective is not sufficient to allow AV to be accepted by the population. Surveys of

B. Vigne (✉) · R. Orjuela · J.-P. Lauffenburger · M. Basset
Université de Haute-Alsace, IRIMAS UR7499, Mulhouse, France
e-mail: Benoit.Vigne@uha.fr

R. Orjuela
e-mail: Rodolfo.Orjuela@uha.fr

J.-P. Lauffenburger
e-mail: Jean-Philippe.Lauffenburger@uha.fr

M. Basset
e-mail: Michel.Basset@uha.fr

D. Theilliol et al. (eds.), *Recent Developments in Model-Based and Data-Driven Methods for Advanced Control and Diagnosis*, Studies in Systems, Decision and Control 467,
https://doi.org/10.1007/978-3-031-27540-1_30

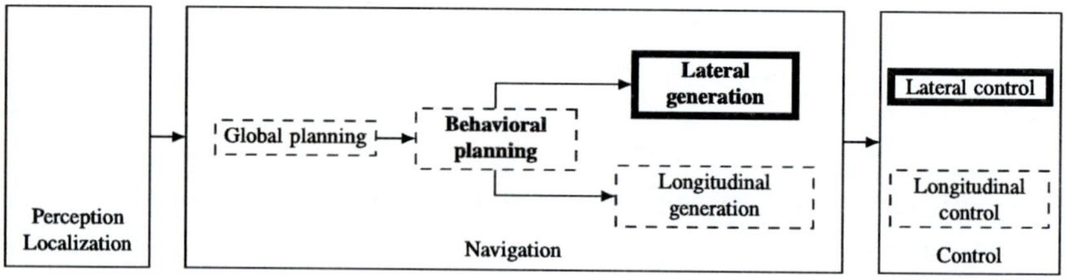

Fig. 1 Global architecture of autonomous driving control

public opinion reveal concerns about the security of vehicles without driver control [13]. To enhance trust in this new technology, one way consists on the development of human-like interfaces between the user and the system [11]. Another aims to adapt the system to the user behavior, i.e. by personalizing the vehicle dynamic to the driving style [4]. In this context of adaptation, a previous study proposed a method to personalize the path shapes with one driving style parameter in an overtaking maneuver [14]. This method showed good path planning results in simulation in different conditions (varying driving styles, car and truck overtaking). This paper assumed an efficient control loop able to track the generated adapted path. The main purpose of this article is to present the control architecture of the dedicated motion planning and highlight the performance of the planning-control loop either in simulation and in real experiments.

Motion planning refers to the Navigation layer in the global architecture of the AV [6]. As shown in Fig. 1, the Navigation integrates a global planner establishing a route from a start to a destination point dealing with mission constraints (road type, time spent, particular points of interest,. . .). This global path has to be adapted to the local driving situation by a behavioral planner which generates dedicated maneuvers (overtaking, road crossing, merging on highway,. . .). Finally, a local planner is used for trajectory generation, i.e. lateral deviation and speed profile generation according to safety and vehicle dynamic constraints. The Navigation layer is preceded by a Perception/Localization layer carrying out the positioning and obstacle detection challenges and followed by a Control layer aiming to provide the vehicle control signals required to follow the defined path, i.e. throttle/braking and steering controls.

This paper focuses on the lateral path planning and control layers described in Fig. 1 with a particular interest on a driving style parametrized lane change maneuver. The outline of the paper is as follows: Sect. 2 presents the parametrized path planner. The lateral controller description is done in Sect. 3. The experimental validation process is detailed in Sect. 4 with results and discussion. Finally, Sect. 5 concludes the paper.

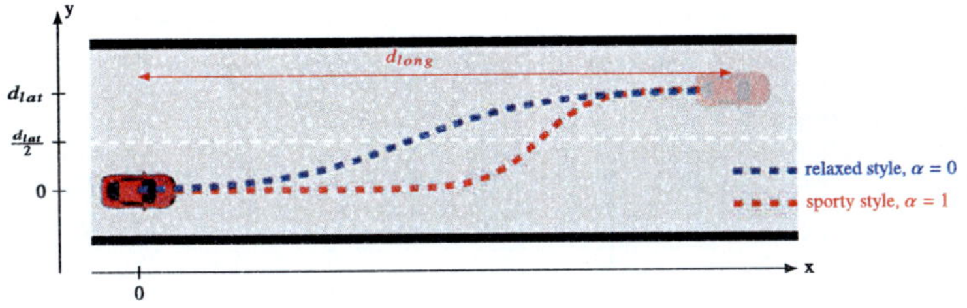

Fig. 2 Lane change path function of driving style parameter α

2 Personalized Path Generation

In robotics and generally speaking in vehicle control applications, parametric function interpolation methods are commonly used to generate lateral deviation paths. The mathematical models are typically 3 to 5 order polynomial functions [12], considering that 5th order functions are optimal with respect to jerk minimization [15]. Clothoid curves have the property to provide a continuous curvature [3]. Bézier curves are known for their low computational cost [7]. This list is not exhaustive and more details could be find in [6]. The sigmoid curve has the benefit to be limited to 4 parameters for its computation. Its definition is:

$$y = \frac{d_{lat}}{1 + \exp^{-a(x - \frac{d_{long}}{2} - b)}} \tag{1}$$

The lateral deviation y depends on the longitudinal vehicle position x and four parameters (Fig. 2): d_{lat} corresponds to the entire lateral deviation from the initial lane to the adjacent lane and depends on the road width. d_{long} is the global length of the lane change maneuver and depends on the time to execute the entire maneuver. So, these two parameters are predefined. a is a form factor representing the curvature. Finally, b is a longitudinal translation allowing a spatial delay. a and b are used to adjust the shape of the lane change path to the driving style. Previous studies [1] and [14] have already shown that for security, dynamic and continuity constraints, both parameters a and b have to be bounded, i.e. $a \in [a_{min},\ a_{max}]$ and $b \in [b_{min},\ b_{max}]$. Independently and intuitively, the form factor a can be associated to a relaxed driving style because the lower a is, the smoother the path is, with low jerk and acceleration rates. So, a cost criterion relative to a and a_{min} is defined in (2) with an optimization function to be minimized:

$$\begin{cases} J_{relaxed} = \min\limits_{a} \left(\dfrac{a - a_{min}}{a_{max} - a_{min}} \right)^2 \\ a_{min} \leq a \leq a_{max} \end{cases} \tag{2}$$

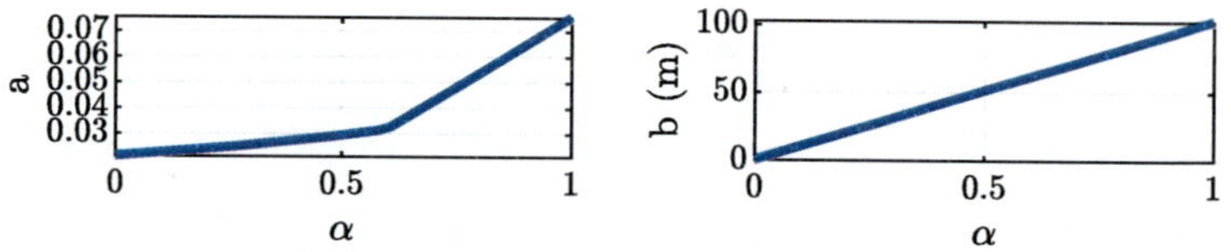

Fig. 3 Sigmoid parameters function of driving style parameter α

The sporty driving style is associated to the b parameter. The higher b is, the tighter the path is, reducing the driving time (and distance) on the adjacent lane which gives a sporty feeling. So, a cost criterion relative to b and b_{max} is defined in (3) with an optimization function to be minimized:

$$\begin{cases} J_{tight} = \min\limits_{b} \left(\dfrac{b_{max} - b}{b_{max} - b_{min}} \right)^2 \\ b_{min} \leq b \leq b_{max} \end{cases} \tag{3}$$

As a conclusion, the cost criterion (2) defines a relaxed style on the one hand and the cost criterion (3) defines a sporty style on the second one. To manage a continuous transition from one to the other, a weighting factor α varying from 0 to 1 links both as follows:

$$\begin{cases} J = \alpha J_{relaxed} + (1 - \alpha) J_{tight} \\ 0 \leq \alpha \leq 1 \end{cases} \tag{4}$$

Finally, the optimization process finds the best couple $[a, b]$ matching with the predefined driving style parameter α as seen in Fig. 3. Both parameters a and b increase with a driving style evolving from relaxed to sporty involving increasingly sharp path shape as seen in Fig. 2. After this trajectory generation, a lateral controller described in the coming section is used to track the reference path.

3 Lateral Guidance

The control design is based on the lateral dynamic bicycle model (Fig. 4) described by the following state space form:

$$\dot{\xi} = A\xi + B\delta_f \tag{5}$$

with the state vector $\xi = [y \ \ \psi]^T$ where y is the lateral deviation in the vehicle local frame, ψ its heading angle and δ_f the front wheel steering angle. The matrices A and B depend on vehicle parameters and their definition can be found in [9].

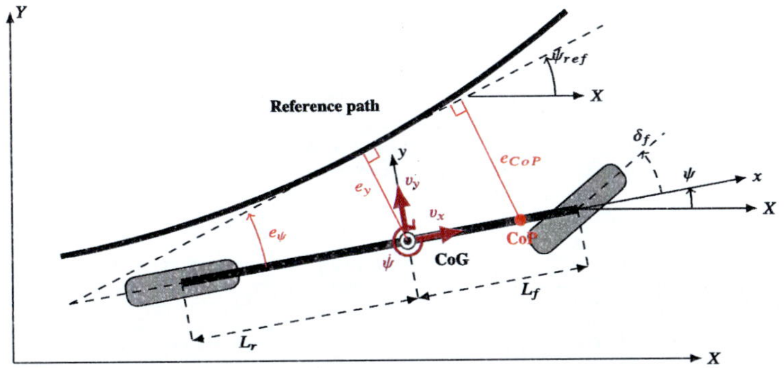

Fig. 4 lateral and orientation error for path following

The controller shown in Fig. 5 provides the appropriate steering angle δ_f in order to follow a reference path. For that purpose, it has to minimize the lateral error e_y and the orientation error e_ψ between the vehicle and the reference path as shown in Fig. 4 and defined by:

$$\dot{e_y} = v_y + v_x e_\psi \qquad e_\psi = \psi - \psi_{ref} \tag{6}$$

Here, the center of percussion (CoP) is preferred to the CoG to define the lateral error (Fig. 4) due to its position in front of the vehicle allowing a better anticipation of the trajectory similarly to a look-ahead control method [2]. The lateral CoP error is defined by:

$$e_{CoP} = e_y + x_{CoP} e_\psi \tag{7}$$

with the CoP position defined by $x_{CoP} = \dfrac{I_z}{m L_f}$ where I_z represents the vehicle inertia moment. From the bicycle model (5), both errors (6) and (7), the tracking error model used to design the lateral guidance controller is:

$$\dot{\xi}_{CoP} = A_{CoP}\xi_{CoP} + B_{CoP}\delta_f + D_{CoP}\omega_{ref} \tag{8}$$

where $\xi = [e_{CoP} \ \dot{e}_{CoP} \ e_\psi \ \dot{e}_\psi]^T$ is the error state vector, $\omega_{ref} = [\dot{\psi}_{ref} \ \ddot{\psi}_{ref}]$ is the disturbance vector with the desired yaw rate and yaw acceleration. The controller used to track the parametrized path is based on [2] using both a feed-forward and a robust state-feedback control:

$$\delta_f = u_{FB} + u_{FF} \tag{9}$$

where u_{FF} represents the feed-forward action and u_{FB} is the feedback action.

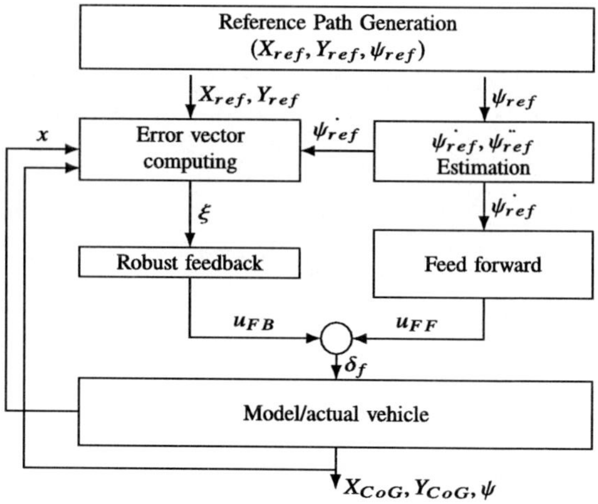

Fig. 5 controller architecture

The feed-forward aim is to reduce the impact of the disturbance $\omega_{ref}(t)$ on the lateral error. From the definition of the matrix D_{CoP} (see [2]), the feed-forward control is obtained in order to remove the action of the $\dot{\psi}_{ref}$ in the disturbance vector:

$$u_{FF} = \frac{m}{2R_l C_f}\left(\frac{2R_l C_f L_f}{m V_x} + V_x\right)\dot{\psi}_{ref} + \frac{m}{2R_l C_f} x_{CoP} \ddot{\psi}_{ref} \tag{10}$$

with $R_l = \frac{L_f + L_r}{L_r}$.

The state feedback action ensures the lateral stability by guarantying the error vector exponential convergence towards zero as well as it attenuates the impact of w_{ref} on the state variables. For this purpose, the following control law is proposed:

$$u_{FB}(t) = -K\xi(t) \tag{11}$$

where K represents the gain of the feedback. The Linear Matrix Inequality (LMI) method is used to formulate this robust control problem and its resolution is done with the vehicle model parameters given in [2]. The obtained gains for K are: [0.0417 − −0.0352 0.9971 0.0408].

4 Experimental Validation Process

Experimental platform

The experimental validation process requires at least a vehicle and an appropriate zone (test track) to carry out trials in secured conditions. The driving platform is usually a commercial vehicle customized with sensors, actuators and embedded computing systems allowing technical skills and time in development. The test zone is generally a private driving track due to the security conditions and the law agreements to be respected.

The employed test vehicle is based on a Renault Grand Scenic (Fig. 6a) with an architecture for autonomous driving shown in Fig. 6c. In the case of this study, the Perception layer provides to the Navigation layer the current vehicle position in the North East Down (NED) frame by using two real time kinematic (RTK) global positioning systems (GPS). Proprioceptive metrics (v_x, v_y, ψ, $\dot{\psi}$) required for the error state vector computation are obtained from an inertial navigation system which provides also the lateral acceleration a_y. The lateral control is performed by a servomotor coupled to the steering column and the constant longitudinal speed by the vehicle cruise control module. All Algorithms for localization, trajectory planning and control are implemented in the middleware RTMAPS (Real Time, Multisensor applications) from Intempora [10]. RTMaps runs on an embedded system integrating an industrial-grade GPU computing platform with an Intel core i7-9700 and 2×16 GB of RAM. It allows also to timestamp, record, synchronize and replay the collected data. Trials are performed on a private track at the National Automobile Museum of Mulhouse [5]. Its is formed by two rings links together and a peripheral track as shown in Fig. 6b. It is about 100 m long and 40 m wide.

Experimental protocol

Experiments are conducted on the straight line of the museum track. First, the vehicle drives on the track from a starting point during 80 m and records position points with a sampling time T_s of 100 ms (Fig. 6b). The point set defines the global path. Independently, a local path is generated for a driving style parameter α and a longitudinal speed v_x. Afterwards, the path is sampled with a spatial period $\Delta x = v_x T_s$. Finally, the local path is merged with the global one to obtain the reference path for the lateral controller. This process is carried out off-line.

From the starting point, the autonomous vehicle performs the trajectory according to the reference path and records metrics relative to the vehicle dynamics and the controller. The experimentation is performed with three different driving style parameter ($\alpha = 0$: relaxed style, $\alpha = 0.5$: daily style and $\alpha = 1$: sporty style) at a longitudinal velocity v_x of 30 km/h. The road width d_{lat} is fixed to 3 m and the total length maneuver d_{long} to 80 m.

Experimental results

Reference trajectories for each driving profile are shown in Fig. 7a. The sportier the driving style, the steeper the path shape, resulting in progressive lateral acceleration. Due to the experimental conditions, the reference and vehicle starting positions are slightly different generating undesired initial lateral and heading errors. But after around 30 meters, the lateral controller corrects the trajectories and the vehicle lane changes match with the expecting references (Fig. 7a). By removing the initial transient trajectories, i.e. the first 30 m (Fig. 7b), lateral errors reveal a maximal deviation of 9 cm between the reference and the trajectory corresponding to 2.5% of the entire lane change deviation (3 m). Heading errors show a maximal deviation of –0.002 rad

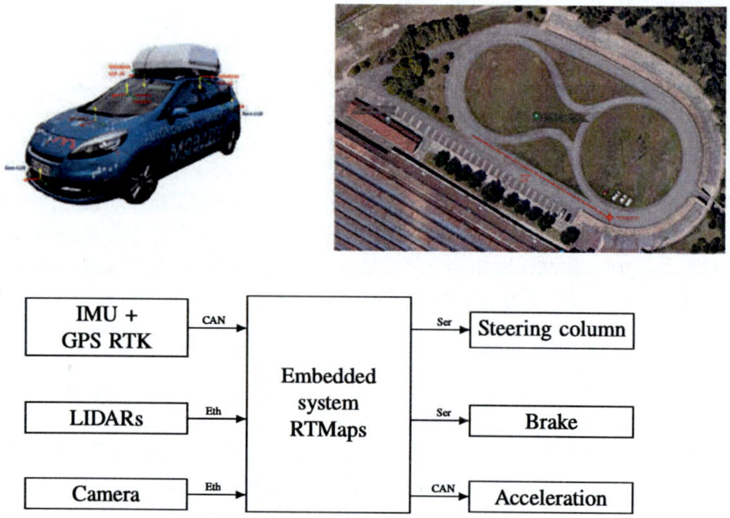

Fig. 6 Artemips platform **a**, test environment **b** and architecture **c**

proving the controller efficiency. After a median filter processing in order to reduce the signal noise, the maximum value of the vehicle lateral acceleration appears to be depending on the driving style parameter but without exceeding the pre-set limits ($a_{max} = \pm 2\ \mathrm{ms}^{-2}$) parametrized in the optimization proces .

Finally, the experimentation proves the feasibility of the lane change customization, as well as the global to local path transform and the efficiency of the lateral controller for low speeds. However, the test track is too short to validate the algorithms at high speeds.

Simulation results

In order to extend the validation with more varying driving scenario, the algorithms are tested by a simulation process requiring to replace the real test vehicle by a mathematical model.

In our case, a 2 Degree Of Freedom (DOF), bicyle model, is adopted. It returns the different state variables allowing localization and correction. The lateral controller is implemented following the functional diagram in Fig. 5. Numerical testing conditions are similar to the experimental process: $v_x = 30$ km/h, $\alpha \in [0; 0.5; 1]$. The time step is 10 ms. Curves are shown in the Fig. 7c and present the same characteristics as the experimental results. At low speed, the different non-linearities of the autonomous vehicle (i.e ground/tire interaction forces, body deformation, ...) are irrelevant and the 2 DOF model is efficient: For the different driving styles, the correlation coefficient r between simulated and real trajectories is near 1 as shown in Fig. 7d.

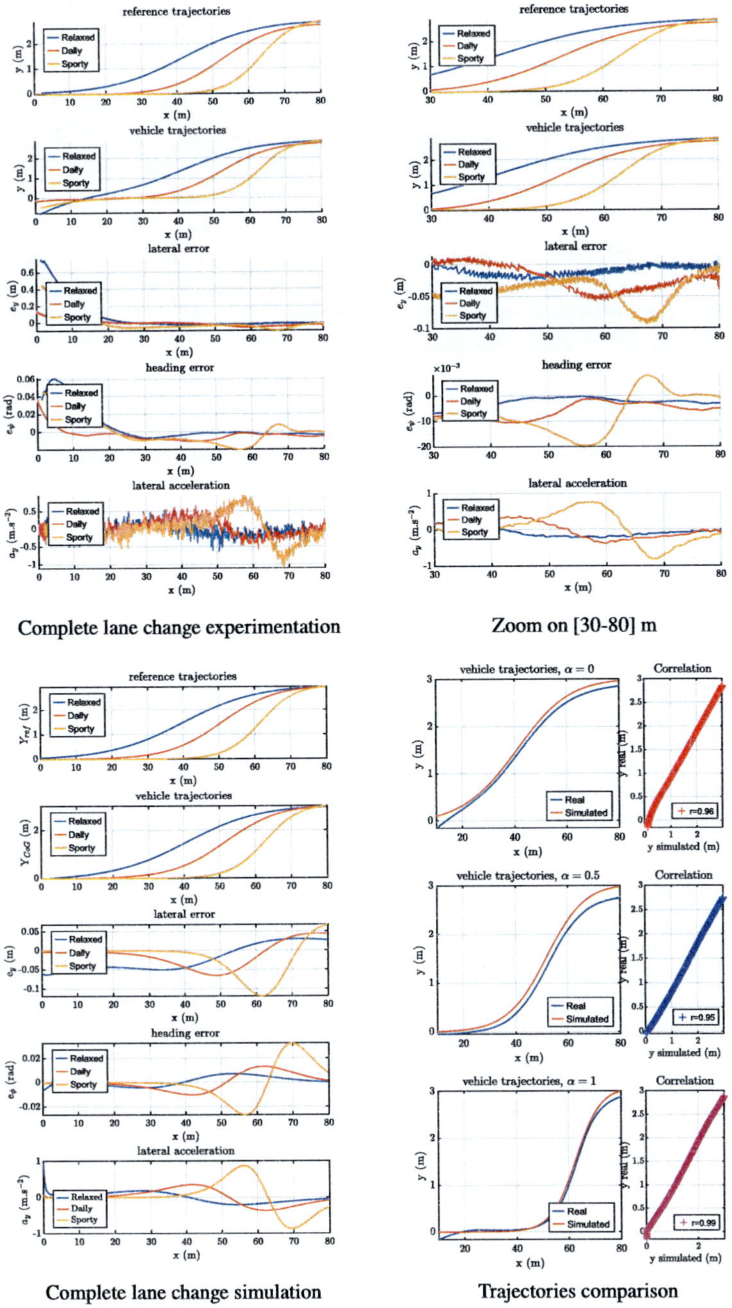

Complete lane change experimentation

Zoom on [30-80] m

Complete lane change simulation

Trajectories comparison

Fig. 7 Lane change maneuver results

5 Conclusion and Perspectives

The shape of a lane change maneuver trajectory is personalized with one sole parameter allowing to provide from a comfortable to a sporty driving path. The theoretical concept has been validated on one side by an experimental process on a real autonomous vehicle and on another side by simulation tests. However, to satisfy a full validation, the reference path generation has to be embedded in the real-time process. Moreover, due to the length of the test track, the maneuver is performed at low speed. To overcome these drawbacks and accelerate the transition between simulations and experimentations, it is aimed in future works to develop a full digital twin vehicle on an open source simulator.

Acknowledgements The authors gratefully acknowledge the help of our team members, Thomas Josso-Laurain and Jonathan Ledy, whose contributions were essential for the experimental process. The authors would also like to thanks the National Automobile Museum of Mulhouse for the access to its test track.

References

1. Ammour, M., Orjuela, R., Basset, M.: Trajectory Reference Generation and Guidance Control for Autonomous Vehicle Lane Change Maneuver. In: 2020 28th Mediterranean Conference on Control and Automation (MED). IEEE, Saint-Raphaël, France (2020)
2. Boudali, M., Orjuela, R., Basset, M.: A comparison of two guidance strategies for autonomous vehicles. In: 20th IFAC World Congress. IFAC, Toulouse, France (2017)
3. Alia, C., Gilles, T., Reine, T., Ali, C.: Local trajectory planning and tracking for autonomous vehicle navigation using clothoid tentacles method. In: 2015 IEEE Intelligent Vehicles Symposium (IV). IEEE, Seoul, Korea (South) (2015)
4. Chu, D., Deng, Z., He, Y., Wu, C., Sun, C., Lu, Z.: Curve speed model for driver assistance based on driving style classification. IET Intell. Transp. Syst. **11**(8), (2017)
5. Musée National de l'Automobile. https://www.musee-automobile.fr (2022)
6. González, D., Pérez, J., Milanés, V., Nashashibi, F.: A review of motion planning techniques for automated vehicles. IEEE Trans. Intell. **17**(4), 1135–1145 (2016)
7. Moreau, J., Melchior, P., Victor, S., Moze, M., Aioun, F., Guillemard, F.: Reactive path planning for autonomous vehicle using bézier curve optimization. In: 2019 IEEE Intelligent Vehicles Symposium (IV). IEEE, Paris, France (2019)
8. ONSIR. Road accidents: final data. Technical Report, ONSIR (2021)
9. Rajamani, Rajesh: Vehicle Dynamics and Control. Springer, US, Bosto, MA, USA (2012)
10. INTEMPORA Rtmaps. https://intempora.com (2022)
11. Ruijten, P.A.M., Terken, J.M.B., Chandramouli, S.N.: Enhancing trust in autonomous vehicles through intelligent user interfaces that mimic human behavior. Multimodal Technol. Interact. **2**(4), (2018)
12. Said, A., Talj, R., Francis, C., Shraim, H.: Local trajectory planning for autonomous vehicle with static and dynamic obstacles avoidance. In: 2021 IEEE International Intelligent Transportation Systems Conference (ITSC). IEEE, Indianapolis, IN, USA (2021)
13. Schoettle, B., Sivak, M.: A survey of public opinion about autonomous and self-driving vehicles in the US, the UK, and Australia. Technical Report (2014)

14. Vigne, B., Orjuela, R., Lauffenburger, J-P., Basset, M.: A personalized path generation for an autonomous vehicle overtaking maneuver. In: 11th IFAC Symposium on intelligent autonomous vehicles, IAV 2022. IFAC, Prague, Czech Republic (2022)
15. Ziegler, J., Stiller, C.: Spatiotemporal state lattices for fast trajectory planning in dynamic on-road driving scenarios. In: 2009 IEEE/RSJ International Conference on Intelligent Robots and Systems. IEEE, St Louis, MO, USA (2009)

Robust $\mathcal{L}_2$ Proportional Integral Observer Based Controller Design with Unmeasurable Premise variables and Input Saturation

Ines Righi, Sabrina Aouaouda, Mohammed Chadli, and Said Mammar

Abstract In this paper, a descriptor Takagi-Sugeno (T-S) fuzzy system is considered, subject to disturbances, sensor and actuator faults, in the presence of unmeasurable premise variables and input saturation, which the later are considered as uncertainties of the system. A robust $\mathcal{L}_2$ Proportional-Integral (PI) observer based controller is designed. This observer based controller, can estimate both the system states, the faults, and stabilize closed-loop states trajectories. An integrated robust controller strategy is adopted by a novel structure of non Parallel Distributed Compensation (non-PDC) control law. Moreover, the stability conditions of controller design are expressed in term of Linear Matrix Inequalities (LMIs) constrained optimization problem, by the proposed one-step design procedure. With the help of LMI TOOLBOX in MATLAB, we can easily design the type PI observer for efficient robust state/fault estimation and solve the $\mathcal{L}_2$ observer based controller design problem of discrete-time T-S fuzzy systems. The applicability of the proposed scheme is illustrated via a numerical example.

I. Righi · S. Aouaouda
University of Souk Ahras, LEER- BP 1553, Souk, Ahras 41000, Algeria
e-mail: i.righi@univ-soukahras.dz

S. Aouaouda
e-mail: sabrina.aouaouda@univ-soukahras.dz

M. Chadli (✉) · S. Mammar
Université Paris-Saclay Evry, IBISC, 91020 Evry, France
e-mail: mchadli20@gmail.com

S. Mammar
e-mail: said.mammar@ibisc.univ-evry.fr

D. Theilliol et al. (eds.), *Recent Developments in Model-Based and Data-Driven Methods for Advanced Control and Diagnosis*, Studies in Systems, Decision and Control 467,
https://doi.org/10.1007/978-3-031-27540-1_31

1 Introduction

Among the systems studied in control theory, systems with unknown inputs (UI) are of particular interest. The challenge is thus to obtain a good estimation of the state without any a priori knowledge on the unknown input. Fortunately, a T-S and LPV fuzzy model provide an effective way to express the complicated nonlinear systems, via a set of local linear models interpolated by membership functions . As a result, the nonlinear control systems theory can be widely exploited to analyze and synthesize the nonlinear systems expressed as multiple model approach [3]. Therefore, there is a rapidly growing interest in FE and FTC problems for nonlinear system based on the T-S fuzzy method and many important results have been reported for the topic in the literature [5–7, 11].This paper focuses on discrete-time T-S fuzzy systems with UI and disturbances, subject to states and input constraints. However, in many real systems, the frequent occurrence of unknown faults often lead to performance degradation and even instability of the system [1, 2]. In order to strengthen the system reliability and guarantee system stability, the FE and FTC have received considerable attention during the past few decades, and plenty of results of these research fields, have been reported in the literature [9, 12]. Using FE/FTC results to design faults observer and FEs for nonlinear systems directly is a challenging issue. For exogenous disturbances, several minimum variance observers have been proposed for such T-S systems [15]. However, in the present paper the disturbances have been assumed to be with finite energy. Hence, the use of $\mathcal{L}_2$ observer is particularly suited, as this kind of observer aims to minimize the transfer of energy between the disturbances and the estimation error [17]. For that, the main contribution of this paper corresponds to the design of an observer in charge of the sensor and actuator fault estimation simultaneously along with the system state, with fewer disturbances than the one presents at the system dynamics.This paper is dedicated to the study of the observer design based on the $\mathcal{L}_2$ approach, for T-S system with unmeasurable premise variables; the main objective is to address the FE problem for a class of constrained T-S fuzzy models subject to sensor and actuator faults, this controller can also be designed in order to maintain stability, acceptable dynamic performance and steady state of the overall system, also reduce the conservatism compared to previous work, such as [6, 17, 21], despite the presence of faults. This paper is organized as follows: Sect. 2 formulates the brief description of some notations and problem statements, for modeling of the T-S structure of studied system. The description of the proposed observer, the modified constrained robust controller, the control problem definitions and the presentation of main results with the LMI based design conditions for robust $\mathcal{L}_2$ PI observer based controller, are presented in Sect. 3. The effectiveness of the proposed methods is clearly demonstrated by means of example in Sect. 4, and finally section V provides some conclusions (Fig. 2).

2 Notations and Problem Statements

2.1 Notations

The following notations are adopted to represent conveniently the different expressions, given a set of nonlinear function: $h_i(.)$, $v_k(.)$, $i \in I_r$, $k \in I_{r_e}$ are the nonlinear scalar functions. This work focuses on unmeasurable premise variables, whose measurements can be obtained from the observer design, which depends on the state vector, and can be equivalently represented by a vector of states, expressed as $h_i(z_k)$ and $v_k(z_k)$, satisfying the convex sum property. For a vector x and z, $x(k)$, $z(k)$ defined by x_k, z_k and $x(k+1)$ defined by x_{k+1}, the same for the other vector I_r denotes the set $(1, 2, \cdots, r)$, r_e denotes the set $(1, 2, \cdots, r_e)$, R^+represents the set of positive real integer. I denote the identity matrix. An asterix $*$ symbolizes the symmetric block matrices. N_n denotes the set $(1, 2, \cdots, n)$.

2.2 Problem Statements

In the following section, the controller is derived using the descriptor form Sufficient LMI constraints are derived from Lyapunov theory. Compared to [21], in the following section a constrained controller is proposed, for this we consider the following class of T-S fuzzy model, subject to input saturation, external disturbances, sensor/actuator fault:

$$\begin{cases} \sum_{k=1}^{r_e} v_k(z_k) E_k x_{k+1} = \sum_{i=1}^{r} h_i(z_k)(A_i x_k + B_i sat(u_k)) + F_a f_k + B_\omega \omega_k \\ y_k = C x_k + F_s f_k \end{cases} \tag{1a}$$

The state-space matrices E_k, A_i, B_i, C, B_ω are of the appropriate dimensions,F_a,F_s, and the matrix C involved in (1a) is assumed to be a full row rank, k is a current samples, where $i \in I_r$ represent the ith linear right hand-side submodel, $k \in I_{r_e}$ represents the kth linear left hand-side submodel of T-S fuzzy model (1a). Besides ΔE, ΔA, ΔB corresponding to the kth/ith subsystem contains the bounded uncertain terms, which can be rewritten as: $\Delta E = H_a D_{e,k} N_e$, $\Delta A = H_a D_{a,k} N_a$, $\Delta B = H_b D_{b,k} N_b$, with H_a, H_b, N_a, N_b, H_e, N_e are known constant matrices, with $D_{a,k}$, $D_{b,k}$ and $D_{e,k}$ are unknown matrices functions bounded, for all index $\epsilon = a, b$ and e, $i \in I_r, k \in I_{r_e}$, a one has $D_{\epsilon,k}^T D_{\epsilon,k} \leq I$.
System (1a) can be represented by a polytopic form:

$$\begin{cases} \sum_{k=1}^{r_e} v_k(\hat{z}_k) \mathbb{E}_k x_{k+1} = \sum_{i=1}^{r} h_i(\hat{z}_k)(\mathbb{A}_i x_k + \mathbb{B}_i sat(u_k)) + F_a f_k + B_\omega \omega_k \\ y_k = C x_k + F_s f_k \end{cases} \tag{1b}$$

where the membership functions are denote $h_i(\hat{z_k})$ and $v_k(\hat{z_k})$ vary within the convex sets Ω_1 and Ω_2:

$$\Omega_1 = (h_i(\hat{z_k}) \in I_r; h_i(\hat{z_k}) = (h_1(\hat{z_k}), ..., h_r(\hat{z_k}))^T \tag{2a}$$

$$\Omega_2 = (v_k(\hat{z_k}) \in I_{r_e}; v_k(\hat{z_k}) = (v_1(\hat{z_k}), ..., v_{r_e}(\hat{z_k}))^T \tag{2b}$$

Note that $h_i(\hat{z_k})$ and $v_k(\hat{z_k})$ depend on the variable $\hat{z_k}$ verifying the convex sum property. The augmented form is adopted, where $x^* = \left[x_k \; x_{k+}\right]^T$. The T-S system (1a) can be equivalently rewritten in the following compact augmented form:

$$\begin{cases} E^* x^*_{k+1} = (\mathbb{A}^* x^*_k + \mathbb{B}^* sat(u_k) + F^*_a f_k + B^*_\omega \omega_k \\ y_k = C^* x^*_k + F_s f_k \end{cases} \tag{3a}$$

$$\mathbb{A}^* = \begin{bmatrix} 0 & I \\ (A_{\hat{h}} + \Delta A) & (E_{\hat{v}} + \Delta E) \end{bmatrix}; \mathbb{B}^* = \begin{bmatrix} 0 \\ (B_{\hat{h}} + \Delta B) \end{bmatrix}; E^* = \begin{bmatrix} I & 0 \\ 0 & 0 \end{bmatrix} \tag{3b}$$

3 Control Problem

3.1 Control Law

In order to satisfy the desired constrained controller performance, the augmented controller design, is defined for system (3a):

$$\begin{cases} u_k = F^*_{\hat{h}\hat{v}} H^{-1*}_{\hat{h}\hat{v}} x^*_k \\ \psi(u_{k(l)}) = u_{k(l)} - sat(u_{k(l)}) \\ \psi(0) = 0 \end{cases} \tag{4a}$$

$$F^*_{\hat{h}\hat{v}} = \left[F_{\hat{h}\hat{v}} \; 0\right]; H^*_{\hat{h}\hat{v}} = \begin{bmatrix} H_{\hat{h}\hat{v}} & 0 \\ 0 & 0 \end{bmatrix}. \tag{4b}$$

The gains $F^*_{\hat{h}\hat{v}}$ and $H^*_{\hat{h}\hat{v}}$ are matrices controllers to be determined, x^*_k is the estimated augmented state variable of x^*_k. The architecture of the proposed constrained robust controller design is based on the scheme depicted in Fig. 1. System (1a) or (3a) achieves observability conditions, as detailed in [23]. To derive the controller laws, a PI observer based controller is synthesized to estimate both faults and states for system (3a) and has the augmented form:

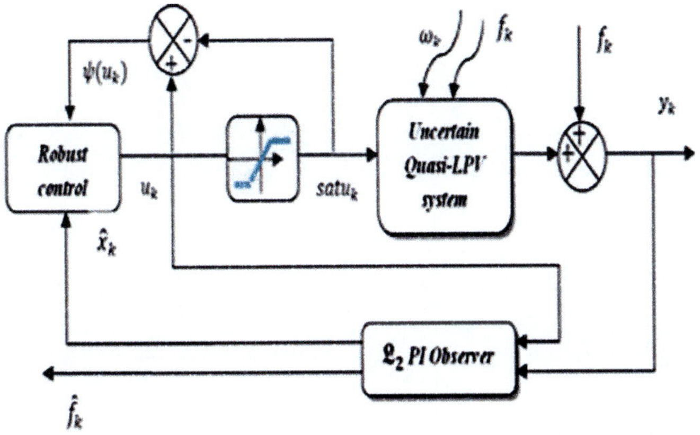

Fig. 1 The proposed robust controller design scheme

$$\begin{cases} E^*\hat{x_{k+1}}^* = A^*\hat{x_k}^* + B^*(u_k) + F_a^*\hat{f}_k + L^*_{P\hat{h}\hat{v}}(y_k - \hat{y}_k) \\ y_k = C^*\hat{x_k}^* + F_s\hat{f}_k \\ \hat{f}_{k+} = \hat{f}_k - L_{Ia\hat{h}\hat{v}}(y_k - \hat{y}_k) \\ u_k = F^*_{\hat{h}\hat{v}}H^{-1*}_{\hat{h}\hat{v}}x^*_k \end{cases} \tag{5a}$$

$$L^*_{P\hat{h}\hat{v}} = \begin{bmatrix} 0 \\ L_{P\hat{h}\hat{v}} \end{bmatrix}. \tag{5b}$$

$L^*_{P\hat{h}\hat{v}}$ is the augmented proportional gain for estimating the augmented variable state. $L_{Ia\hat{h}\hat{v}}$ is the integral gain for estimating the fault (Fig. 3).

The estimation error between the system (4a) and the observer (5a) is given by:

$$e^*_{0,k} = x^*_k - \hat{x}^*_k \tag{6}$$

The fault error is defined by:

$$e_{f,k} = f_k - \hat{f}_k \tag{7}$$

In the following, substituting u_k of (4a) into (3a) we consider the augmented form as follows:

$$\begin{cases} E^*x^*_{k+1} = (A^* + B^*F^*_{\hat{h}\hat{v}}H^{-1*}_{\hat{h}\hat{v}})x^*_k + F^*_a f_k + B^*_\omega \omega_k - B^*(F_{\hat{h}\hat{v}}H^{-1*}_{\hat{h}\hat{v}}e^*_{0,k} - \psi(u_k)) \\ y_k = C^*x^*_k + F_s f_k \end{cases} \tag{8}$$

Fig. 2 The control inputs

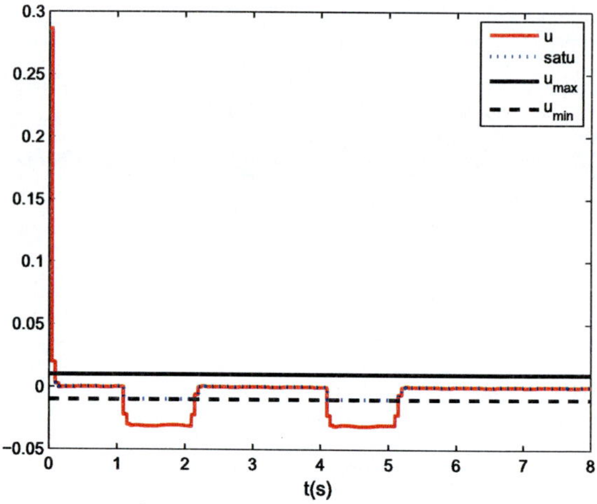

Fig. 3 States variables and their estimate

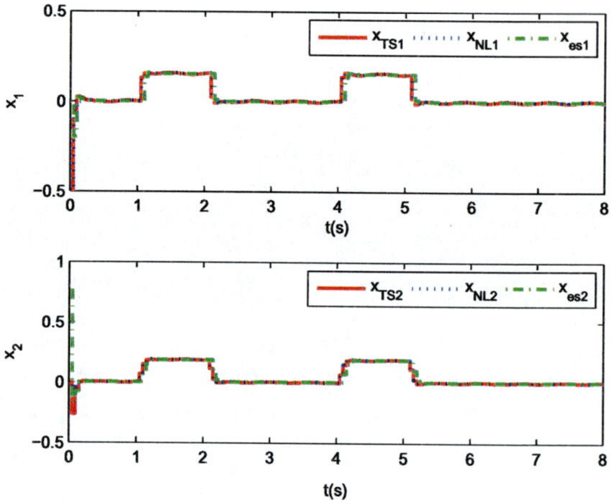

Now, using the augmented state vector is defined by

$$X_k = \left[x_k^* \ f_k\right]^T \tag{9}$$

(9), and system (8) becomes:

$$\begin{cases} \epsilon^* X_{k+1}^* = \mathcal{A}_{\hat{h}\hat{h}\hat{v}} X_k - \mathcal{B}^* \psi(u_k) + \mathcal{B}_\omega \omega_k - \mathbb{B}^* F_{\hat{h}\hat{v}}^* H_{\hat{h}\hat{v}}^{-1*} e_{0,k}^* \\ y_k = \mathcal{C} X_k \\ u_k = \mathcal{F}_{\hat{h}\hat{v}}^* \mathcal{H}_{\hat{h}\hat{v}}^{-1*} X_k - F_{\hat{h}\hat{v}}^* H_{\hat{h}\hat{v}}^{-1*} e_{0,k}^* \end{cases} \tag{10a}$$

$$\begin{cases} \mathcal{A}_{\hat{h}\hat{h}\hat{v}} = \begin{bmatrix} (\mathbb{A}^* + \mathbb{B}^* F^*_{\hat{h}\hat{v}} H^{-1*}_{\hat{h}\hat{v}}) & F^*_a \\ 0 & I_f \end{bmatrix}; \epsilon^* = \begin{bmatrix} E_* & 0 \\ 0 & I_f \end{bmatrix}; \mathcal{F}^*_{\hat{h}\hat{v}} = \begin{bmatrix} F^*_{\hat{h}\hat{v}} & 0 \end{bmatrix}; \mathcal{H}^*_{\hat{h}\hat{v}} = \begin{bmatrix} H^*_{\hat{h}\hat{v}} & 0 \\ 0 & 0 \end{bmatrix} \\ \mathcal{B}^*_\omega = \begin{bmatrix} B^*_\omega \\ 0 \end{bmatrix}; \mathcal{B}^* = \begin{bmatrix} \mathbb{B}^* \\ 0 \end{bmatrix}; \mathcal{C} = \begin{bmatrix} C^* & 0 \end{bmatrix} \end{cases} \tag{10b}$$

The dynamic error is defined by:

$$E^* e^*_{0,k+} = E^* x^*_{k+} - E^* \hat{x}^*_{k+} = (A_{\hat{h}\hat{v}} - L_{P\hat{h}\hat{v}} C) e^*_{0,k} + (F^*_a - L_{I\hat{h}\hat{v}} F_s) - B^*_{\hat{h}} \psi(u_k) + B_\omega \omega + \Delta_1 \tag{11a}$$

with $\Delta_1 = (\Delta A^* + \Delta B^* F^*_{\hat{h}\hat{v}} H^{-1*}_{\hat{h}\hat{v}}) x^*_k - \Delta B^* F^*_{\hat{h}\hat{v}} H^{-1*}_{\hat{h}\hat{v}} e^*_{0,k} - \Delta B^* \psi(u_k)$

The dynamics of the fault estimation, with considering the fault constant: $f_{k+} = f_k$ [16]. Considering the dynamics (11a) and (11b), the augmented error dynamics is obtained as follows:

$$\begin{cases} e_{f,k+} = f_k - \hat{f}_k = L_{I\hat{h}\hat{v}} C e^*_{0,k} + (L_{I\hat{h}\hat{v}} F_s + I_{e,f}) e_{f,k} \end{cases} \tag{11b}$$

Now, we consider the augmented system, by combining the system (8) and the error dynamics (11), the closed-loop T-S system is obtained as follows:

$$\begin{cases} \epsilon e_{k+} = \mathcal{M}_{\hat{h}\hat{h}\hat{v}} e_k - B_{\hat{h}} \psi(u_k) + B_\omega \omega_k \\ u_k = \mathcal{F}^*_{\hat{h}\hat{v}} \mathcal{H}^{-1*}_{\hat{h}\hat{v}} X_k - \mathcal{F}^*_{\hat{h}\hat{v}} \mathcal{H}^{-1*}_{\hat{h}\hat{v}} e_k \end{cases} \tag{12a}$$

$$\mathcal{M}_{\hat{h}\hat{h}\hat{v}} = \begin{bmatrix} (A_{\hat{h}}\hat{v} - L_{P\hat{h}\hat{v}} C) - & (F^*_a - L_{I\hat{h}\hat{v}} F_s) \\ L_{I\hat{h}\hat{v}} C & (L_{I_{\hat{h}\hat{v}}} F_s + I_{ef}) \end{bmatrix}; \epsilon = \begin{bmatrix} E^* & 0 \\ 0 & I_{ef} \end{bmatrix}. \tag{12b}$$

Now, we consider the augmented system $\overline{x}_k = [X_k e_k]^T$, by combining the system (8) and the error dynamics (11), the closed-loop T-S system is obtained as follows:

$$\begin{cases} E\overline{x}_{k+1} = \mathcal{A}_{\hat{h}\hat{h}\hat{v}} \overline{x}_k - \mathcal{B}\psi(u_k) + \mathcal{B}_\omega \omega_k \\ y_k = \mathcal{C}\overline{x}_k \\ u_k = \mathcal{K}^*_{\hat{h}\hat{v}} \mathcal{L}^{-1*}_{\hat{h}\hat{v}} \overline{x}_k \end{cases} \tag{13a}$$

$$\mathcal{K}^*_{\hat{h}\hat{v}} = \begin{bmatrix} \mathcal{F}_{\hat{h}\hat{v}} & -\mathcal{F}_{\hat{h}\hat{v}} \end{bmatrix} \mathcal{L}^*_{\hat{h}\hat{v}} = \begin{bmatrix} \mathcal{H}_{\hat{h}\hat{v}} & 0 \\ 0 & 0 \end{bmatrix} \tag{13b}$$

Moreover, to handle the non-linearity $sat(u_k)$, the dead-zone function ψ_{u_k} will be employed [15], where:

$$\psi(u_{k(l)}) = u_{k(l)} - sat(u_{k(l)}) \tag{14}$$

we use the generalized sector condition proposed by [15] to deal with the dead zone function. Also, the set: $\mathcal{D}_u = (\overline{x}_{k(l)} \in R^{n_x}; |(u_{k(l)} - v_{k(l)})| < u^2_{max(l)}; l \in I_{n_l})$, with the auxiliary signal: $v_k = \mathcal{W}^*_{\hat{h}\hat{v}} \mathcal{L}^{-1*}_{\hat{h}\hat{v}} \overline{x}_k$ used as a degree of freedom in the design conditions, and the condition: $\psi(u_{k(l)}) S^{-1}_{\hat{h}\hat{v}} (\psi(u_k) - v_{k(l)})$ holds. One way to construct the estimate DoA is to employ level sets taken from the Lyapunov function associated with the closed-loop system. To this end, a non quadratic Lyapunov function is considered

$$V(\overline{x}_k) = \overline{x}_k^T \overline{E}^T \overline{P}_h^{-1} \overline{E} \overline{x}_k \tag{15}$$

a level set associated with the Lyapunov function can be defined as in the following lemma:

Lemma 3.1 *Suppose that* $V(\overline{x}_k)$ *given in* (15) *is a Lyapunov function for system* (13a). *Then, a possible level set is given by:*

$$\Big\{ \mathcal{L}_v = \varepsilon(\overline{E}^{-T} \overline{P}_h^{-1} \overline{E}, 1) = \bigcap_{i=1}^{r} \varepsilon(\overline{E}^T \overline{P}_i^{-1} \overline{E}, 1) \subset \mathcal{D}_{\overline{x}} \tag{16}$$

$$\Big\{ \rho > 0, \varepsilon(\overline{E}^T \overline{P}_h^{-1} \overline{E}, \rho) = (\overline{x}_k \in R^{n_x} : \overline{x}_k^T \overline{E}^T \overline{P}_h^{-1} \overline{E} \overline{x}_k < \rho) \tag{17}$$

Definition 1 The state trajectories of T-S descriptor system (13a) are contained within the following polyhedral set (validity domain):

$$\mathcal{D}_{\overline{x}} = (|\mathcal{N}_m^T \overline{x}_k| < 1; m \in I_{n_m}) \tag{18}$$

Where the given matrix $\mathcal{N}_m$ represents the state constraints of system (13a), for all $\overline{x}_k \in R^{n_{\overline{x}}}$. □

3.2 *LMI-Based Design Conditions of Constrained Descriptor System*

Property 1 [local stability]. Given a scalar α', the initial condition x(0) belong to a specific set in the state-space, which $\varepsilon(\overline{E}^T \overline{P}_h^{-1} \overline{E}, \rho)$ is a region of asymptotic stability (RAS) for the saturated system. In the presence of disturbances, the controller guarantees that the trajectories of (13a) are bounded, there exist a matrix $\overline{P}_i^T > 0$ and a positive scalar $\rho > 0$ such that, for any $\overline{x}(0) \in \varepsilon(\overline{E}^T \overline{P}_h^{-1} \overline{E}, \rho)$ and in the presence of ω, the trajectories of the saturated system remains inside the polyhedral set $\mathcal{D}_{\overline{x}_k}$, and do not leave the ellipsoid $\varepsilon(\overline{E}^T \overline{P}_h^{-1} \overline{E}, \rho)$ and converges exponentially to the equilibrium point with a decay rate less than α' , satisfying the following property.
Property 2 [$\mathcal{L}_2$ gain-performance]. Given vector $\mathcal{N}_m$ defined in assumption 1, and a positive scalar δ depending in the type of disturbances involved in the dynamics of system (13a), see [15] we distinguish the following control problem.

Control problem. In the presence of ω_k. There exist positive scalar ρ and γ such that for all $\overline{x}_k \in \mathcal{L}_v$, the corresponding closed-loop trajectory (13a) remains inside the validity domain $\mathcal{D}_{\overline{x}_k}$ in (18). Moreover the $\mathcal{L}_2$-gain of the state vector $\overline{x}_k$ follows:

$$||\overline{x}_k||_2^2 \leq \gamma^2 ||\omega_k||_2^2 + \rho, k > 0 \tag{19}$$

Where the objective is to attenuate the effects of exogenous input ω_k on the augmented state space by minimizing γ and ρ.

The objective now is to compute the gains of observer based controller (5a), to ensure the stability of the closed-loop system (13a), guarantying the trajectories performance for all ω_k not nul, sufficient conditions to achieve this objective are given through the following theorem:

Theorem 1 *For a given the discrete-time T-S descriptor system (3a) with a non-linearities parameter $z_k \in \Omega_1$ and Ω_2, under input saturation with the proposed observer based controller (5a), whose validity domain is defined by $\mathcal{D}_{\overline{x}}$,is locally exponentially stable if there exist a positive definite matrices $P_{i11} = P_{i11}^T > 0$, $P_{i12} = P_{i12}^T > 0$, $P_{i2} = P_{i2}^T > 0$, $P_{i4} = P_{i4}^T > 0$, $(H_{11}, H_{12}, H_{21}, H_{22}) \in R^{n_x \times n_x}$, $(P_{i11}, P_{i12}) \in R^{n_x \times n_x}$, $P_{i2} \in R^{n_f \times n_f}$, $P_{i4} \in R^{n_{ef} \times n_{ef}}$, an observer gains $(L_{I11}, L_{I12}, L_{I21}, L_{I22}) \in R^{n_u \times n_x}$, $(L_{P11}, L_{P12}, L_{P21}, L_{P22}) \in R^{n_y \times n_x}$, a positive diagonal matrices $S_{jk} \in R^{n_u \times n_u}$ a positive scalars $\tau_{11}, \tau_{12}, \tau_{31}, \tau_{32}, \tau_b, \gamma = \sqrt{\gamma^2}, \rho$, η where $(i, j) \in (I_r \times I_r)$, $k \in I_{r_e}$, such that:*

$$\begin{cases} \min \gamma, \rho, \eta \\ \overline{E}^T \overline{P}_h^{-1} \overline{E} \geq 0 \end{cases} \quad \textit{such that}$$

$$\rho + \gamma^2 \delta < 1 \tag{20}$$

$$\begin{bmatrix} \Psi^{(1,1)} & * & * & * & * \\ 0 & P_{i2} & * & * & * \\ 0 & 0 & \Psi^{(1,1)} & * & * \\ 0 & 0 & 0 & P_{i4} & * \\ T_{jk}^* & -W_{2jk}^* & T_{jk}^* & -W_{4jk}^* & u_{max}^2/\rho \end{bmatrix} \geq 0, m \in N_m, i, j \in N_r \tag{21}$$

$$\begin{bmatrix} \Psi^{(1,1)} & * & * & * & * \\ 0 & P_{i2} & * & * & * \\ 0 & 0 & \Psi^{(1,1)} & * & * \\ 0 & 0 & 0 & P_{i4} & * \\ N_m^* H_{jk}^* & 0 & 0 & 0 & 1/\rho \end{bmatrix} \geq 0, m \in N_m, i, j \in N_r \tag{22}$$

$$\begin{cases} T_{ij}^k < 0 \\ \frac{2}{r-1} T_{ii}^k + T_{ij}^k + T_{ji}^T < 0, \end{cases} \tag{23}$$

$$T_{ij}^k = \begin{bmatrix} \Psi + X^T \mathcal{G} X & * & * \\ Z^T & -\mathcal{G} I & * \\ \theta_{12}^T & 0 & -\theta_{22} \end{bmatrix} \tag{24}$$

4 Illustrative Example

This section gives design examples for the following nonlinear model:

$$\begin{cases} 2x_{1,k+} - (1 + x_{1,k}^2)^{-1} x_{2,k+} = x_{2,k}^2 x_{1,k} - 0.5\beta x_{2,k} + 0.1 u_k \\ (1 + x_{1,k}^2)^{-1} x_{1,k+} + 2x_{2,k+} = 0.7x_{1,k} + sin(x_{1,k}) x_{2,k} \\ y_k = x_{1,k} \end{cases} \tag{25}$$

Then, the nonlinearities are chosen as:

$$\left\{ z_k = [z_1, z_2, z_3]^T ;\, z_1 = (1 + x_{1,k}^2)^{-1}, z_2 = x_{2,k}^2, z_3 = sin(x_{1,k}) \right. \tag{26}$$

Solving the LMIs of Theorem 1, the unknown gains of the PI observer based controller are obtained. The constants α', δ and ε are selected as 1, 0.2 and 1, and since the main objective is to estimate the state variables and the faults.

5 Conclusion

It was considered that the descriptor T-S fuzzy system was affected by external disturbance, sensor and actuator faults, with unmeasurable premise variables and input saturation. The used strategy was based on the $\mathcal{L}_2$ performance criteria to be robust against disturbance and faults. Furthermore, it was demonstrated that the proposed approach is suitable to estimate system states and sensor/actuator faults by a Proportional-Integral PI observer based controller, and stabilize the states and controller into equilibrium point. Finally, a numerical example was presented to show the effectiveness and applicability of the proposed approach.

References

1. Edwards, C., Spurgeon, S.K., Patton, R.J.: Sliding mode observers for fault detection and isolation. Autom. **36**(4), 541–553 (2000)
2. Su, X.J., Liu, X.X., Song, Y.D.: Event-triggered sliding mode control for multi-area power systems. IEEE Trans. Ind. Electron. **64**(8), 6732–6741 (2017)
3. Chadli, M.: «Stabilité et Commande de Systemes décrits par des Structures Multimodeles: Approche LMI »Thèse INPL-CRAN, Nancy, France (2002)

4. Li, X.J., Yang, G.H.: Dynamic observer-based robust control and fault detection for linear systems. IET Control. Theory Appl. **6**(17), 2657–2666 (2012)
5. Zhang, K., Jiang, B., Staroswiecki, M.: Dynamic output feedback fault tolerant controller design for Takagi-Sugeno fuzzy systems with actuator faults. IEEE Trans. Fuzzy Syst. **18**(1), 194–201 (2010)
6. Zhang, K., Jiang, B., Shi, P.: Analysis and design of robust H fault estimation observer with finite-frequency specifications for discrete-time fuzzy systems. IEEE Trans. Cybern. **45**(7), 1225–1235 (2015)
7. Dong, J.X., Hou, J.T.: Output feedback fault-tolerant control by a set theoretic description of T-S fuzzy systems. Appl. Math. Comput. **301**, 117–134 (2017)
8. Jia, Q.X., Chen, W., Wang, P.: Design of a PD-type learning observer for reconstruction of actuator faults in descriptor systems. IET Control. Theory Appl. **11**(1), 17–24 (2017)
9. Huang, S.J., Yang, G.H.: Fault tolerant controller design for T-S fuzzy systems with time-varying delay and actuator faults: a K-step fault-estimation approach. IEEE Trans. Fuzzy Syst. **22**(6), 1526–1540 (2014)
10. Chadli, M., Abdo, A., Ding, S.X.: H2/H fault detection filter design for discrete-time Takagi Sugeno fuzzy system. Autom. **49**, 1996–2005 (2013)
11. Liu, M., Cao, X.B., Shi, P.: Fault estimation and tolerant control for fuzzy stochastic systems. IEEE Trans. Fuzzy Syst. **21**(2), 221–229 (2013)
12. Liu, M., Cao, X.B., Shi, P.: Fuzzy-model-based fault tolerant design for nonlinear stochastic systems against simultaneous sensor and actuator faults. IEEE Trans. Fuzzy Syst. **21**(5), 789–799 (2013)
13. Aouaouda, S., Chadli, M., Boukhnifer, M., Karimi, H.R.: Robust fault tolerant tracking controller design for vehicle dynamics: a descriptor approach. mechatronics. **30**, 316–326 (2015)
14. Aouaouda, S., Chadli, M.: Model reference tracking control for uncertain takagi-sugeno systems subject to sensor faults. IFAC-PapersOnLine **48**(21), 1250–1255 (2015)
15. Tarbouriech, S., Garcia, G., da Silva Jr, J.M.G., Queinnec, I.: Stability and Stabilization of Linear Systems With Saturating Actuators. Springer Science & Business Media (2011)
16. Escobet, T., Bregon, A., Pulido, B., Puig, V.: Fault Diagnosis of Dynamic Systems. Springer International Publishing (2019)
17. Do, M.H., Koenig, D., Theilliol, D.: Robust H proportional-integral observer-based controller for uncertain LPV system. J. Frankl. Inst. **357**(4), 2099–2130 (2020)
18. Alwi, H., Edwards, C., Marcos, A.: Fault reconstruction using a LPV sliding mode observer for a class of LPV systems. J. Frankl. Inst. **349**(2), 510–530 (2012)
19. Zhen-Hua, W.A.N.G., Rodrigues, M., Theilliol, D., Yi, S.H.E.N.: Sensor fault estimation filter design for discrete-time linear time-varying systems. Acta Autom. Sin. **40**(10), 2364–2369 (2014)
20. Tanaka, K., Wang, H.O.: Fuzzy Control Systems Design and Analysis: a Linear Matrix Inequality Approach. John Wiley & Sons (2004)
21. Guzman, J., López-Estrada, F.R., Estrada-Manzo, V., Valencia-Palomo, G.: Actuator fault estimation based on a proportional integral observer with nonquadratic Lyapunov functions. Int. J. Syst. Sci. 1–14 (2021)
22. Boyd, S., Ghaoui, L., Feron, E.: Linear Matrix Inequalities in Systems and Control Theory. SIAM, Philadelphia, PA, USA (1994)
23. Zhang, K., Jiang, B., Shi, P., Cocquempot, V.: Observer-Based Fault Estimation Techniques. Springer International Publishing (2018)
24. Righi, I., Aouaouda, S., Chadli, M., Khelil, K.: Robust controllers design for constrained nonlinear parameter varying descriptor systems. Int. J. Robust Nonlinear Control. **31**(17), 8295–8328 (2021)
25. Righi, I., Aouaouda, S., Chadli, M., Khelil, K.: Robust $\mathcal{L}_2$ control for uncertain TS descriptor model with input saturation. In: 2020 28th Mediterranean Conference on Control and Automation (MED), pp. 428–433. IEEE (2020)